The Urban Book Series

AF800204

Aims and Scope

The Urban Book Series is a resource for urban studies and geography research worldwide. It provides a unique and innovative resource for the latest developments in the field, nurturing a comprehensive and encompassing publication venue for urban studies, urban geography, planning and regional development.

The series publishes peer-reviewed volumes related to urbanization, sustainability, urban environments, sustainable urbanism, governance, globalization, urban and sustainable development, spatial and area studies, urban management, urban infrastructure, urban dynamics, green cities and urban landscapes. It also invites research which documents urbanization processes and urban dynamics on a national, regional and local level, welcoming case studies, as well as comparative and applied research.

The series will appeal to urbanists, geographers, planners, engineers, architects, policy makers, and to all of those interested in a wide-ranging overview of contemporary urban studies and innovations in the field. It accepts monographs, edited volumes and textbooks.

More information about this series at http://www.springer.com/series/14773

K. V. Raju • A. Ravindra • S. Manasi
K. C. Smitha • Ravindra Srinivas

Urban Environmental Governance in India

Browsing Bengaluru

Springer

K. V. Raju
ICRISAT Development Centre
International Crops Research Institute for Semi Arid Tropics
Hyderabad, India

S. Manasi
Centre for Research in Urban Affairs
Institute for Social and Economic Change
Bengaluru, Karnataka, India

Ravindra Srinivas
Centre for Sustainable Development
Bengaluru, Karnataka, India

A. Ravindra
Institute for Social and Economic Change
Bengaluru, Karnataka, India

K. C. Smitha
Centre for Research in Urban Affairs
Institute for Social and Economic Change
Bengaluru, Karnataka, India

ISSN 2365-757X ISSN 2365-7588 (electronic)
The Urban Book Series
ISBN 978-3-030-08794-4 ISBN 978-3-319-73468-2 (eBook)
https://doi.org/10.1007/978-3-319-73468-2

Printed on acid-free paper

This Springer imprint is published by the registered company Springer International Publishing AG part of Springer Nature.
The registered company address is: Gewerbestrasse 11, 6330 Cham, Switzerland

Preface

Rapid urbanisation is a reality. It is estimated that by 2050, 64% of the world and 86% of the developed world will be urbanised and three billion people will be living in cities. UN projections indicate that 1.1 billion people will move into cities between 2016 and 2030. More specifically, the pace of urbanisation has been quick in recent years in Asia, Africa and Latin America. In the context of India, 590 million people live in Indian cities, and it has been a challenge to meet the infrastructure demand of the growing populace. India has 68 cities with population greater than one million. Cities in India contribute a significant percentage (58%) of the country's GDP. Thus, urban governance is important because urbanisation has become a powerful force in developing countries. Additionally, environmental governance is all the more relevant given the link between supply and demand for resources to meet growing population in the context of resource scarcity. Concomitantly, growing megacities are increasingly showing signs of deterioration in human quality of life and environmental degradation.

Since the 1980s, the importance of urban areas has been re-evaluated. Throughout the last decade, three elements have become significant: (a) shift in world population from rural to urban areas, (b) re-evaluation of the economic importance of cities and (c) reaffirmation of the significance of local institution building for development. Therefore, it is important to rethink ways of bringing about improvement in the quality of life. The exercise of political, economic and administrative authority in the management of country's affairs at all levels needs to incorporate collective decision-making, include formal and informal sectors, promote participation and representation and integrate involvement at both national and local levels.

In this backdrop, environmental governance becomes most crucial. It is comprised of complex set of mechanisms, processes and institutions through which citizens and groups articulate their interests, mediate their differences and exercise their legal rights and obligations. It is a continuous process through which conflicting and diverse interests may be accommodated and cooperative action may be taken. For instance, as part of environmental governance at the local level, we can see neighbourhood cooperatives have installed and maintained waste recycling plans, regional initiatives of state agencies and the regulation of deforestation.

Several authors have specified that 'good governance' has to be participatory, transparent, accountable, effective and equitable and must promote the rule of law. It should also assure political, social and economic equity by prioritisation based on broad consensus in society. It should entail several debates about decentralisation, focusing on the administrative and political dimensions of devolving power to individuals/groups in civil society. However, there is a gradual shift towards a notion of the 'guiding state' indicating Constructive interaction between the state, the private sector and civil society. Such guiding state is important for openness to new ideas, and responsiveness to citizens enhancing the quality of environmental governance. Responsiveness improves when citizens are well informed and collectively seek better performance from these agencies.

It is therefore, important to delineate and understand 'urban governance' given the distinct challenges of cities and also in the context of the book, since the terminology is used quite broadly. The United Nations Human Settlements Programme (UN-HABITAT) provides the following definition: 'Urban governance is the sum of the many ways individuals and institutions, public and private, plan and manage the common affairs of the city. It is a continuing process through which conflicting or diverse interests may be accommodated and cooperative action can be taken. It includes formal institutions as well as informal arrangements and the social capital of citizens' (UN-Habitat 2001:14). Thus new urban challenges have triggered the evolution of new institutions, of which civil society organisations (non-governmental organisations, resident welfare associations, religious institutions) involved in governance have gained prominence. Further to this urban background, in this book we discuss 'environmental governance' in detail.

Hyderabad, India — K. V. Raju
Bengaluru, Karnataka, India — A. Ravindra
Bengaluru, Karnataka, India — S. Manasi
Bengaluru, Karnataka, India — K. C. Smitha
Bengaluru, Karnataka, India — Ravindra Srinivas

Contents

Chapter 1
Introduction

Abstract Environmental governance has emerged as one of the important dimensions of today's world, including India. In fact, it has become more complex and challenging for both humanity and its governing institution. This may be attributed to various implications caused by human activities across the landscape, both rural and urban. The growing conflict between use and conservation of environment and economic development is not confined to mere debates and academic research. In this respect, the Stockholm Conference on the Human Environment held in 1972 turned out to be a defining moment in terms of the recognition of challenges involved in environmental governance by the global community. Ever since, environmental governance issues got increasingly internalised in the global agenda at all levels. The role of democratic institutions in environmental governance is becoming increasingly recognised in the context of changing urban environment. Hence, policy responses and institutions can take place at different scales, i.e. from global to the national and local or subnational level.

1.1 Environmental Governance

Environmental governance has emerged as one of the important dimensions of today's world, including India. In fact, it has become more complex and challenging for humanity and its governing institution. This may be attributed to various implications caused by human activities across the landscape, both rural and urban. The growing conflict between use and conservation of environment and economic development is not confined to mere debates and academic research. In this respect, the Stockholm Conference on the Human Environment held in 1972 turned out to be a defining moment in terms of the recognition of challenges involved in environmental governance by the global community. Ever since, environmental governance issues got increasingly internalised in the global agenda at all levels. The role of democratic institutions in environmental governance is becoming increasingly recognised in the context of changing urban environment. Hence, policy responses and institutions can take place at different scales, i.e. from global to the national and local or subnational level. We in this book address the issues based on the research that was carried out as a part of a larger study on 'Urban Environmental Governance in India: Browsing Bengaluru' initiated by the Centre for Ecological Economics

K. V. Raju et al., *Urban Environmental Governance in India*, The Urban Book Series, https://doi.org/10.1007/978-3-319-73468-2_1

and Natural Resources (CEENR) and Centre for Research in Urban Affairs at the Institute for Social and Economic Change (ISEC), Bangalore. Hence, urban environmental governance in the context of cities has to be deconstructed to suit the varied environmental issues in city contexts. The central objectives of this book are the following:

1. To understand the new actors, institutional complexity and challenges encountered are influencing urban environmental governance.
2. To capture the need for and evolution of urban environmental governance from global and Indian perspective.
3. To examine the best practices and sustainable city/urban models and approaches from global and Indian perspectives.
4. To explore the role of institutions and governance structures and their influence on the outcomes of urban environmental governance in India.
5. To explore the relationship between urban environment and environmental laws in India
6. To study the trends and status of environmental resource use in Bangalore Metropolitan Area (BMA) and to demonstrate how the local-level institutions and governance structures impact the regulation of urban environmental governance in India.

Thematic areas selected for the study include the following:

1. Governance (history, organisations/agencies, local elected bodies like city corporation in respect of delivery of services)
2. Spatial-demographic dimensions (population growth, density patterns, spatial growth)
3. Basic amenities/services (water supply, sanitation, solid waste management, urban lakes, urban transportation, energy/power supply, etc.)
4. Urban poverty (slums and access to basic amenities)
5. Quality of urban life (air quality, drinking water quality, ground water, etc.)

1.2 The Study Area

The city of Bangalore is one of the fastest-growing million plus cities experiencing a steady increase in population concomitantly spatial growth. The city's growth is driven by cluster of 'IT/BT' industries, favourable climate, high quality of life, cosmopolitan culture and intense international activities. The city has rightly come to be known as 'IT hub of Asia' and 'Silicon Valley of India'. Besides, Bangalore City is reaching threshold status of 'international city' or 'world city'. Unlike other megacities, the city of Bangalore has its share of concerns besides being faced with an immense pressure of serving the immediate day-to-day needs/services of the people such as access to water supply, sanitation, management of solid waste and conservation of fast-diminishing/depleting ecology and biodiversity.

1.3 Methodology

Our study is largely based on extensive review of literature and information, collection of data and holding discussions with officials across local departments in Bangalore City. The data required for the study was gathered from several sources. Firstly, we obtained and collated secondary data on various thematic areas such as governance, institutions and spatial-demographic indicators from available reports and documents such as City Development Report (2006) by BBMP, Revised City Development Plan (2009) by JNNRUM, Revised Master Plan (2015) by BDA, Bangalore City Structural Plan (2030) by BMRDA, ABIDe Plan (2020), etc. Secondly, we collected data from relevant local-level departments such as Bruhat Bengaluru Mahanagara Palike (BBMP), Bangalore Development Authority (BDA), Karnataka State Pollution Control Board (KSPCB), Central Ground Water Control Board (CGWB), BESCOM, Bangalore Water Supply and Sewerage Board (BWSSB) and Lake Development Authority (LDA). The data included compiled sources of annual reports, pilot study reports, yearly statistics and related data. Thirdly, personal interviews were conducted with the officials of the concerned local department to elicit their views on the present trends and status of environmental resource use in Bangalore Metropolitan Area (BMA). In addition, relevant data was collected from the local bodies across sectors on various indicators (such as access, adequacy and quality) and thematic areas for filling the gaps for which information was not available from the above two sources. Finally, the data available on internet with regard to departments at local levels including information on history and origin was collected and collated for the study.

We collected data on governance, spatial and demographic growth of the city from the websites of the city local-level departments such as Bruhat Bengaluru Mahanagara Palike (BBMP), BDA and KUIDFC. The data on drinking water supply was obtained from BWSSB. Similarly, data on land use changes and management of urban lakes was collected from BDA, BBMP and Lake Development Authority (LDA). With respect to power supply for the city, data was collected from BESCOM, Corporate Office, Bangalore. Information regarding solid waste management was obtained from BBMP, Solid Waste Unit.

With the exception of institutions and governance structures in the city, the rest of the data gathered was processed for analysing the critical and not so critical situations impacting environmental governance. The final output is spread across eight chapters based on thematic areas. In most cases, the available data used in the study relates to post-2000. The next section provides details of each chapter and key findings.

1.4 Book Chapters

The second chapter examines different contextual issues such as urbanisation, urban environmental threats and crisis-like situations at both global and domestic levels. The purpose here is to explore how the process of urbanisation and concomitant environmental threats has contributed to the evolution of global and urban environmental governance. The third chapter deals with evidence regarding sustainable models, methods and approaches which are followed and practised across the global/world cities. The fourth chapter is primarily concerned with how the political settings in terms of institutional regimes (rules, regulations, norms and standards) and governance structures influence the environmental outcomes. Further, this chapter addresses linkages between urban environment and the evolution of environmental legislations and legal framework. The legal framework enumerates the gap between legislation and actual implementation. The fifth chapter summarises how the growth patterns of both spatial and demography have carried with them enormous environmental problems inevitably leading to the verge of environmental crisis-like situation for the city. The sixth chapter discusses the current trends and status of environmental resource use in Bangalore Metropolitan Area (BMA). The seventh chapter examines how the dynamics of local institutions and governance structures triangulate and interface affecting the management of environmental resources which is crucial for regulating environmental governance at the local level. Finally, the eighth chapter summarises the study of environmental governance in the Bengaluru metropolis. Further, this chapter presents an action plan for promoting effective environmental governance for the city.

Chapter 2
Urban Environmental Governance: Global Experience

Abstract This chapter is intended to present strategic approaches and an overview of urban environmental governance from global perspective. It provides a brief overview of the evolution of global environmental governance concomitantly the interface with urban transformation. The chapter briefly discusses the complex issues associated with environment from a global economic growth and urbanisation lens. This evolution is related to environmental governance in terms of practice with a broader context of sustainable development which was initially associated with developed countries and later by developing countries. Further, the chapter proceeds to highlight the nexus between 'urbanisation' and 'environment' in terms of how it has drastically changed the quality of urban life in India and argues that environmental issues are inextricably linked to urbanisation.

Keywords Urbanisation · Environmental governance · City · Environmental challenges · Urban poverty · Urban development · Ecological hazards

2.1 Introduction

The world is getting rapidly urbanised. By 2030, particularly, developing countries of Asia and Africa will have more people living in urban areas than rural (UN-HABITAT 2008). Cities are reconfiguring on a massive scale at both spatial and demographic levels in the form of clusters, urban corridors, suburban sprawl and agglomeration. Such an urban transformation drastically impacts the territorial, economic, political, sociocultural and ecological systems in terms of generating negative externalities. Similarly, cities in the Asia-Pacific region constitute not only engines of economic growth but also reflect interconnectivity between global and regional economies. Certainly, they reflect the emerging links between urbanisation and interdependence among cities of the globe besides symbolising new patterns of increased economic activities. Such global and regional convergence can trigger imbalance for urban regional development besides leading to a diffused spatial development by way of prioritising economic centres rather than environmental and ecological concerns.

K. V. Raju et al., *Urban Environmental Governance in India*, The Urban Book Series, https://doi.org/10.1007/978-3-319-73468-2_2

2.2 Need for Urban Environmental Governance: The Global Context

The unprecedented urbanisation process is a remarkable phenomenon of the twenty-first century (UNEP 2002) as reflected in the fact that nearly 82% of the urban population live in the developing countries and that the aggregate annual population increase in the six developing countries' cities like New Delhi and Mumbai (India), Dhaka (Bangladesh), Lagos (Nigeria), Kinshasa (Democratic Republic of Congo) and Karachi (Pakistan) is higher than the entire population of Europe (UNHSP 2013: 25). Some of the fastest-growing cities have emerged in Asia and Africa (Bangalore, Mumbai, Seoul and Tokyo). Not only they differ in size and density, but also their urban development which is mostly unplanned and uncoordinated, hence facing severe social, environmental and ecological crisis. Thus, urban development in the developing countries is most ambivalent underlying the dynamic process of diversification, capital accumulation, specialisation and spatial expansion (Nas and Veenman 1998: 102).

The process of globalisation includes two distinct broad processes: (i) movement of commodities, capital, people and information technology through space referred to as 'deterritorialisation' and (ii) spatial reconfiguration which allows these flows to happen referred to as 'reterritorialisation' (Brenner 1999; Marcotullio 2003: 226). Concomitantly, the cities' physical transformation, as a part of reterritorialisation process, embarks on changes in land use pattern such as creation of (i) 'industrial parks' in major metropolitan cities like Singapore, Taipei, Bangkok and Seoul; (ii) urban corridors like Bangalore-Mysore corridor, Bangalore-Tumkur corridor and IT corridor in Bangalore; and (iii) building special zones or industrial enclaves like special economic zones (SEZs) around Bangalore or export process zones (SPZs) located next to the megacities (Shanghai, Singapore, Hong Kong and Tokyo). Such a territorial convergence inevitably leads to the overlapping of ecological and environmental crisis (Nas and Veenman 1998: 102) like the destruction of ecosystems, spread of diseases, pollution, concentration of energy supplies, climate change and so on. The question is how does the 'global processes' impact the local ecology and environment.

Post-globalisation studies have, in fact, elaborated on the relationship between environmental issues and urban development in the context of developing countries (McGranahan et al. 2001; Marcotullio 2003; UN-HABITAT 2008). As the global economic process deepens and identifies with the 'world-city' formation (Harvey 1985; Marcotullio 2003), such process emphasises international standard of infrastructure development. The cities, therefore, often act as administrative/financial/economic as well as cultural high points with high value-added activities (like business, communication, services, information technology, research, etc.). For instance, the table below (see Table 2.1) provides top 20 cities of the world evenly spread across Asia-Pacific, Europe and the America. These cities are ranked by Global Cities Index (GCI) based on their global engagement in five different areas, i.e.

Table 2.1 Top global cities (ranking for 2014)

Cities	Ranking – 2014	Asia-Pacific cities	Ranking – 2014
New York	1	Jakarta	1
London	2	Manila	2
Paris	3	Addis Ababa	3
Tokyo	4	Sao Paulo	4
Hong Kong	5	New Delhi	5
Los Angeles	6	Rio de Janeiro	6
Chicago	7	Bogota	7
Beijing	8	Mumbai	8
Singapore	9	Nairobi	9
Washington	10	Kuala Lumpur	10
Brussels	11	Bangalore	11
Seoul	12	Beijing	12
Toronto	13	Johannesburg	13
Sydney	14	Kolkata	14
Madrid	15	Istanbul	15
Vienna	16	Cape Town	16
Moscow	17	Chennai	17
Shanghai	18	Tunis	18
Berlin	19	Dhaka	19
Buenos Aires	20	Caracas	20

Source: A.T. Kearney, Global Cities Index and Emerging Cities Outlook (2014)

business activity, human capital, information exchange, cultural exchange and political engagement.

For Asia-Pacific region, three cities are ranked as global cities, i.e. Mumbai (ranked 8th), Bangalore (ranked 11th) and Chennai (ranked 17th). With the emergence of such functional city system, it is clear that global cities are not uniform and well equipped with respect of 'innovation' and 'competition'. Due to the functional uniqueness of global/world cities, they differ in their physical, spatial density and overlapping infrastructure. With the overlapping multilevel jurisdiction (legal, administrative, political, etc.) of the cities, the available socioeconomic opportunities further deeply influence the urban development. Arguing further on the interlinkage between urban development and urban environment, environmental problems are categorised into (i) green and (ii) brown agendas clearly suggesting a shift in the environmental problems from the local to global further threatening the ecosystem (Marcotullio 2003; UNHSP 2009). The categorisation of environmental agendas signifies (i) citizen challenges and (ii) scale of environmental problems. Certainly, the process of global integration of cities has not only increased inequalities and levels of poverty but has severely impacted on the relationship between urban development and environment (UNEP 2002). But shifts in the environmental agendas of developing countries differ from the experience of Western countries. The environmental problems are 'sequential' over a long period of time in Western countries, while they are overlapping and accumulating unfinished agendas in the developing countries (Marcotullio 2003). The reason is technological intervention that has changed the relationship pattern of uneven urban development and environment. Thus, the debate has shifted away from the conceptual and theoretical models of environment towards outcomes such as sustainability across a number of sectors and spatial dimensions.

2.3 Global Environmental Threats and Challenges

The Brundtland Commission report, 'Our Common Future', critically remarked that 'the future will be predominantly urban and the most immediate environmental concerns of most people will be urban one's' (UN 1987: 255; Satterthwaite 2003: 74). Nearly two-thirds of the world population will be living in the cities[1] consuming 80–90% of the global energy for immediate use while contributing 70–80% of greenhouse gas emissions (Otto-Zimmermann 2011). Similarly, world population, now at over six billion, is expected to reach around nine billion by 2050.[2] Nearly half of the earth's population is living in cities (Goffman 2005). Population growth is mostly concentrated in the developing world, particularly in Asian region which is estimated to home to more than 50% of the global urban population by 2050 (State and Outlook 2010).

Rapid urbanisation puts severe pressure on access and quality of services available to its population. For example, it is estimated that 80% of the global greenhouse gas emissions originate from cities (UNEP City Alliance Programme). The interface between cities and urban development produces and reproduces environmental and ecological impact at global, national and regional/local scale which include climate change, different kinds of pollution, loss of biodiversity and destruction of sensitive ecological systems altering not only natural features such as erosion and loss of habitat of species but also straining access to and quality of urban services like water, sanitation, land, fuel, transportation and other resources for their survival (see Table 2.4). Such phenomenon is generally referred to as 'urban footprint' (UNESCAP 2005; Satterthwaite 2003). In addition, the urban sprawl has a damaging effect on environment and ecology at the local level. Lack of effective urban management and planning initiatives and incoherent urban policies accentuate the environmental problems. Governing institutions, legal systems and political will are seldom reformed to the changing urban scenario (Hardoy et al. 2001).

Such global environmental challenges are critically linked to urban poverty and sustainability of cities. This ultimately further increases the pressure on the local

[1] 3.5 billion people at present live in cities, and by 2055 an estimated 75% of the world population will live in urban areas. Cities occupy 2% of the earth land accounting for over 70% of both energy consumption and carbon emission (http://www.sustainablecities2013.com/).

[2] By 2025, there will be 37 megacities with populations of over ten million; 22 of those cities will be in Asia (www.sustainablecities2013.com).

Table 2.2 Environmental disasters across major cities of the world

	Population in millions	Earthquake	Volcano	Storms	Tornado	Flood	Storm surge
Tokyo	35.2	X	–	X	X	X	X
Mexico City	19.4	X	X	X	–	–	–
New York	18.7	X	–	X	–	–	X
São Paulo	18.3	–	–	X	–	X	–
Mumbai	18.2	X	–	X	–	X	X
Delhi	15.0	X	–	X	–	X	–
Shanghai	14.5	X	–	X	–	X	X
Kolkata	14.3	X	–	X	X	X	X
Jakarta	13.2	X	–	–	–	X	–
Buenos Aires	12.6	–	–	X	–	X	X

Source: UNHSP (2009: 39)

environment and ecosystems, especially their capacity to promote access[3] to basic amenities (State and outlook 2010). Cities are often prone to various kinds of environmental and ecological disasters like earthquake, volcanic eruptions, storms, tornados and flood and storm surge (see Table 2.2). There has been considerable threat to climate in terms of 60% of degraded ecosystems, loss of 35% of mangroves and destruction of 20% of world's precious coral reefs. Similarly, a potential threat to the loss of natural species has increased by 100%. It is estimated that each year 150,000 sq. kms of forest is lost. Similarly, one out of every four households is living in poverty especially in Asia and Africa (see Table 2.3). Nearly 65% of cities in the developing countries do not treat their waste water resulting in various environmental and ecological hazards within household, at neighbourhood level, even at workplace, at the city/municipality levels, at city region or periphery region as well as in connected linkages between city and global levels.[4]

With increasing urbanisation, cities in Asia are increasingly prone to threats from climate change. Similarly, exploring the impacts of climate change on metropolitan cities is gaining prominence due to rapid changing urban environment. Unmet demands of growing urban population in terms of lack of access to water, sanitation, energy has eventually exposed urban poor being the most exposed and hence, highly vulnerable to the present unique challenges. The impact can manifest in terms of direct or indirect physical, social, economic and health-related problems. Some include (i) rise in sea levels, (ii) tropical cyclones and (iii) heavy precipitation leading to urban floods and landslides, heavy heat islands and drought (Table 2.4).

[3] As per the Food and Agriculture Organisation (FAO) of UN, demand for food supplies, feed and fibre would grow by 70% by 2050 (State and Outlook 2010: 142).

[4] Please see Annexure Table 2.7 for more details on the classification of environmental and ecological hazards suffered by cities of the world.

Table 2.3 Global environmental issues/threats

Global environmental threats	Environmental threats
Global climate change	60% of the ecosystems are either being degraded or used unsustainably 35% of the world's mangroves have been lost 20% of the world's precious coral reefs have been destroyed Species extinction rate is still 1000 times higher than what would be occurring naturally An estimated 90% of the total weight of large predators in the oceans – such as tuna, sharks and swordfish – have disappeared Loss of 150,000 square kilometres of forest each year In India, the rise in air pollution in cities such as Bangalore outpaced even those in China between 2002 and 2010, according to a report by *Time* magazine citing a Tel Aviv University study (www.sustainablecities2013.com) The *Wall Street Journal* reported that in the Indonesian capital of Jakarta, particulates, carbon monoxide and nitrogen dioxide increased by between 40% and 85% in 2011 (www.sustainablecities2013)
Urban poverty	One out of every four households live in poverty 40% of African urban households and 25% of Latin American urban households live below poverty line
Urban waste water	65% of cities in the developing countries do not treat their waste water
Greenhouse gas emission	While large cities of the world consume 75% of the world energy contribute, 75–80% of the heat-trapping greenhouse gases are released into our atmosphere affecting the climate Carbon dioxide from fossil fuel use accounted for only 57% of the global anthropogenic greenhouse gas emissions in 2004 In 2004, agriculture and gases released from land use changes and forestry accounted for 31% of greenhouse gas emissions In 2004, carbon dioxide emissions contributed around 77% of the global anthropogenic greenhouse gas emissions (methane contributed 14.3%, nitrous oxide 7.9% and fluorinated gases 1.1%) 20–25% of carbon dioxide emission is caused by small and rural areas from developed countries Cities probably emit between 30% and 40% of all anthropogenic greenhouse gas emissions While lower-middle income nations contribute 53.2 and high-income nations contribute 29.1% of GHE between 1980 and 2005

Source: Satterthwaite (2008: 239–240), and Satterthwaite (2009: 258)

Table 2.4 Impact of climate change upon urban areas

Ranking by population exposure	Ranking by value of property and infrastructure assets exposure
Kolkata (India)	Miami (USA)
Mumbai (India)	Guangzhou (China)
Dhaka (Bangladesh)	New York (USA)
Guangzhou (China)	Kolkata (India)
Ho Chi Minh City (Vietnam)	Shanghai (China)
Shanghai (China)	Mumbai (India)
Bangkok (Thailand)	Tianjin (China)
Rangoon (Myanmar)	Tokyo (Japan)
Miami (USA)	Hong Kong (China)
Haiphong (Vietnam)	Bangkok (Thailand)

Source: UN-HABITAT (2011: 71)

2.4 Evolution of Global Environmental Governance (GEG)

Broadly, 'environmental governance' has been defined in terms of (i) institutions and 'processes' and (ii) outcomes. From an institutional perspective, 'environmental governance' has been defined as 'the establishment, reaffirmation or change of institutions to resolve conflicts over environmental resources' (Paavola 2005). Similarly, Najam et al. (2006) define global environment governance 'to the sum of organisations, policy instruments, financing mechanism, rules, procedures and norms that regulate global environmental protection'. A study by deLoe et al. (2009) defines 'environmental governance' as a specific form of broader 'governance' and refers to processes and institutions through which societies make decisions that affect the state of environment. Thus, environmental governance has been understood as offering institutional solutions for solving environmental crisis or challenges. The institutions can be in the form of 'policies, financial mechanisms, rules, norms, regulations, distributive and procedural justice' which influences the governance outcomes. Diverse studies have looked into environmental governance in the context of 'process' issues such as community participation, participation of stakeholders including the indigenous communities and minorities in the natural resource management, access to information, capacity building, transparency, accountability and so on. Hardin's (1968) analysis of 'tragedy of commons' highlights appropriation and overexploitation of natural resources due to increasing interdependence. Ostrom (1990) and Ostrom et al. (1994) highlight the successful community governance of common-pool resources (such as fisheries, pastures and groundwater reservoirs) by agreed-upon rules and regulations without government policy intervention to overcome the 'tragedy of commons'. The study by Ostrom (2000) provides an overview of the evolution of social norms and how they enable collective action. Similarly, 'good environmental governance' not only leads to environmental goals, such as conservation and sustainable development, but also how decisions are taken to achieve environmental goals (Jeffrey 2005).

Through the years, Global Environmental Governance has grown both in size and scope by evolving through a number of global/international/multilevel conventions/ treaties/ agreements that guide the process of governance with regard to environmental and ecosystem decline/threats.[5] A series of multilateral environmental agreements (MEAs) have provided the basis for an international environmental regime and regulation. These conventions and treaties are implemented with the help of small organisations called secretariats. Major international organisations such as World Bank, UN-HABITAT, UNEP and ADB and many regional associations/organisations have made substantial efforts towards promoting sustainable development in urban areas. The collective international environmental institutions are composed of organisations/NGOs/civil society and regional organisations, including private sector, multi-national corporation and business community. Major institutional decisions on environmental policy came out during the United Nations Conference on the Human Environment (Stockholm 1972 conference), constituted as a part of the United Nations Environment Programme (UNEP). Some of the major global environmental initiatives[6] such as 'The Rio Earth Summit 1992 and Johannesburg Summit on Sustainable Development 2002' are considered landmark summits on the evolution of GEG system (Najam et al. 2006). Thus, the evolution of global environmental governance is located and traced within a broader context of sustainable development.

The international conventions, treaties and agreements manifest into 900 multilateral environmental agreements (MEAs) (Wingqvist et al. 2012). The question of implementation of these mandates and its effectiveness is measured not simply in terms of ecology and environment but also in terms of 'development' and outcomes like 'equity' which is embedded within the concept of sustainable development (Nanjam 2005). Yet, most of the agreements and treaties are declaratory in nature. And many environmental regimes are in terms of rules and norms created by non-environmental organisations like World Trade Organization, the Millennium Development Goals (MDGs), lending policy safeguards of the International Finance Corporation (IFC), major private banks and so on. In addition to the agreements and treaties, multiple sources of funding through donor aid flow, international financial organisations, UN agencies and international NGOs for environmental governance were given to the national governments across developing countries (Najam et al. 2006). Too many agreements, actors and resources have inevitably led to fragmentation and inefficiency (Najam et al. 2006). Similarly, many global-level reform[7] initiatives were undertaken for the promotion of global environmental governance, which were launched under the guidance of UN and international NGOs.

[5] See Annexure Table 2.10 on environmental threats across different regions of the world.

[6] Refer Annexure for Table 2.7 which provides a glimpse of the global environmental governance through treaties, conventions and organisations.

[7] Refer Annexure for Table 2.8 on the evolution of global reform initiatives with respect to global environmental governance.

Despite the pervasive nature of GEG encompassing various dimensions of environmental governance, it is ineffective in controlling or alleviating global environmental concerns and damages. GEG is yet to get standardised and is still evolving in nature. Some of the defects identified with respect to GEG are (i) complex international regimes[8] which lack cooperation and coordination for effective implementation or enforcement, (ii) ineffective use of resources, (iii) multiplication of civil societies and NGOs both at the national and international levels, (iv) institutional fragmentation and, (v) finally, international courts and tribunals. Such an institutional fragmentation leads to conflicting agendas, duplication of work, geographical dispersion and inconsistency in rules and norms (Najam et al. 2006). One of the strongest criticisms voiced with respect to GEG has been its high dependency on voluntary funding and a short-term implementation of projects. In addition, the signatory nations have completely failed to integrate environmental stipulations into their planning and governance processes. Therefore, a plethora of multilateral and bilateral funding and private funds coupled with a variety of financing instruments (grants, loans, guarantees, technology transfer, etc.) have inevitably led to administrative and institutional fragmentation for the recipient countries (Wingqvist et al. 2012).

2.5 Urban Environmental Governance in India: Issues and Challenges

Indian cities[9] have for long witnessed a steady increase in urban[10] population,[11] i.e. from 27% in 1901 to 38% in 2001 (MoEF 2011; Vishwanath et al. 2013) and, similarly, from 13% in 1900 to 49% in 2005 which is projected to escalate to 60% by 2030 (Bharath et al. 2012). There are 48 urban agglomerations/cities having a population of more than one million in India (in 2011) (ibid). Thus, urbanisation[12] in India is characterised by 'unplanned and unmanageable growth' leading to urban sprawl (Pandey et al. 2006) and an exponential growth of informal or slum[13] settle-

[8]There are more than 500 MEAs registered with the UN, including 61 atmosphere-related; 155 biodiversity-related; 179 related to chemicals, hazardous substances and waste; 46 land conventions; and 196 conventions that are broadly related to issues dealing with water (Najam et al. 2006: 30).

[9]According to 2011 census, 90 million people have been added to Indian urban areas since the previous census in 2001 (Vishwanath et al. 2013: 15).

[10]Indian cities as drivers of economic growth contribute 60% of national income (Pandey et al. 2006: 208).

[11]Indian urbanisation pattern is often referred to as 'agglomeration economies'. As per the Agglomeration Index, Indian urbanisation has reached to 52% (ibid: 24).

[12]Total urban population has increased more than ten times from 26 million to 285 million, an increase of 28% by 2001. An increase from 23% in 1991 to 65% in 1991 is found in respect of Class I cities in India (MoEF 2009: 134). Similarly, as per 2001 census, there are 35 million plus cities in India (ibid 136).

[13]Total slum population in the country is 40.3 million comprising 22.6% of the total urban popula-

ments.[14] Such settlements face a high risk of health hazardous such as tuberculosis, malaria, dengue, cholera, typhoid and plague adversely impacting the environment (MoEF 2011). The trends of urbanisation such as 'agglomeration, conurbation, sub-urbanisation, peri-urbanisation and urban sprawl' have adverse ramifications such as spatial and vehicular density and growing demand for energy and food. The changing land use pattern, particularly, in metropolitan cities of India has a considerable impact on local environment. Thus, since the last three decades, the interface between the 'process' of urbanisation and environment has seriously impacted the quality of urban (Maiti and Aggrawal 2005). Moreover, environmental governance in India suffers from poor urban planning and command-and-control type of environmental management (Pandey et al. 2006).

Environmental deterioration has been closely linked to unsustainable production and consumption patterns (MoEF 2011). Consequently, access to and quality of basic urban services are very poor and do not match the rapid economic growth scenario of India cities (ibid). Various environmental problems plague India,[15] particularly metro cities (coastal), which are vulnerable to cyclones and annual monsoon floods, rapid population growth, increase in household consumption, industrialisation, poor access to infrastructure and unequal distribution of resources (Anand 2013). Most often, unplanned urbanisation in India has led to specific environmental and ecological impacts such as shortage of housing, worsening of water quality, various types of pollution, problems associated with disposal of waste and hazardous wastes most common in metro cities like Mumbai, Kolkata, Chennai, Delhi, Bangalore, Kanpur, Hyderabad and so on (Maiti and Aggrawal 2005; MoEF 2009; Sridhar and Kashyap 2012).

2.5.1 *Major Urban Challenges in India*

Indian urbanisation is often referred to as an 'agglomeration economy' pushed by various external and internal factors such as urban-rural migration, concentration of industrial/IT/BT clusters, creation and implementation of large-scale infrastructure projects and creation of special economic zones (SEZs) that have eventually spurred a spatial and territorial transformation seriously posing threats to environment and ecology. Some of the challenges of Indian urbanisation include the following:

---

tion of cities or towns (MoEF 2009: 140).

[14] The United Nations Human Settlements Programme (UNHSP) states that 43% of urban residents in the developing countries like Brazil and India live in slums.

[15] On 2.4% of land area, India sustains 16.7 of the world population exerting a tremendous pressure on its natural resources (MoEF 2011).

2.5.1.1 Cities as Engines of Economic Growth

More than 90% of the world's urban population growth is taking place in the developing countries coupled with increasing number of largest cities[16] (UN-HABITAT 2010: 4; UNHSP 2011: 2). It is estimated that 'half of the world's megacities (12 out 21) are now in Asia. Similarly, seven out of ten most populous cities of the world are now in Asia (Tokyo, New Delhi, Mumbai, Shanghai, Kolkata, Dhaka and Karachi). Irrespective of the nature of countries (high/middle/low income), cities in Asia[17] have recorded the highest growth rate for the past two decades, contributing as much as 30% of the global economic output (in 2008). Asian cities are most often referred to as 'agglomeration economies', 'factory of the world' for international financial centres and 'knowledge economies' (ibid). In Asia two most significant reasons for registering a high economic growth are (i) migration (rural-urban or urban-urban) and (ii) reclassification of 'rural' to 'urban' (UN-HABITAT 2010). A high economic growth rate has led to an increased demand for better infrastructure and resources (physical and basic amenities) particularly land. In fact, the entire process is a defining feature of the rapid urbanisation pattern in Asia. One of the most threatening features of India's urbanisation process is population/demographic explosion in cities[18] (Maiti and Aggrawal 2005). The total urban population has increased from 26 to 285 million.[19] Concomitantly, a vast majority of migrant population has settled in slums and informal settlements of India's metro cities, such as New Delhi, Kolkata, Mumbai and Chennai (see Table 2.5). The total slum population in India has recorded an increase of 41% in million plus cities.

This unabated urbanisation has unleashed serious environmental problems in terms of inadequate housing, spatial density, lack of access to basic services,[20] excessive pollution, degeneration of nonrenewable resources and an increase in informal activities. Indeed, urbanisation has propelled the urban territorial restructuring in terms of 'agglomeration, urban corridors, conurbation, special zones and suburbanisation' instigating transition in the land use management. Besides, megacities or million plus cities do not adequately practise sustainable measures, thus grossly affecting the local environment. Therefore, the capacity of local governance

[16] The number of million plus cities has drastically increased from 75 in 1950 to 447 in 2011, while simultaneously there is an increase in the average size of the world's 100 largest cities from 2.0 to 7.6 million. By 2020, it is projected that there would be 527 cities with a population of more than one million (UNHSP 2011: 2).

[17] Asia constitutes the second largest urbanised region with 42.2% of the population living in urban areas which is slightly more than Africa's 40% (UN-HABITAT 2010: 6). It is further estimated that between 2010 and 2020, a total of 411 million people will be added to Asian cities or 60% of the growth in the world's urban population (ibid).

[18] The number of million plus cities has increased from 23 in 1991 to 35 as per 2001 census. Population growth is recorded from 19% in 1951 to 33% in 1991 (Maiti and Aggrawal 2005: 279).

[19] There is a continuous increase in urban population from 11% (in 1901) to 17% in 1951 to 28% in 2001. Similarly, nearly 60% of the urban population live in Class I cities (Maiti and Aggrawal 2005: 278).

[20] See Annexure Table.

Table 2.5 Growth of slum population in the four metropolitan cities in India

Metropolitan cities	1981(%)	1991(%)	2001(%)
Greater Mumbai	30.8	43.2	48.9
Kolkata	30.3	36.3	32.6
New Delhi	18.0	22.5	18.9
Chennai	13.8	15.3	17.7

Source: Maiti and Aggrawal (2005): 281

in terms of planning, management and governance is part of an urban growth crisis triggering 'negative externalities' on environment and ecology.

2.5.1.2 Rural-Urban Migration

One of the chief characteristics of India's urbanisation is rural-urban migration.[21] Some of the major cities or million plus cities of India like New Delhi, Mumbai, Bangalore,[22] Chennai and Kolkata are experiencing an increase in migration from rural to metropolitan cities. The percentage of urban population in India[23] (million plus cities) have risen from 6% in 1901 to 19% in 1951 and further to 33% in 1991 (Maiti and Aggrawal 2005: 280; Sridhar and Wan 2014). A high percentage of population (68.7%) is concentrated in Class I cities[24] of India leading to the deficiency of urban basic services (Kundu 2006). A vast migrant population[25] is attracted to urban centres in search of jobs or income-generating opportunities due to expanding infrastructure facilities and access to better basic services. They inevitably settle in temporary or informal dwellings resulting in the swelling of slums in metro cities. According to 1991 census, slum population in India swelled to 41% residing in the million plus cities (ibid: 281; MoEF 2009: 138; Khan et al. 2011). Such a dramatic increase in migrant population[26] has exerted an adverse impact on the environment as this increase causes varying degrees of land degradation and inappropriate land management and unsustainable practices such as land shortage, encroachment, insecure land tenure and poverty. The impact is irreversible in terms of both spatial and temporal, particularly the loss of local ecosystems (in the form of physical

[21] As per Census 2001, the share of rural-urban migration population constituted 16.4% in India. Especially, Karnataka accounts for 11.9% of rural-urban migration.

[22] As per Census 2001, there is an upward trend of migration in Karnataka between 1991 and 2001 which is 34%. Rural-urban migration is basically labour migration (Roychowdhury et al. 2012: 13–16).

[23] India constitutes one of the ten megacentres of biodiversity (Singh 1995: 57).

[24] The number of Class I cities has increased from 24 in 1901 to 393 in 2001 (Kundu 2006: 29).

[25] 'Migration can be seen as livelihood and income diversifying criteria'. Economic reasons constitute one of the major factors for rural-urban migration in search of income-generating opportunities in metro cities (Roychowdhury et al. 2012: 31–33).

[26] As per Census 2001, 36 million intrastate migrations to urban centres have been observed (MoEF 2009: 139).

changes like pollution/encroachment of lakes, tanks, urban floods, urban heat islands, climate change and so on.).

2.5.1.3 Unplanned Urban Development

Urban centres constitute hub of 'economic activities', because of rapid industrialisation and migration resulting in a huge population. Nearly 50% of the urban population comprises migrants (MoEF 2009) who resort to excessive consumption of untapped natural resources inadvertently producing pollution and illegal waste in the process. However, such an increase in per capita resource consumption makes India highly susceptible to environmental degradation (Singh 1995). These problems are associated with unplanned development and unabated urban growth.[27] Such an unregulated high urban growth rate in India, particularly over the last two decades, has led to a skewed access to and quality of urban basic services.[28] The problem of poor management is usually associated with a contagious outgrowth of cities' high urban spatial density and proliferation of unplanned settlements/slums[29] which largely do not have access to an adequate water supply, sanitation, housing, waste disposal or electricity. The problem is further compounded by inadequate resources/finances of urban local bodies to cater to the growing demand for services and infrastructure. Any combination of these factors gives rise to urban health problems and new disease patterns mainly due to unhygienic living conditions, pollution, inadequate access and malnutrition. The concentration of unplanned settlements like urban slums leads to the spread of communicable and infectious diseases such as tuberculosis, hepatitis, dengue, malaria, pneumonia, etc. (Pandey et al. 2006: 211; Bhandari 2006; MoEF 2009).

2.5.1.4 Changing Land Use Pattern

An indiscriminate urbanisation process has a tremendous impact on the urban land use pattern. While promoting 'world-city' infrastructure in Indian metropolitan cities, infrastructure projects on a massive scale have been implemented (such as metrorail, urban corridors, ring-roads, IT/BT industrial zones, special economic zones (SEZs) and so on in addition to huge commercial and residential complexes bringing about tremendous changes in the land use pattern. In addition, vast tracts of vacant land are occupied or absorbed by the slums (essentially migrant population) while further getting relocated to urban peripheries or marginalised land

[27] As per Census 2001, 27.8% of Indian population lives in urban centres (MoEF 2009: 134).

[28] Refer to Table 2.11 for more details on the status of urban basic services in India.

[29] The estimated slum population has increased from 46.26 million in 1991 to 61.82 million in 2001 showing a growth of 15.56% (CSO 2011: 123). The NSSO data for July 2008 to June 2009 reveal that 49,000 slums exist in urban areas of India (including both notified and non-notified) and they have increased to 50.6% for 2002 (ibid: 123).

contributing to city segmentation (Kundu 2006). For example, half of the city population in New Delhi and Mumbai lives in unauthorised areas (Toutain and Gopiprasad 2006). Such development and redevelopment projects cut across geographical boundaries in the form of continuous and discontinuous urban sprawls which pass through administrative jurisdictions of many municipalities and villages. The increasing population and competing demands for land have resulted in a significant decline in the per capita availability of land from 0.89 ha in 1951 to 0.3 ha in 2001 (MoEF 2001, 2011).

Delivery of services is also affected by the segmentation of urban areas across metropolitan cities in India more than 70% of non-notified slums do not have access to basic amenities (CSO 2011). The proliferation of multiple agencies for land use through conflicting legislations, without clear defined role complicates the protection of environment. Moreover, the emergence of master plans (such as city development plan, comprehensive development plan, JNNURM and so on) for addressing the urban issues has never prioritised local environment and ecology. Urban land is often tagged as 'real estate value' with a high priority given to the 'economic' value of land while improving their current or future infrastructure and functioning capacity but completely neglecting the environment and biodiversity of the city (Toutain and Gopiprasad 2006). The city-level zonal regulations, by-laws and norms are barely followed for all types of development projects, subsequently resulting in land encroachments or illegal occupation. There is no coherent policy addressing the issue of interlinkage of urban land use and local development which presuppose the integration with environment parameters (such as water bodies, spatial, biosphere, climate factors, power, solid waste, waste management, transport). This offers a fresh challenge to the ecosystem and nonrenewable resources in terms of disintegrating the balance between urbanisation and preservation of environment in India.

2.5.1.5 Urban Informal Economy

'Informal economy'[30] has been the key feature of Asian cities (UN-HABITAT 2010: 87). In India 86% of the total workforce is employed in the informal sector (between 2004 and 2005) (Naik 2009; Chandrasekhar and Ghosh 2013). The presence of the informal sector is characterised by the dynamics of urbanisation process (UN-HABITAT 2010). Particularly, the share of the informal workers is highest in the states of Andhra Pradesh, Himachal Pradesh, Karnataka and Madhya Pradesh in India (Naik 2009). The participation of women in informal jobs has increased substantially in Asian cities, particularly, in the form of 'invisible' jobs, i.e. menial jobs like domestic labourers, piece-rate workers and assisting to small

[30] Often 'informal sector' or 'unorganised sector' has been used synonymously. Those employed in the informal sector do not have job security or social security benefits.

family enterprises with low wage rates without social security to the agony of women besides contributing to the deterioration of health and poverty (Kundu 1999; Roychowdhury 2004; UN-HABITAT 2009). As informal workers[31] do not enjoy job or income security (low income) and lack of secure property rights, as such they are excluded from or have inadequate access to basic amenities or infrastructure facilities in cities. Most of them live in extremely unhygienic conditions, and the condition has further deteriorated with the segmentation of cities into formal and informal settlements further accentuating the situation (Kundu 1999). Therefore, there is a significant correlation between informal sector employment and an increase in the incidence of urban poverty (Kundu 1999; Naik 2009; UN-HABITAT 2009).

2.5.1.6 Urban Poverty and Environment

Urban poverty[32] is one of the major problems in Asian cities[33] (UN-HABITAT 2009). It is estimated that at least a billion urban dwellers[34] have a very poor access daily needs, often in temporary shelters/shacks and overcrowded houses, often leading a life of very poor quality termed as 'slums/informal settlements' (Satterthwaite 2003; Satterthwaite et al. 2011). The conditions of slums in metropolitan cities of India are very deplorable (see Table 2.6) as these settlements generally occupy either vacant or private lands without a secured tenancy, again termed as 'illegal', which does not authorise them to access basic services (such as water, toilets, drainage, waste collection, electricity supply or housing) (see Table 2.6). Between 1990 and 2008, the shares of urban population in terms of access to safe drinking water[35]

[31] Informal sector jobs constitute (i) daily wage in construction, rickshaw pulling, hawking and street vending, jobs in textile/garment sector, carpet making, agarbathi/cigarette/beedi making or garbage collection.

[32] UN-HABITAT (2009: 109) defines poverty as one of the social exclusionary approaches which refer to the 'phenomenon whereby individuals or groups are unable to fully participate in political processes'. In India, particularly, poverty is measured in terms of consumption and levels of income. The study by Satterthwaite (2003) provides a broader definition of poverty by including eight parameters which include (i) inadequate income, (ii) risky assets, (iii) inadequate shelter, (iv) inadequate provision of public infrastructure, (v) inadequate provision of basic amenities, (vi) no safety net to ensure basic consumption, (vii) inadequate protection of rights of the poor and (viii) powerlessness and voicelessness of the poor and lack of means to ensure accountability from donors, public agencies and NGOs.

[33] Satterthwaite et al. (2011). Engaging with the urban poor and their organisations for poverty, reduction and urban governance, an issue paper for the United Nations Development Programme.

[34] Most of the slum dwellers/informal settlements are located near drainage channels, under bridges, near open waste sites or low-lying areas.

[35] It is estimated that over the next 20 years, the global demand for water will increase by 40% while by more than 50% in the developing countries (State and Outlook 2010: 143). Similarly, an average annual investment would amount to USD 772 billion for water and waste water management around the world by 2050 (ibid).

Table 2.6 Environmental conditions in urban slums of India

Environmental conditions	Urban poor NFHS* 2	Urban poor	Urban nonpoor	Overall urban	Overall rural	All India
	2000	2005–2006				
Households with access to piped water supply at home (%)	13.2	18.5	62.2	50.7	11.8	24.5
Households with access to public tap/ hand pump for drinking water (%)	72.4	72.4	30.7	41.6	69.3	42.0
Household using a sanitary facility for the disposal of excreta (flush/pit toilet) (%)	40.5	47.2	95.9	83.2	26.0	44.7
Median number of household members per sleeping room	3.5	4.0	3.0	3.3	4.0	3.5
Infectious diseases						
Prevalence of medically treated TB (per 100,000)	535	461	258	307	469	418
Prevalence of HIV among adult population (age 15–49) (%)	–	0.47	0.31	0.35	0.25	0.28

*National Family Health Survey
Source: UNHSP (2013: 128)

declined by between 3% and 12% in Bangladesh, Indonesia, Myanmar and Nepal (UNHPS 2013: 143). A study by Sridhar and Kashyap (2012) reveals that Kolkata has the highest density of slums followed by Chennai. Although New Delhi has the highest number of slums in absolute numbers, the city of Mumbai supports large number of slum dwellers (a staggering 57.7%).

Studies clearly point to the fact that there exists a strong evidence to prove that the urban environmental crisis is a major contributory factor to urban poverty in Asia and African cities (Satterthwaite 2003). Although the positive role of slum or urban poor in the protection of local ecology is recognised (as waste pickers, recyclers or reclaimers of waste from domestic or commercial/industries), they suffer from multiple deprivations that include unsure jobs vs low income, unsure housing, lack of access to amenities and infrastructure and so on. Most often the urban poor residing in informal settlements face serious environmental hazards like urban floods, homelessness and health-related issues that significantly contribute to poverty. Dimensions of the urban poverty-environmental nexus include (Satterthwaite n.d.; 2003):

(i) Inadequate access to basic amenities: Most of the urban poor living in informal settlements do not have access to drinking water and sanitation facilities; they consequently suffer from innumerable health issues and diseases such as dengue, malaria, cholera, TB and so on.
(ii) Occupational hazards: As most of the urban poor are employed in the informal sector, they are mostly exposed to various kinds of pollution, and particularly

those employed in industrial/energy or transport sectors suffer from various kinds of health problems.

(iii) Urban vulnerability to natural disasters: Since unabsorbed populations from slums live mostly in unhygienic or congested or substandard settlements, they are often victims of natural disasters like floods, rising heat/temperatures and earthquakes and subsequently are exposed to epidemics, premature deaths and injuries due to accidents in cities.

The problem is further accentuated by global warming (Satterthwaite 2003), structural issues and apathetic institutions underpinning poverty, for instance, who fight against polluter industries, donors and sometimes government. Therefore, a range of environmental crisis faced by the urban poor in urban areas overlap socio-economic and political factors (ibid). In addition, four kinds of environmental degradation have been identified with respect to urban development including (i) high use or waste of nonrenewable resources, (ii) high use of renewable resources, (iii) high levels of biodegradable waste generation and (iv) generation of high levels of non-biodegradable emissions (Satterthwaite 2003).

2.6 Conclusion

The obvious pressure and the resultant devastating effect on the environment are most pronounced due to unabated urbanisation in the twenty-first century not only in respect of the developed but also developing countries. The inevitable linkages between the urbanisation process and environment and ecology continue to bother the present generation as well as the generations to come. Therefore, the system of environmental governance has gained significance within a broader framework of sustainability.

Due to rapid urbanisation, the metropolitan cities of India are facing daunting task of accommodating the needs of people on the one hand and addressing the challenges posed by environmental degradation on the other. In these conditions, how do the world cities cope up with environmental crisis unleashed by urbanisation? Are there any tangible efforts made towards promoting sustainable cities? Therefore, Chap. 3 provides a glimpse of sustainable city models across the world including India (Tables 2.7, 2.8, 2.9, 2.10, and 2.11).

Table 2.7 Global environmental governance: treaties/conventions/organisations

Conference reports	
Early 1970s	*UN Conference on the Human Environment, 1972, Stockholm* Discussed environmental degradation and transboundary pollution, began United Nations Environment Programme *Vancouver Declaration on Human Settlements, 1976* Recognises the serious condition of many human settlements and recommends strengthening international cooperation, particularly regarding the basic needs in developing countries
1980s	The Brundtland Report (Our Common Future), 1987 Defined sustainable development as 'development that meets the needs of the present without compromising the ability of future generations to meet their own needs'
1990s	*United Nations Conference on Environment and Development (UNCED): Earth Summit*, June 1992, Rio de Janeiro Established Agenda 21, established the term 'Sustainable Development' *Agenda 21*, 1992 A global blueprint for sustainable development on national, regional and local levels that has become the basis for many plans *International Conference on Population and Development*, 1994, Cairo Discussed relationships between population, sustained economic growth and sustainable development; affirmed right to education *Earth Summit +5, New York*, 1997 Reviewed and appraised the implementation of agenda
2000s	*The World Summit on Sustainable Development (Rio + 10), Johannesburg 2002* Refocused attention on Rio commitments and Agenda 21. Worked on access to safe water, proper sanitation and clean energy, as well as reversing ecosystem decline *International Treaty on Plant Genetic* Resources *for Food and Agriculture, 2004* Provides for conservation and sustainable use of plant genetic resources for food and agriculture and the sharing of derived benefit
Treaties, Conventions and Organisations	
1970s	*Antarctic Treaty, 1959* Ensures 'in the interests of all mankind that Antarctica shall continue forever to be used exclusively for peaceful purposes and shall not become the scene or object of international discord' *Ramsar Convention on Wetlands, 1971, Iran* Provides an international framework for the conservation and use of wetlands and their resources; emphasises wildfowl habitat
	Convention Concerning the Protection of the World Cultural and Natural Heritage, 1972, Paris Sets up an international committee to protect historical and natural sites, requires an inventory of endangered world heritage sites. Recognises that nature and culture are complementary *Convention on International Trade in Endangered Species of Wild Fauna and Flora (CITES), 1973, Washington* Enforces international trade in wild animals and plants; establishes global list of endangered species *Convention on Long-Range Transboundary Air Pollution, 1979, Geneva* Combats acidification on a broad regional basis, brings together research and policy. Has been extended by eight new protocols

(continued)

Table 2.7 (continued)

1980s	*UN Convention on the Law of the Sea, 1982, Montego Bay (not fully ratified)* Develops principles from the 1970 resolution that the seabed and ocean floor, beyond the limits of national jurisdiction, are the common heritage of mankind *Convention on the Protection of the Ozone Layer, 1985, Vienna* Encourages research and cooperation; set a precedent for early response to environmental problems *Montreal Protocol on Substances that Deplete the Ozone Layer, 1987* Protects the ozone layer by controlling total global emissions of substances that deplete it, particularly chlorofluorocarbons *Convention on the Control of Transboundary Movements of Hazardous Wastes and Their Disposal, 1989, Basel* Encourages disposal of hazardous wastes within country of origin; provides for waste reduction and disposal
1990s	*United Nations Framework Convention on Climate Change, 1992, New York* Recognises that global warming is a problem and sets an objective of stabilising greenhouse gas emissions, requires regular inventories of such emissions and places heaviest burden on industrialised countries *Convention on Biological Diversity, 1992, Rio de Janeiro* Regulates the conservation and sustainable use of biological diversity, the equitable sharing of genetic resource benefits and appropriate transfer of relevant technologies *Kyoto Protocol to the United Nations Framework on Climate Change, 1997, Kyoto (entered into force 2004 but never ratified by the USA)* Sets targets on greenhouse gas emissions. See CSA's Global Warming and the Kyoto Protocol
2000s	*Cartagena Biosafety Protocol, 2000 (implemented 2003)* Ensures protection in the transfer and use of living modified organisms that may have adverse effects on conservation and biological diversity and on human health *Doha Declaration, 2001* Links international trade, development and the environment within the context of the World Trade Organization
Global organisations working on environment	
1970s	*Intergovernmental Oceanographic Commission (IOC), 1960* Provides for global cooperation in the study of the ocean. Coordinates national programmes and knowledge sharing *United Nations Environment Programme (UNEP), 1973, Nairobi, Kenya* Provides leadership and encourages partnership in caring for the environment; coordinates information and programmes *United Nations Human Settlements Programme, UN-Habitat, 1978* Ameliorates problems stemming from urban growth, especially in the developing world; promotes sustainable development *GEMS/Water Programme, 1977* Provides data and information on inland water quality
1980s	*World Commission on Environment and Development (WCED)*, 1983 Promotes sustainable development; developed Brundtland Report *Intergovernmental Panel on Climate Change (IPCC)*, 1988 Assesses information relevant to understanding the scientific basis of risk of human-induced climate change, its potential impacts and options for adaptation and mitigation

(continued)

Table 2.7 (continued)

1990s	*Global Environment Facility* (*GEF*), 1991 Helps developing countries fund environmental programmes *Commission on Sustainable Development, 1992* Ensures effective follow-up of UNCED; monitors and reports on implementation of the Earth Summit agreements *Committee on Trade and Environment, 1994* The part of the World Trade Organization that concerns itself with environmental issues, including ruling on trade disputes over the environment
2000s	*Pew Oceans Commission, 2000* Assessed policies on marine resources and an array of problems facing oceans

Source: http://www.csa.com/discoveryguides/ern/05aug/chart.php#unced

Table 2.8 Global reform initiatives on global environmental governance

Global reform initiatives	Aims
UN Secretary General, Kofi Annan, launched a UN-wide reform initiative (1997)	To improve the coordination and effectiveness of environmental institutions by releasing 1997 programme for reform
The Nairobi Declaration on the Role and Mandate of UNEP	The declaration was adopted by the UNEP Governing Council and endorsed by the UN General Assembly to revive UNEP
The UN Task Force on Environment and Human Settlements (1997)	Created two new coordinating bodies: the Environmental Management Group (EMG) and the Global Ministerial Environment Forum (GMEF)
The Inter-agency Environment Management Group (1999)	To provide UNEP with an effective and strong coordinating role within the UN system on environmental matters
The Malmo Declaration (2000) was adopted by the GMEF	To strengthen UNEP and broaden its financial base and how to better incorporate non-state actors into the GEG system
The Cartagena Process (2000–2002)	To improve international policymaking coherence
The Johannesburg Plan of Implementation (2002)	For the full implementation of the Cartagena decision
The Eighth Special Session of the UNEP Governing Council/Global Ministerial Environment Forum Jeju, Republic of Korea	To discuss progress on the Cartagena decision
French President, Jacques Chirac, calls for creation of a United Nations Environmental Organization (UNEO) at the UN General Assembly (2003)	Informal working group was set up to facilitate dialogue among governments on UNEP reform
The Bali Strategic Plan for Technical Support and Capacity-Building was adopted by the GC/GMEF (2004)	The Bali Plan outlined proposals for improving the capacity of developing countries and economies in transition to implement MEAs
The UN Summit (2005)	To strengthen coordination within the framework of international environmental governance and for the integration of environmental activities at the operational level into a broader sustainable development framework

(continued)

Table 2.8 (continued)

Global reform initiatives	Aims
A High-Level Panel on UN-Wide Coherence in the Areas of Humanitarian Assistance, the Environment and Development (2006)	It was created after the World Summit in New York (2005)
International NGOs	The World Resources Institute (WRI) – USA The Institute of Sustainable Development and International Relations (IDDRI) – France The Ecologic Institute: Science and Policy for Sustainable World – Germany The Global Environmental Governance Project – Yale University

Source: Najam et al. (2006): 22–23

Table 2.9 Environmental and ecological hazards in cities

Scale	Type of hazard	Specific hazards
Within household and its plot	Biological pathogens	Water-borne, water-washed (or water-scarce), airborne, food-borne and vector-borne, including some water-related vectors (e.g. *Aedes* mosquitoes breeding in water containers where households lack reliable piped supply)
	Chemical pollutants	Indoor air pollution from fires, stoves or heaters. Accidental poisoning from household chemicals. Occupational exposure of home workers
	Physical hazards	Household accidents – burns and scalds, cuts and falls. Physical hazards from home-based economic activities. Inadequate protection from rain, extreme temperatures
Neighbourhood	Biological pathogens	Pathogens in waste water, solid waste (if not removed from the site) and local water bodies. Disease vectors, e.g. malaria-spreading *Anopheles* mosquitoes breeding in stagnant water or filariasis-spreading *Culex* mosquitoes breeding in blocked drains, latrines or septic tanks
	Chemical pollutants	Ambient air pollution from fires, stoves, etc.; also perhaps from burning garbage if there is no regular garbage collection service. Air and water pollution and wastes from 'cottage' industries and motor vehicles
	Physical hazards	Site-related hazards, e.g. housing on slopes with risks of landslides; sites regularly flooded, sites at risk from earthquakes
Workplace	Biological pathogens	Overcrowding/poor ventilation aids transmission of infectious diseases
	Chemical pollutants	Toxic chemicals, dust, etc.
	Physical hazards	Dangerous machinery, noise, etc.

(continued)

Table 2.9 (continued)

Scale	Type of hazard	Specific hazards
City (or municipality within a larger city)	Biological pathogens	Pathogens in the open water bodies (often from sewerage), also at municipal dumps, contaminated water in piped system
	Chemical pollutants	Ambient air pollution (mostly from industry and motor vehicles, motor vehicles' role generally growing), water pollution, hazardous wastes
	Physical hazards	Traffic hazards. Violence. 'Natural' disasters and their 'unnaturally large' impact because of an inadequate attention given to prevention and mitigation
	Citizens' access to land for housing	Insecure land tenure experienced by the urban poor and migrant settlers in cities
	Heat island effect and thermal inversions	Raised temperatures a health risk, especially for vulnerable groups (e.g. elderly, very young). Air pollutants may become trapped, increasing their concentration and the length of people's exposure to them
City region (or city periphery)	Resource degradation	Soil erosion from poor watershed management or land development or clearance, deforestation, water pollution, ecological damage from acid precipitation and ozone plumes, loss of biodiversity
	Land or water pollution from waste dumping	Pollution of land from dumping of conventional household, industrial and commercial solid wastes and toxic/hazardous wastes. Leaching of toxic chemicals from waste dumps into water. Contaminated industrial sites. Pollution of surface water and groundwater from sewage and surface run-offs
	Pre-emption or loss of resources	Freshwater for city pre-empting its use for agriculture; expansion of paved area over good-quality agricultural land
Links between city and global issues	Nonrenewable resource use	Fossil fuel use, use of other mineral resources, loss of biodiversity, loss of nonrenewable resources in urban waste streams
	Nonrenewable sink use	Persistent chemicals in urban waste streams, greenhouse gas emissions, stratospheric ozone-depleting chemicals
	Overuse of 'finite' renewable Resources	Scale of consumption that is incompatible with global limits for soil, forests, freshwater, etc.

Source: Satterthwaite (1999), The Links Between Poverty and the Environment in Urban Areas of Africa, Asia and Latin America, New York: United Nations Development Programme (UNDP) and the European Commission (EC).

Table 2.10 Environmental threats across different regions of the world

	Land	Forests	Biodiversity	Freshwater	Coastal and marine	Atmosphere	Urban areas	Disasters
Africa	Land use Soil degradation, sealing and contamination Soil erosion	Loss of natural forests Forest degradation Sustainable forest management	Agricultural intensification Genetically modified organisms	Water quantity and quality Policy and legislative framework	Coastal erosion Pollution	Air pollution Stratospheric ozone depletion Greenhouse gas emissions	Air quality Noise pollution Solid waste	Storms and floods Earthquakes Human-caused disasters
Asia and the Pacific	Land degradation Land tenure	Deforestation Forest degradation	Habitat loss and degradation Overexploitation of resources and illegal trade	Decreasing water available per capita Water quality	Habitat conversion and destruction Pollution Overexploitation of fisheries	Air pollution Ozone depletion Air quality	Solid waste Water supply and sanitation Air quality	Drought Hurricanes Floods Earthquakes Spills of hazardous substances
Europe	Land use Soil degradation, sealing and contamination Soil erosion	Loss of natural forests Forest degradation Sustainable forest management	Agricultural intensification Genetically modified organisms	Water quantity and quality Policy and legislative framework	Coastal erosion Pollution	Air pollution Stratospheric ozone depletion Greenhouse gas emission	Air quality Noise pollution Solid waste	Storms and floods Earthquakes Human-caused disasters
Latin America and Caribbean	Land degradation Land tenure	Deforestation Forest degradation	Habitat loss and degradation Overexploitation of resources and illegal trade	Decreasing water available per capita Water quality	Habitat conversion and destruction Pollution Overexploitation of fisheries	Air pollution Ozone depletion Air quality	Solid waste Water supply and sanitation Air quality	Drought Hurricanes Floods Earthquakes Spills of hazardous substances

(continued)

Table 2.10 (continued)

	Land	Forests	Biodiversity	Freshwater	Coastal and marine	Atmosphere	Urban areas	Disasters
North America	Land degradation Pesticides	Forest health Old growth forests	Habitat destruction and degradation Bioinvasion	Groundwater Great Lakes water quality	Conversion of fragile ecosystems Overexploitation of marine resources Pollution	Stratospheric ozone depletion Greenhouse gases and climate change	Urban sprawl Ecological footprint	Floods and climate change Forest fires
West Asia	Land degradation Rangeland deterioration	Degradation Overexploitation Sustainable forest management	Habitat Degradation and loss Overexploitation of species	Increasing water demand Overexploitation of groundwater Water quality	Coastal development and urbanisation Overexploitation of resources Marine pollution	Air pollution Ozone-depleting substances Climate change	Land conversion Solid waste	Drought Oil discharges Armed conflict
Polar region	Degradation Erosion Climate change	Boreal forest issues Threats to forest tundra	Climate change Ozone depletion Overexploitation	Alien species Pollution	Overexploitation of fisheries Pollution Climate change	Stratospheric ozone depletion Long-range air pollution Climate change	Sanitation and waste	Floods Oil discharges Pest invasion

Source: UNEP (2002:31)

Table 2.11 Status of urban environment in India (Kamyotra and Bharadwaj 2011)

Urban services	Access and quality	Environmental impact
Water	Water availability on a per capita cubic metre basis is estimated to decline from 1730 to 1240 in India The average availability of water is less than 4 h a day, some areas receive water only for 1 h on alternate days The per capita water supply ranges from 9 lpcd to 584 lpcd across urban India Poor quality of operation and maintenance costs leads to wastage of water to the extent of 25–50% Low pressure and intermittent supply lead to the contamination in the distribution network Increase in domestic consumption of water over the next 20 years will double from 25 billion m3 to 52 billion m3 As per the 54th NSS round (National Sample Survey), 59% of households share public water and 15% do not have access to drinking water As per the World Bank (WB) report, 27 Asian cities, with over 1 million population including Chennai, Delhi, Kolkata and Mumbai are ranked worst performing in terms of hours of water availability per day The capacity utilisation has been reported to be less than 50% in 40% of the towns and less than 75% in a further 20% of towns Due to old and rusted pipes or poor maintenance of the system, these losses sometimes go up to 50% A study by Sridhar and Kashyap (2012) points out that the availability of water is below the specified norms. While water supply coverage in the city of Kolkata is very low with just 27.3% of the city households being connected	Wastage of water Contamination of water in the distribution network 90% of water supplied is polluted in class two cities of India Water-borne diseases are the major cause for mortality rate in India Sanitation-related diseases are responsible for 60% of the environmental diseases Decline in access and quality of water seriously affects food and biomass in the country In India almost 70% of its surface water resources and a growing percentage of its groundwater reserves are contaminated by biological, toxic, organic and un-organic pollutants In 1995, the Central Pollution Control Board identified severely polluted stretches across 18 major rivers in India Long-term intake of fluoride can cause tooth decay and crippled bones. Arsenic can cause skin cancer and skin pigmentation The overexploitation of aquifers, depletion of water resources and pollution by urban human wastes are causing serious health problems

(continued)

Table 2.11 (continued)

Urban services	Access and quality	Environmental impact
Sewerage	Combined, the 22 largest cities in the country produce over 7267 million litres of domestic waste water per day 72 of 4400 towns in India have partial sewerage facilities, and 17 have some form of primary treatment facilities before disposal While waste generation in Class 1 cities more than doubled from 1978 to 1995 The treatment capacity decreased from 39% to 24% during 1995 Of the total waste water generated in the metropolitan cities in India, barely 30% or 30% is treated before disposal Out of 345 towns, 95% do not have waste water treatment plants Cities and townships of Karnataka state generate approximately 2260 million litres of sewage per day. Only 80% is collected and treated less Only 36 out of 218 urban local bodies in Karnataka have underground drainage system (UGD) 65% of urban households do not have closed drainage facility Collection systems exist for only about 30% of the waste water through sewer lines and treatment capacity exists for about 7000 million litres/day In Class I cities of India, more than 71% of waste water is not treated, while in Class II cities, more than 97% of waste water is not treated (up to 2009) Out of 300 Class I cities, about 70 have partial sewerage systems and sewage treatment facilities Only 54% of New Delhi has sewerage coverage	In 118 cities, it is discharged indirectly into rivers, lakes, ponds or creeks, while in 63 cities it is used for agriculture Pollution of urban water bodies (like tanks, lakes and groundwater) It is estimated that 75 to 80% of water pollution by volume is caused by domestic sewage Large urban population is at risk of being exposed to water-borne diseases of infectious (bacterial, viral or animal infections) or chemical nature (due to fluoride or arsenic). Water-borne diseases are still a great health concern in India Hazardous wastes are a source of groundwater pollution It has been assessed that 80% of pollution is caused by sewage alone

(continued)

Table 2.11 (continued)

Urban services	Access and quality	Environmental impact
Solid waste[a]	About 48 million tonnes of solid waste are generated in the urban areas every day, an eight-fold increase since independence Only 72% of waste is collected daily Non-degradable waste is increasing in an alarming proportion, and the production and consumption of plastic have increased more than 70 times between 1960 and 1995 Seventy per cent of Indian cities have inadequate waste transportation facilities Lack of common disposal sites Absence of secure landfills E-waste worth US$1.5 billion was generated in India in the year 2003 Cities' generated overall quantity of solid waste amounts to about 5% Indian industries are generating nearly 7.66 million metric tonnes (MT) of hazardous waste. Only 73 hazardous dumping sites are available for Indian cities For the year 2005, 1.46 lakh tonnes of e-waste were generated and were expected to reach 8.0 lakh tonnes by 2012. About 65 cities in India generate more than 60% of total e-waste in India Out of 115 sewage treatment plants in major Indian States, 35% continued to discharge polluted water beyond the stipulated norms 49% (11% (Urban) and 65% (Rural) do not have access to toilets in India	Accumulation and decomposition of waste on streets and public places with adverse effect on public health Public littering Exacerbating unhygienic environmental conditions are leading to both physical and health problems of the urban poor. Often they suffer from respiratory disorders, diarrhoea, fungal and other skin infections, transient loss of memory and depression Rampant illegal dumping of industrial hazardous waste leads to biological contamination of rivers, lakes and canal pits and groundwater source with high pollution loads Uncontrolled release of chromium-contaminated waste and sludge contaminates aquifers E- and plastic waste containing lead, cadmium and mercury are negatively affecting India's ecological systems besides posing challenges to sustainable development
Health care	0.1–1.5 kg per bed per day of health-care waste (HCW) is generated. Total HCW generated increased from 890 tonnes to 920 tonnes per day between 1997 and 2002 25% of health waste generated is hazardous[b] About 42% of health-care workers do not have knowledge on classification and segregation of biomedical waste Open dumping and burning of waste is a common practice	25% of HCW is infectious

(continued)

Table 2.11 (continued)

Urban services	Access and quality	Environmental impact
Air pollution	Data from 36 cities indicate that ambient air pollution far exceeds the WHO guidelines The percentage of cities with 'dangerous' air has increased from 15% to 21% More than 35% of urban households are exposed to high levels of indoor air pollution During 2007, the highest concentration of NO among all two residential areas was observed at Town Hall, Delhi Death due to air pollution in Indian cities increased by 30% between 1992 and 1995 Conformity to the RSPM standards well above the National Ambient Air Quality (ABAQ) in New Delhi is the worst and closely followed by Mumbai	Respiratory and lung problems Standards regarding the main air pollutants of public health concern were violated at most of the monitoring stations The health of over 900 million urban population around the world is deteriorating on daily basis due to high levels of air pollution SO, NO and suspended 2 x particulate matter (SPM) damage the human respiratory and cardiorespiratory systems in various ways contributing to respiratory-related morbidity In Mumbai alone, the prevalence of respiratory diseases amounts to 22.2% Urban air pollution is estimated to cause over 250,000 deaths and billions of cases of respiratory illnesses every year
Urban transportation	The total number of motor vehicles increased from 0.3 million in 1951 to 67 million in 2003 There is a drastic increase in the number of two-wheelers from 8% in 1951 to 70.9% in 2003 India's total SO2 and NOx emission increased from 7.12 million metric tonnes to 9.82 million metric tonnes between 1992 and 2005 with a CAGR of 3.63% Total vehicle population of India is more than 85 million (about 1% share of the world) An average 10% increase has been found in each year, which is a serious concern for air pollution Urban transport constitutes 60 to 80% of motor vehicles	Traffic congestion, increase in accident rates, wastage of fuel and environmental pollution (emission of carbon monoxide, hydrocarbons, nitrogen oxides and other toxic substances) Rate of accidents has gone up from 16,000 in 1981 to 80,000, respectively, post-2001 Respirable suspended particulate matter (RSPM) levels are fairly high, and SPM exceeds national standards in many cities Vehicular emissions containing pollutants such as sulphur dioxide, nitrogen oxides, carbon monoxide, lead, ozone, benzene and hydrocarbons pollute urban areas Encroachment of footpaths/ roads

(continued)

Table 2.11 (continued)

Urban services	Access and quality	Environmental impact
Land	Per capita availability of land declined from 0.89 ha in 1951 to 0.3 ha in 2001 Land degradation is between 16 and 57% of total geographical area	High storm intensity, soil erosion, climate change
Energy	35.5% of the population still lives without access to electricity in India Transmission and distribution loss of electricity in New Delhi is as high as 19.64%	Use of wood fuel and kerosene Increase in gas emissions

Source: Planning Commission (2002), Pandey et al. (2006), Planning Commission (2002), Environmental Management and Policy Research Institute (2012), MoEF (2009, 2011), CSO (2011: 111), Sridhar and Kashyap (2012)
India produces about 42 million tonnes of urban solid waste annually. The per capita waste generation varies between 0.2 kg. and 0.6 kg. per day, and the current municipal solid waste generation is estimated to be approximately 0.4 kg per capita per day (Planning Commission 2002: 652). Every year eight million tonnes of plastic waste are generated in India (Annual Report 2011: 138)

References

Anand, V. S. (2013). Global environmental issues, open access scientific reports, Accessed on 2nd Dec 2013. http://www.omicsonline.org/scientific-reports/2157-7617-SR-632.pdf

Annual Report (2011/12). *Abatement of pollution, annual report 2011/12*. New Delhi: Ministry of Environment and Forestry.

Bandari, L. (2006). Social infrastructure: Urban health and education. In*Infrastructure development finance company 'Indian infrastructure report 2011: Water: Policy and performance for sustainable development'*. New Delhi: Oxford University Press.

Bharath, H. A., Durgappa, V. S., & Ramachandra T. V. (2012). Modeling and simulation of urbanisation in greater Bangalore, India. http://wgbis.ces.iisc.ernet.in/energy/water/paper/iitm_urbanisation/results.htm

Brenner, N. (1999). Globalization as reterritorialisation: The re-scaling of urban governance in the European Union. *Urban Studies, 36*(3), 431–451.

Chandrasekhar, C.P., Ghosh, J. (2013). *India still a vast informal economy*, Business Line, The Hindu.

CSO. (2011). *Millennium development goals india country report 2011, Central Statistical Organisation*. New Delhi: Ministry of Statistics and Programme, Implementation, Government of India.

Environmental Management and Policy Research Institute. (2012). *State of environment report, Karnataka 2011*. Bangalore: Environmental Management & Policy Research Institute.

Goffman, E. (2005). Global environmental governance, environmental policy issue, access http://www.csa.com/discoveryguides/ern/05aug/overview.php

Governance Research, *CSERGE Working Paper EDM 05-01*, Centre for Social and Economic Research on the Global Environment, University of East Anglia: UK.

Hardin, G. (1968). The tragedy of the commons. *Science, 162*(3859), 1243–1248.

Hardoy, J. E., Mitlin, D., & Satterthwaite, D. (2001). *Environmental problems in an urbanizing world*. London: Earthscan.

Harvey, D. (1985). The marxian theory of the state. *A Radical Journal of Geography, 17*(2–3), 174–181.

Jeffrey, M. I. (2005). Environmental governance: A comparative analysis of public participation and access to justice. *Journal of South Pacific Law, 9*(2), 1–31.

Kearney, A. T. (2014). Global cities index and emerging cities outlook. http://www.atkearney.in/researchstudies/global-cities-index/2014

Kamyotra, J.S., and R.M Bhardwaj (2011). Municipal Waste Managment in India. In Infrastructure Development Finance Company, 'Indian Infrastructure Report 2011: Water: Policy and Performance for Sustainable Development', New Delhi: Oxford University Press.

Khan, J. H., Hassan, T., & Shamshad. (2011). Socio-economic causes of rural to urban migration in India. *Asia-Pacific Journal of Social Sciences*, (*3*, 2), 138–158.

Kundu, A. (1999). *Urban informal sector in India: Macro trends and policy perspectives, discussant paper 25, Issue in Development, Development Policy Department*. Geneva: ILO.

Kundu, A. (2006). Trends and patterns of urbanization and their economic implications. In*Indian infrastructure report 2006: Urban infrastructure*. New Delhi: Oxford University Press.

de Loë, R. C., Armitage, D., Plummer, R., Davidson, S., & Moraru, L. (2009). *From government to governance: A state-of-the-art review of environmental governance. Final report. Prepared for Alberta Environment, Environmental Stewardship, Environmental Relations*. Guelph, ON: Rob de Loë Consulting Services.

Maiti, S., & Agrawal, P. K. (2005). Environmental degradation in the context of growing urbanization: A focus on the metropolitan cities of India. *Journal of Human Ecology, 17*(4), 277–287.

Marcotullio, P. J. (2003). Globalisation, urban form and environmental conditions in Asia-Pacific cities. *Urban Studies, 40*(2), 219–247.

McGranahan, G. et al. (2001). *The citizens at risk: from urban sanitation to sustainable cities*. London: Earthscan.

MoEF. (2001). *India state of environment 2001*. New Delhi: Ministry of Environment and Forests, Government of India.

MoEF. (2009). *State of environment report, 2009 India*. New Delhi: Ministry of Environment and Forestry, Government of India.

MoEF. (2011). *Sustainable development in India: Stocktaking in the run up to Rio+20*. New Delhi: Ministry of Environment and Forestry, Government of India.

Naik, A.K. (2009). Informal sector and informal workers in India, *Paper prepared for the* special IARIW-SAIM Conference on "Measuring the Informal Economy in Developing Countries".

Najam, A., Papa, M., & Taiyab, N. (2006). *Global environmental governance: A reform agenda*. Manitoba: International Institute for Sustainable Development.

Nanjam, A. (2005). Developing countries and global environmental governance: From contestation to participation to engagement. *International Environmental Agreements, 5*, 303–321.

Nas, P. J. M., & Veenman, M. (1998). Towards sustainable cities: Urban community and environment in the third world. *International Journal of Anthropology, 13*(2), 101–115.

Ostrom, E. (1990). *Governing commons: The evolution of institutions for collective action*. Cambridge: Cambridge University Press.

Ostrom, E. (2000). Collective action and the evolution of social norms. *Journal of Economic Perspectives, 14*(3), 137–158.

Ostrom, E., Gardner, R., & Walker, J. (1994). *Rules, games, and common-pool resources*. Michigan: University of Michigan.

Otto-Zimmermann, K. (2011). Embarking on global environmental governance, ICLEI Paper 2011–1 vs,2b, Local Governments for Sustainability.

Paavola, J. (2005). Environmental conflicts and institutions as conceptual cornerstones of environmental governance research, CSERGE Working Paper EDM 05-01, Centre for Social and Economic Research on the Global Environment, University of East Anglia, UK.

Pandey, S., Singhal, S., Jaswal, P., & Guliani, M. (2006). Urban environment. In 3inetwork (Ed.), *Indian infrastructure report 2006: Urban infrastructure*. New Delhi: Oxford University Press.

Planning Commission. (2002). *Civic amenities in urban areas, tenth plan document*. New Delhi: Planning Commission.

Roychowdhury, S. (2004). Globalisation and labour. *Economic Political Weekly, XXXIX*(1), 105–108.

Roychowdhury, S., James, K. S., Sivanna, N., & Vani, B. P. (2012). *Migration, informal work and welfare: A policy perpective on Karnataka cities, project report*. Bangalore: Institute for Social and Economic Change.

Satterthwaite, D. (1999). Development in an urban context, The Environment Times. http://www.grida.no/publications/et/ep2/page/2483.aspx

Satterthwaite, D. (2003). The links between poverty and the environment in urban areas of Africa, Asia and Latin America. *Annals of the American Academy of Political and Social Science, 590*, Rethinking Sustainable Development, pp. 73–92.

Satterthwaite, D. (2008). Cities' contribution to global warming: Notes on the allocation of greenhouse gas emissions. *Environment and Urbanization, 20*(2), 539–549.

Satterthwaite, D. (2009). The implications of population growth and urbanization for climate change. *Environment and Urbanization, 21*(2), 545–567.

Satterthwaite, D. (n.d.). Development in an urban context, The Environment Times, http://www.grida.no/publications/et/ep2/page/2483.aspx

Satterthwaite, D., Mitlin, D., Patel, S. (2011). Engaging with the urban poor and their organisations for poverty, reduction and urban governance, An issues paper for the United Nations Development Programme.

Singh, S. P. (1995). India's population problem, a global issue of environment and sustainability: An appeal for cooperation. *Bulletin of the Ecological Society of America, 76*(1), 55–59.

Sridhar, K. S., & Kashyap, N. (2012). *Stat of India's cities: An assessment of urban conditions in four mega cities*. Bangalore: Public Affairs Centre.

Sridhar, K. S., & Wan, G. (2014). *Urbanisation in Asia: Governance, infrastructure and the environment*. New Delhi: Springer.

State and Outlook (2010). The European environment. Access.

Stockholm. (1972). Stockholm declaration of 972. http://staging.unep.org/Documents.Multilingual/Default.Print.asp?DocumentID=97&ArticleID=1503&l=en

Toutain, O., & Gopiprasad, S. (2006). Planning for urban infrastructure. In*Indian infrastructure report 2006: Urban infrastructure*. New Delhi: Oxford University Press.

UNEP. (2002). *Global environmental outlook3: Past, present and future perspectives*. London: Earthscan.

UN-HABITAT. (2008). *State of the world's cities 2010/2011: Bridging the urban divide*. London: United Nations Human Settlement Programme.

UN-HABITAT. (2010). State of Asian cities, Japan, United Nations Human Settlement Programme (UN-HABITAT).

UN-HABITAT. (2011). *Global report on human settlements 2011: Cities and climate change, United Nations Human Settlements*. London: Earthscan.

UNESCAP. (2005). *The state of the environment in Asia and the Pacific 2005: economic growth and sustainability*. UN: Thailand.

UNHSP. (2009). *Planning sustainable cities: Global report on human settlements 2009*. London: UNHSP and Earthscan.

UNHSP. (2011). *Cities and climate change: Global report on human settlements 2011*. London: Earthscan.

UNHSP. (2013). *State of world cities 2012/2013: Prosperity of cities*. New York: Routledge.

Vishwanath, T., Lall, S. V., & Dowall, D. (2013). *Urbanization beyond municipal boundaries: Nuturing metropolitan economies and connecting peri-urban areas in India*. Washington, DC: The World Bank.

Wingqvist, G. Ö., Drakenberg, O., Slunge, D., Sjöstedt, M., & Ekbom, A. (2012). *The role of governance for improved environmental outcomes, report 6514*. Sweden: Swedish Environmental Protection Agency.

World Commission on Environment and Development. (1987). *Our common future: World commission on environment and development*. New York: Oxford University Press.

Chapter 3
Emerging Urban Environment Governance Models/Approaches

Abstract The process of urbanisation has been the most striking global phenomenon of the twenty-first century for both developed and developing countries. The process has inevitably unveiled a series of environmental challenges and vulnerabilities due to lack of urban policies, inadequate resources and services to meet growing demands. The major concern for both developed and developing countries is how to promote urban sustainability and evolve a coping mechanism to alleviate environmental stress and disasters. But over the years, amidst barriers posed by contemporary urban challenges, pragmatic efforts are made conceiving different urban sustainable approaches and models as a framework offering effective solutions to the environmental threats and thereby improve the quality of urban life. These sustainable models are practised at different levels among varied sectors spread across the world cities including India. This study, therefore, captures such different sets of sustainable models and approaches used by the cities across the world.

Keywords Urban growth · Sustainable city · Environmental models · Green agenda · Global city

3.1 Introduction

The process of urbanisation has been an inevitable trend of the twenty-first century, for both developed countries and developing countries, posing demography, environment, ecology, economic and socioeconomic challenges to cities (Grimm et al. 2008; UNHSP 2009). Particularly, cities are reeling under globalisation process and concomitant economic restructuring, and major concerns have been urban sustainability and coping mechanism to alleviate environmental stress and disasters. Some of the most visible environmental challenges have been growing urban density and lack of resources to meet their growing demands; many cities[1] are designed with no

[1] The city of London needs 125 times its own area to provide the resources to consume, and Japan's capital city Tokyo has grown to just under 13 million, and its population is more than 35 million (PECC 2010: 6).

K. V. Raju et al., *Urban Environmental Governance in India*, The Urban Book Series, https://doi.org/10.1007/978-3-319-73468-2_3

regard to resource (protection, conservation and sustainability) and disregard or negligence towards various kinds of urban environmental pollution. Therefore, cities adopt various models and approaches according to the size of the population, urban planning and management and available financial resources to alleviate environmental and ecological constraints. This chapter examines different urban models and approaches practised by the cities in both developed and developing countries to promote sustainability.

3.2 Contemporary Urban Challenges of Global Cities

The process of urbanisation and city growth are accompanied by a multitude of environment problems and challenges. Some of them are presented in the table (see Table 3.1) below.

Table 3.1 Urban challenges in global cities

Areas	Key challenges	Impact
Urban sprawl	Inclusion of peri-urban and villages to the administrative boundaries of the core city Complex institutional and regulatory system	Lack of access to basic amenities, depletion of 'commons' vs natural resources that include depletion of nonrenewable resources, changes in land use pattern, depletion of energy, increase in waste production, reduction of 'ecological footprint' of the city, insecure land tenure, increase in demographic and spatial density, deforestation, increase in slums
Greenhouse gas emission	Climate change	Natural hazards like rise in temperature, heavy rainfall, urban floods, rise in water/sea level, increase in cost of fossil fuels
Economic restructuring	Rural-urban migration Impact on urban labour markets Gender dimension to restructuring	Polarisation/immobility in social and economic ladders, particularly, poor-income jobs, rapid growth of informal sector leading to the informalisation/casualisation of jobs and income, industrial relocation to developing countries' 'cities', increase in service sector industries, dramatic spatial restructuring, thereby dramatic increase in income inequality, reduction in public spending on 'infrastructure' facilities, increase in proportion of poverty
Institutional dynamics	Expansion of political system from 'government to governance'	-Multilevel involvement of non-state actors like civil society, NGOs, international funding agencies, transnational corporation, privatisation, deregulation and decentralisation of services concomitantly

Source: Compiled from different sources

3.3 Urban Environmental/Sustainable City Models/Approaches: Across Global Cities

Cities act as important economic catalyst, to condition the dynamics of the prevailing socioeconomic conditions and environmental interest which have adopted various approaches and models to promote healthy eco- and social systems. Various urban models and approaches that were developed across the globe's world cities are drawn towards 'sustainability' as a model, set of enduring principles, approach, methodology, ideology or framework for offering effective solutions to the environmental threats and improving the quality of urban life. Indeed, the concept of sustainability has been analysed in global, national, regional, corporate and local/urban settings. Effective urban governance is linked to the effects of institutions, and management of environment and local ecology eventually promotes sustainability. Cities around the world have implemented a number of sustainable initiatives across sectors like water, land, solid waste, urban landscape and so on. But the standard set for achieving sustainability differs from city to city.

The need for promoting sustainable city model appeared first two decades ago, i.e. the UN report entitled 'Our Common Future' (WCED 1987) points to the fact that depletion of natural resources, poverty and pollution destroying not only the planet for present but also for future generation. To save the earth from inevitable destruction, the landmark 'Earth Summit' of 1992, by UN-HABITAT and UNEP, undertook serious efforts towards promoting sustainable city models amidst global economic growth scenario. These efforts took the shape of 'Sustainable City Programme' initially covering 66 cities. According to the UN Development Programme's Human Development Report 2011, the model of 'sustainability' is inextricably linked to the promotion of equity, i.e. fairness, social justice and greater access to the better quality life.

To overcome the critical challenges posed by urbanisation and make up for loss of environment and ecological resources, various sustainable city models are practised and implemented across the world cities. The table above (see Table 3.2) provides a glimpse of various sustainable city models adopted by the world cities. These models are conceptualised based on the partnership, available resources, priority of the sector, community participation and networking. Some of the sustainable city models include (i) resilient city, (ii) eco-city, (iii) low-carbon city, (iv) biodiverse city, (v) inclusive city, (vi) green city, (vii) green economy and (viii) smart cities. Various cities across the global have emerged as successful models (see Table 3.3) implementing approaches and principles of sustainability. These principles vary across the sectors like water, waste management, energy, urban transportation and so on.

Unfortunately, under the compiled list of sustainable cities, only Auroville from India figures, but the city of Bengaluru does not fall under any of the above mentioned categories (see Table 3.3) which are undertaking effective sustainable initiatives. More often though, complex institutional regimes, weak urban planning,

Table 3.2 Urban environmental models and approaches across globe

Models	Year	Model cities	Initiatives
Sustainable[a] Cities – Landmark 1992 'Earth Summit' in Rio de Janeiro, Brazil, UN-HABITAT (UNCHS) and the UNEP	1992	City of Moreland in Australia Porto Alegre, Goiânia and Curitiba in Brazil Bafut, Cameroon China, Nanjing Vancouver, Canada San Francisco, USA Oslo, Norway Curitiba, Brazil Copenhagen, Denmark Masdar City (Near Abu Dhabi), Las Vegas	More compact, energy-efficient, clean and less polluted, more accessible and offer transport choices. They are vibrant and attract more entrepreneurial skill opportunities necessary for growth and prosperity thereby enhancing the cities' competitiveness, livability and environmental performance Ability to strike healthy balance between economic growth and environmental preservation thereby contributing to healthy living standards Action to slow climate change, prevent further degradation and reduce inequalities, as environmental deterioration threatens to reverse recent progress in human development for the world's poorest
Urban Ecological Footprint[b] – the connection between city dwellers and the bio-geophysical environment in which they reside	1992		More compact, greater heterogeneity and functionality, effective provisioning of public goods, more ecological space for imagination and social interaction
Resilient city[c]: city's ability to respond to a natural resource shortage and the recognition of the human impact on climate change			Capacity to adapt to inner systems to withstand shocks and natural disaster. While enhancing each subsystem, the policies increase the degree of collaboration between urban subsystems (social, environmental-infrastructure, economic and institutional system). Reduces its ecological footprint while simultaneously improving the quality
Eco-city first coined by Richard Register – Eco-city Berkeley: Building Cities for a Healthy Future. The UNDP (2010) defines eco-city as 'a city that provides an acceptable standard of living for its human occupants without depleting the ecosystems and biochemical cycles, on which it depends'.	1987	Canada – Calgary and Ottawa Australia – Melbourne India – Auroville Hong Kong	More green space, preservation of rich biodiversity, legally protect to safeguard key indigenous ecosystems, cleaner environment, protection of landscape and high quality of life. The eco-city[d] is tied up with three goals; they are (i) eco-industry (resource conservation, use of renewable energy and life cycle production), (ii) eco-scape (built environment, open spaces and maximising accessibility) and (iii) eco-culture (deriving balance between human and nature, environmental ethics to maintain high-quality ethics)

Low-carbon city – Low-carbon, low-emission or even carbon-neutral cities are signposts to sustainability and global climate change mitigation			
Biodiverse city – A biodiverse city considers its relationship with nature both within and beyond its administrative boundaries and delivers sustainable solutions to natural risks and urban development challenges			
Inclusive city[e] – an inclusive city provides the opportunities and supportive mechanism that enable the residents to develop their full potential and gain their share of urban advantage	2009 and 2010/2011-UN-HABITAT	Morocco; Kharkov, Ukraine Portland, USA	(i) Assessing the past and measuring the progress, (ii) establishing new and effective institutions, (iii) building new linkages and alliances among various tiers of the government, (iv) developing sustained comprehensive vision and (v) ensuring equitable redistribution of opportunities
Green city – ADB – how cities succeed or fail to achieve quality environmentally to their residents and or capacity to address the environmental challenges		Bogotá, Colombia Hamburg, Germany Stockholm, Sweden Copenhagen, Denmark Curitiba, Brazil Portland, Oregon Malmö, Sweden San Francisco, California Reykjavik, Iceland	Globally contribute to environmental sustainability. How city undertakes collective action and the outcomes of these actions thus form the criteria for determining how 'green' is the city. Certain investments must be made on infrastructure to make the city 'green' such as planning, transport, energy efficiency, industrial metabolism, water supply and sanitation facilities and, finally, effective governance and institutional measures
Green economy[f] is a new approach to bring together sound economic decisions and actions with sustainable patterns of consumption and production.	2012		

(continued)

Table 3.2 (continued)

Models	Year	Model cities	Initiatives
Smart city	2014	To derive principles of planning for promoting sustainable developmental initiatives inspired by European experience involving stakeholders like administration, business and professionals	

Source: Compiled from different sources

[a]The term sustainability was first defined in the Brundtland Commission (UNCHSP 2009).

[b]The 'ecological footprint' is an average measure of the amount of land required to sustain an individual (UN-HABITAT 2010/11: 17). The concept of 'urban ecological footprint' was developed in 1992 by Dr. William Rees and M. Wackernagel from the University of Columbia. It refers to the area of land needed to provide the resources used and absorb the wastes produced by the community.

[c]Zhao et al. (2013). Understanding Resilient Urban Futures: A Systematic Modelling Approach, Sustainability, 5, pp. 3202–3223

[d]Lindfield and Steninberg (2012). *Green Cities: Urban Development Series*, Philippines: ADB

[e]UN-HABITAT (2010/11). *State of the World Cities 2010/2011: Bridging the Urban Divide*, London: Earthscan

[f]Green urban economy was elaborated on by the ICLEI World Secretariat on behalf of the German Ministry for Economic Cooperation and Development (BMZ) in 2012. It is used as a conceptual tool for implementation at local levels and provides input to the further development of integrated approaches for sustainable urban development.

Table 3.3 Sustainable city models across the world

Country	City	Area	Sustainable initiatives	Approach/best practice
Canada	Vancouver Calgary Ottawa	Energy efficiency Eco-city	Hydroelectric power generation (90 per cent) Mass transit, bike lanes (248 kilometres) Ride-sharing programmes and greenways Strict building codes Reduction in gas emissions Vancouver has the lowest per capita carbon emissions of any major city Calgary, Alberta, ranked as the top eco-city in the world in 2010 for its 'excellent level of service on waste removal, sewage systems and water drinkability and availability, coupled with relatively low air pollution' Ottawa, Ontario, ranked as third in the top eco-cities in the world. The survey was performed in conjunction with the reputable Mercer Quality of Living Survey	Green city of the world – 2020
USA	San Francisco	Waste management	Air quality, waste management and commitment to eco-friendly commuting options 77 per cent waste recycled Eco-friendly-commuting options An incredibly effective 11-year-old zero waste program, by 2020; the city hopes to bring that up to 100%	Electric vehicle capital of the USA Waste management award
Norway	Oslo	Renewable energy	Biomethane from waste to power mass transit and heating and an 'eco-certification' City's heating system is currently powered by 80 per cent renewable energy, mainly from biomass from residual waste	Leading 'sustainable city of Europe'
Brazil	Curitiba	Green space and clean environment	99 per cent of population recycle their waste Over 1.5 million trees have been planted along city streets, and a network of 28 parks and forests is developed	Best living place in Brazil

(continued)

Table 3.3 (continued)

Country	City	Area	Sustainable initiatives	Approach/best practice
Copenhagen	Denmark	Renewable energy Carbon measurement and planning	More than a third of the city's 1.2 million people regularly cycle to work and focus on city's carbon footprint Wind production – supplying roughly 19 per cent of the country's power needs World's first carbon-neutral capital by 2025 Mandatory green-roof policy Installation of 'pocket parks'	Shining green jewel the first carbon-neutral capital city by 2025
Latin America	Bogota	Urban transportation	70% of the city's population who travel daily by bus (electric buses)	
Australia	Melbourne Moreland	Energy-efficient built environment Carbon-neutral free city	Sustainable building programme innovative city designed and managed property-tax-based financing to improve the energy and water efficiency of private commercial buildings in the city The city of Moreland in Melbourne's north has programmes for becoming carbon neutral, one of which is 'Zero Carbon Moreland', among other existing sustainable implementations and proposals	Sustainable building programmes Carbon-free initiatives
Mexico	Mexico	Air quality	Improved reduction in air pollution and emission Urban sprawl containment, to public awareness campaigns	Mexico City took the air Quality award for ProAire
Germany	Munich	Green energy	Green electricity at its own plants by 2025 to meet the power requirements of the entire municipality of Munich	Green power by 2025 Green energy award
USA	New York	Adaptation and resilient infrastructure	Promotes resilient infrastructure 250 ambitious infrastructure resilience initiatives across a number of categories	Resilient city
Brazil	Rio de Janeiro	Morar Carioca Program – Urban Revitalization Plan	Comprehensive sustainable urban revitalisation with a aim to formalising all of the city's favelas by 2020 A combination of better landscaping, infrastructure, educational tools	Sustainable communities
Singapore	Singapore	Intelligent city infrastructure	Incorporates a range of smart transportation technologies, including one of the world's first Electronic Road Pricing Systems, real-time traffic information delivered through GPS-enabled taxis and a highly integrated public transportation system	Integrated Transport System Intelligent City Infrastructure recipient

Japan	Tokyo	Finance and economic development	Reduction of gas emission Cut emissions by a total of 13% in the city and prevented over seven million tonnes of CO_2 from being released	Finance and economic development
India	Auroville Chennai, Delhi, Kolkata and Mumbai, Bengaluru	Vibrant community culture Environment-friendly fuel transport system[a]	Expertise in renewable energy systems, habitat restoration, ecology skills, mindfulness practices Phasing out lead in petrol in the four metropolitan cities of India in 1994 followed by introduction of CNG and LPG (for autos) in Delhi, Mumbai and Surat and Bengaluru	Vibrant community culture
Kenya	Hacienda – Mombasa	Largest development of eco-friendly residential properties in East Africa	Construction is currently ongoing, and it will eventually be one of Africa's first self-sustaining estates	Eco-friendly residential properties in East Africa
New Zealand	City of Waitakere	Greater Auckland urban region was New Zealand's first eco-city	Working from the Greenprint, a guiding document that the city council developed in the early 1990s	First eco-city
Sweden	Gothenburg	Sustainable city in Sweden	Low environmental impact, contain passive houses, good recycling system for waste, etc.	Especially Älvstaden (central city by the river Göta Älv), which is a good example of sustainable city in Sweden

Source: (http://www.planetizen.com/node/55504) (www.ecoimagination.com) (http://www.fastcoexist.com/3016816/the-10-cities-that-are-leading-the-way-in-urban-sustainability)

[a]The MoEF notified fuel specifications in 1996, and in April 2002 the Supreme Court of India directed for phasing out diesel buses and priority for CNG allocation. It also ordered for drawing and supplying CNG for other polluted cities such as Agra, Faridabad, Jharia, Jodhapur, Kanpur, Lucknow, Patna, Pune and Varanasi (Pandey et al. 2006: 214).

top-down policies and ineffective management can be attributed as prime reasons for lack of efforts towards sustainable initiatives in India. Further, at practical level, mounting population, economic growth regimes, inadequate access to basic amenities, resultant varied forms of pollution and greenhouse gas emission, etc. can be few barriers against sustainability efforts.

Similarly, based on the current environment-related stress and ecological strains in the cities of global or India, environmental agendas have been classified as (i) blue, (ii) green, (iii) brown and (iv) purple agendas. Such classification is done to effectively analyse the environmental and ecological pressures, underlying causes and impact, to undertake legislations and to suggest a framework for remedial measures (Table 3.4).

Table 3.4 Characteristics of green and brown agendas in urban environment

Green agenda	Brown agenda
Natural systems of the local, bioregional and global ecosystem. The green agenda for ecological health – in terms of protection of biodiversity and management of natural ecosystem	The brown agenda is essential for making a city work, for a healthy and liveable environment and for creating the human and economic opportunities that have driven cities throughout their history. Thus, the brown agenda is about the optimisation of land use, engineering of waste systems, the minimising of energy consumption and transport, the reduction in use of materials and the creation of an efficient built environment
Ecosystems that provide green open space used by the city for biodiversity protection and recreation (e.g. greenhouse depletion, ozone depletion)	Waste systems to recycle and remove wastes from cities, including solid, water and air waste
Protection and conservation of water systems that cities use to tap the natural flow for water supply and waste disposal	Energy systems to provide power, heating, cooling and lighting for all city functions
Protection of climate and air systems that provide cities with the requirements for healthy life	Transport systems to enable mobility in the city, including the fuel
Protection and conservation of other ecological services, including agricultural forestry systems providing food and fibre for cities	Building and materials systems that provide the physical basis of life in cities

Source: UNHSP (2009: 114) and Marcotullio (2003: 229)

3.3.1 *Critical Issues in the Way for Promoting Sustainable City Model in India*

Some of the critical issues which need to be taken care of for promoting sustainable efforts as far as India is concerned are as follows:

1. The empirical relationship of sustainable city goals and aims for ecological and environmental protection within the broader framework of governance has received limited attention.
2. Community participation, empowerment, equity, social justice and transparency are not fully defined for analysis of sustainability.
3. Relationship between effective governance and the various economic and social dimension of sustainability is yet to be tapped in India.

3.4 Conclusion

This chapter briefly summarised the sustainable city models practised and implemented amidst environmental and ecological degradation around the world. However, today, many metropolitan cities of India are yet to implement and follow a sustainable pathway. The cities in India are facing accumulated effects of urbanisation and growing demand for basic amenities and infrastructure. Environmental challenges are further accentuated by the growing governance crisis like weak urban planning, proliferation of institutions and abysmal delivery of services. It is high time that cities of India, therefore, introspect their own potential and foster economic, social, environmental and ecological sustainability. In this context, the system of institutions and governance structures plays a critical role in promoting environmental governance which is discussed in detail in Chaps. 4 and 5.

References

Grimm, N. B., Faeth, S. H., Golubiewski, N. E., Redman, C. L., Wu, J., Bai, X., & Briggs, J. M. (2008). Global change and the ecology of cities. *Science, 319*, 756–760.

Lindfield, M., & Florian, S. (2012). *Green Cities: Urban Development Series*, Philippines: ADB.

Marcotullio, P. (2003). Globalisation, urban form and environmental conditions in Asia-Pacific cities, *Urban Studies, 40*(2), 219–247.

Pandey, S., Singhal, S., Jaswal, P., Guliani M. (2006). Urban environment, chapter 10 in indian infrastructure report (2006).

PECC (2010). A better city for tomorrow: Environmental sustainability urban centres: Towards a good governance approach to eco-cities, access www.pecc.com

UNCHSP. (2009). *Planning sustainable cities: Global report on human settlement 2009*. London: UN-HABITAT and Earthscan.

World Commission on Environment and Development (1987). *Our Common Future: World Commission on Environment and Development. New York*: Oxford University Press.
UN-HABITAT (2010/11). *State of the World cities 2010/2011: Bridging the Urban Divide*. London: Earthscan.
Zhao, P., Chapman, R., Randal, E., & Howden, P. (2013). Understanding resilient urban futures: A systematic modelling appraoch, *Sustainability, 57*: 3202–3223.

Chapter 4
Urban Environment Laws and Governance

Abstract The Bhopal gas tragedy in 1984 brought to the fore the lacunae in the environmental laws and the need for a more robust and comprehensive legislation, which resulted in the Environment Protection (EP) Act 1986. There are a number of legislations enacted under the EP Act that are relevant to urban environment – solid waste, air pollution, e-waste in groundwater and the most challenging issue of urban land use laws. Environmental governance should also be analysed in the context of prevalent laws and institutions administering the environmental policies and programmes. Compliance, enforcement, violations and dispute settlement are the key components of environmental governance. This chapter examines these components in the context of urban environmental laws. Four specific cases in which compliance has eluded the law have been discussed as well as the role of other factors such as technology, stakeholder participation, capacity building and a defined financial mechanism in effective environmental governance.

Keywords Environmental laws · Environmental governance · Urban laws · Environmental challenges · Environmental acts

4.1 Introduction

Under the Constitution of India, land, housing, urban development and provision of civic services are a state subject and, hence, under the purview of the state government. The state government can enact and enforce laws and policies related to urban governance. The central government plays an important role through allocation of resources under the national Five-Year Plans and devolution of funds to states.

The State Urban Development Department is in charge of Town Planning Department, Urban Development Authority, water supply, sanitation and sewerage. The Housing Department is in charge of housing policy and land ceiling. The main agency for urban governance is the Bruhat Bengaluru Mahanagara Palike (BBMP). It is responsible for the provision of all municipal services except public transport, supply of electricity and water supply, which are handled by separate agencies. Broadly the municipal functions include public health, regulatory, infrastructure, public safety, welfare and developmental activities. The 12th Schedule

K. V. Raju et al., *Urban Environmental Governance in India*, The Urban Book Series, https://doi.org/10.1007/978-3-319-73468-2_4

of the Constitution has given a list of 18 functions which apart from the traditional functions has included planning for socio-economic development, urban forestry, protection of ecology and environment, poverty alleviation, cultural heritage and primary health care. The incorporation of these functions into the municipal laws of each state varies. Karnataka has included all the functions mentioned in the 12th Schedule to its municipal laws. However, when it comes to the discretionary and obligatory distribution of the functions, Karnataka has kept urban forestry, environment and planning as discretionary.

The Ministry of Environment, GOI, has enacted a number of laws for the protection and management of environment. The earliest were the Water and Air Acts that attempted to deal with environmental issues by prohibiting the discharge of pollutants into the atmosphere and water so that people could breathe clean air and drink uncontaminated water. The CPCB and SPCBs were established under these acts to measure, test, inspect and take action against those causing pollution. Each state has also set up an Environment Department to carry out the functions of environment protection at the state level. The SPCBs are placed under the administrative control of the state department.

The Bhopal gas tragedy in 1984 brought to the fore the lacunae in the environmental laws and the need for a more robust and comprehensive legislation, which resulted in the Environment Protection (EP) Act 1986. This act gave the central government immense powers to formulate any rules for the protection of environment that could supersede other legislations which contravened the provisions of the EP Act. The EP Act does not specifically mention the term 'city' or 'town' or 'urban agglomeration', but the provisions in the EP Rules apply to 'any area' where emissions or discharges from any source exceed the environmental standards.[1] The powers are vested in the state government to prohibit/restrict any activity in an area taking into consideration standards of quality of environment laid down in that area, environmentally compatible land use and the biodiversity.[2]

In spite of the EP Act, the judiciary did take a view that some projects of national importance cannot be overshadowed by environmental hurdles. A writ petition was filed in the Bombay High Court by Goa Foundation (1992) asking for a directive to the Konkan Railway Corporation to get a clearance from the Ministry of Environment as per the EP Act. The court however, after examining the arguments, said that the damage to environment was negligible compared to the public importance of the project which will benefit the long standing aspirations of the people of 'west coast'. The Ministry of Environment subsequently in 1996 introduced the EIA statute, making it mandatory for major projects to get environmental clearance through an EIA and an EMP. The EIA guidelines were amended in 2006 to include construction projects having greater than 20,000sqm. This covers most big urban buildings such as shopping malls, commercial complexes, tech parks and so on. It also mandates a common municipal solid waste management facility in these buildings. All project proponents who conduct an EIA are required to provide information on

[1] Section 3 (b)EP Rules 1987.

[2] Section 5 EP Rules 1987.

change in land use, use of natural resources and nonrenewable resources, extent of solid waste and release of pollutants from fossil fuel, equipment and so on. They also have to declare the risk of land contamination from different sources of pollution. In 2009, the EIA made it compulsory that all information must be made public after obtaining environmental clearance through newspapers. Such information need have to be given to ULBs and also display the information for 30 days. In 2009, all hotels (<20 or >20 beds) through a notification were directed to discharge their effluents into the municipal sewer. In the same year the NAAQ standards for various parameters were revised, the most significant being the RSPM which was revised to 2.5 microns from 10 microns. Given the fact that air pollution levels were rising significantly in urban areas, leading to plausibly more related health disorders, the CPCB found the need to revise these standards.[3]

4.2 Environmental Laws

There are a number of legislations enacted under the EP Act that are relevant to urban environment. One of the first is the *Hazardous Waste Management Rules 1989*. This is mainly applicable to hazardous waste-generating industries, some of which were located within city limits in the 1980s and early 1990s. The Supreme Court in July 1996 ordered 168 polluting industries located in Delhi to be shifted out and prohibited any hazardous industry being set up in residential areas. While many industry owners wanted to shut down their industries and sell off the prime land in Delhi at exorbitant prices, the SC in its ruling said that the owners can sell only 32% of their land, and in case of shutdown, the labourers should be paid 6 years of annual wages. This compelled many to relocate their industries to neighbouring states of Haryana and Rajasthan. However, the Master Plan of Delhi 1990–2001 allowed for 'light' industry to be set up in residential areas. This created a leeway for many industries which were not in the hazardous category (H) to start in the areas vacated by the 'H' category industries without the necessity for any permission from the Delhi Pollution Control Committee (DPCC). Since there was confusion prevailing about the location of light industries in these areas, the matter was again cleared by the HC of Delhi in its judgement given in 2004 that the DPC has the mandate to take any action if the industry was in violation of any law, but if it was permitted under the Master Plan, then the industry had the right to carry on its activities.

In 1999 the Plastic Recycling Rules were notified, prohibiting the use of recycled plastics for storage of food and restricting the manufacture of carry bags and containers to virgin plastic not less than 20 microns in thickness. The rules were

[3]According to the *Journal of the American Medical Association*, 2.5 PM can cause hardening of arteries leading to cardiovascular diseases. As recently as January 2014, the British Medical Association has said that long-term exposure to ambient air pollution leads to coronary events. WHO estimates that about 3–5% of mortality from cancer and cardiovascular diseases is from fine particulate matter.

further amended in 2003 and 2011 which prohibits the use of virgin or recycled plastic bags/containers of specific size in addition to thickness and also assigned the responsibility of collection of plastic waste to the municipal authorities along with a mechanism of extended producer responsibility (EPR). A number of states took initiative to implement the Plastic Rules. HP was one of the first states to ban the use of plastic, while Delhi in 2009 had allowed the use of only virgin plastic but in 2012 completely banned all plastic. Maharashtra has constituted a Special Cell to monitor compliance of Plastic Rules. The Burdwan municipality in WB took a proactive approach to impose 'pollution cost' on unauthorised use of carry bags of Rs 500 per bag. Bengaluru has not banned the use of plastic bags/containers but restricted it to recycled or virgin plastics in accordance with the Plastic Rules.

The *Biomedical Waste Rules were first enacted in 1998* to manage the storage, treatment and disposal of biomedical waste in a manner that is not harmful to the environment and health. These rules are applicable to any institution generating biomedical waste such as hospitals, nursing homes, laboratories, veterinary institutions and animal houses. The rules have provided for different treatment systems such as incineration, autoclaving, shredding, etc. and specific coloured bags for storage of different categories of waste. The rule also prohibits storing untreated waste for more than 48 hrs, and all waste must be disposed in authorised landfill sites. An amendment made in 2000 obligated the municipal authority to provide the common disposal site for biomedical waste. In case the waste generator is outside the municipal area, it is the responsibility of the generator to locate the site and ensure safe disposal of its waste.

In January 2013, the High Court of Delhi directed the civic authorities of Delhi to shift the biomedical waste incinerator from the residential area in the city to outside. The court cited the CPCB guidelines that incinerators are hazardous industry and harmful to the health of people and should therefore not be located within or near residential areas. The Master Plan of Delhi 2021 has also prohibited such industries within the National Capital Territory. For the first time, a judicial magistrate in Orissa, Jagatsinghpur, has convicted a nursing home operator for 2 years of imprisonment for failing to comply with the BMW Rules 1998.

Visibly, in a city, the most important environmental hazard is the solid waste generated in the households and commercial establishments. The *MSW Rules 2000* were notified in the year 2000 for the proper collection, segregation, transportation and disposal of the waste. While the main responsibility of implementing the rules is with the municipal authority, the SPCB shall monitor the compliance especially related to air quality, groundwater, leachate and compost quality. They are also authorised to grant permissions for setting up waste processing/disposal facilities. The rules mandate that biomedical, construction and horticulture waste should not be mixed with the MSW. Detailed guidelines on the standards for a landfill site and the permissible levels to maintain water and air quality are mentioned. Only non-biodegradable waste is allowed to the landfill sites, while all other wastes should be composted or incinerated. Any new waste processing technology requires the prior approval of the CPCB which will evaluate and set the standards for that technology. Not much has been said about segregation, except that awareness programmes must be conducted by the municipal authorities with involvement of local NGOs.

The *Biodiversity Rules 2004* regulate the access, use and sharing of biological resources and traditional knowledge. They mandate that every local body must create a Biodiversity Management Committee (BMC) for the local area with an elected chairperson and other members. The BMC has to prepare a Peoples' Biodiversity Register in consultation with the local people containing information on the biological resources, medicinal and traditional knowledge and its uses. The committee is also expected to maintain a register that will contain information on the access granted for use of resources and benefits shared. The act does not explicitly mention about conservation of bioresources or biodiversity in urban areas. It is more focussed on access, use and sharing of benefits. The BMCs are not initiated in most states except AP, Chhattisgarh and MP.

One of the recent rules enacted by MoEF is the *E-waste Rules 2011*. The rules have fixed specific responsibilities on producers, consumers, collection centres, dismantlers and recyclers of e-waste. The producers of electronic and electric equipment/components have to ensure that the e-waste is collected and managed in a proper manner by sending the waste to authorised collection centres or registered recyclers. The rules also mention that the producers should finance and organise a system for environmentally sound management of e-waste in line with the principle of EPR. The consumers too have to ensure that the waste is given only to authorised recyclers or dismantlers or given back to the producers. Subrule 13 (1) mandates that every producer of electronic and electric equipment will ensure there is no use of certain hazardous substances such as cadmium, mercury, etc. in the manufacture of new equipment. No one is allowed to store e-waste for more than 6 months unless it is being used for recycling.

These rules are having certain loopholes which make it convenient for the informal recyclers to continue with the collection and recycle of e-waste in a hazardous and environmentally harmful manner. Though the rules insist that all recyclers should be registered, there is no way to ensure that it happens, since it is difficult to locate informal recyclers. Many people find it more attractive to dispose their e-waste to local dealers who offer a good price rather than give it back to the manufacturers as required by the law. The biggest impediment to implementing these rules is that it is closely related to the municipal solid waste management. In most places e-waste is mixed with MSW, and proper segregation practices are needed to separate e-waste from MSW. So far in most cities including Bengaluru, segregation of waste remains improper and scarcely implemented.

4.3 Urban Laws and Environment

Land Laws One of the most challenging issues of urban governance is the implementation of land use laws, especially in huge metros like Bengaluru. Many a times land use laws are violated ending in illegal constructions, which means the purpose for which the land was permitted is misused. The most common misuse is where land earmarked for residential purpose is used for commercial activities or permitting certain type of commercial activities which are prohibited. This type of violation needs to be distinguished from building by-law violations, where the land use so permitted may be in the conformation zone but the technical specifications of construction are not met. For instance, the by-laws specify certain floor area ratios and extent of setbacks for different zones but may not have been adhered during the time of construction. From an environmental impact perspective, both violations are important, but the former has greater significance than the latter. However, neither the Town and Country Planning Act (1961) nor the Municipal Corporations Act (Karnataka) has addressed the issue of potential environmental degradation caused by violations of land use. The term environmental degradation is not narrowly confined to air and water pollution but would encompass stress on groundwater levels and water quality, setting up of towers that emit electromagnetic radiations, increased demand for energy and use of LT power for commercial activities, hazardous/toxic waste generated and its disposal, loss of ecology and increased noise and stress levels. The TCP Act of 1963 does mention of that industries in RED category shall not be allowed in non-conformation zones. Interestingly, when the KSPCB issued a closure order to an industry based on the complaints from the residents causing air pollution, the Karnataka High Court in 1987 held that an industry cannot be simply closed on the grounds that it did not obtain consent from the PCB to operate. The PCB then started to file criminal cases against those violating environmental laws which would be kept pending for years in the court.

The TCP in 2007 made an amendment regularising all illegal constructions and violations by making a hefty payment as fine. What impact this could have on the city's environment has to be studied. It is generally believed that mixed land use yields socio-economic benefits and therefore has a positive effect on housing and commercial values. However, when residential areas are altered to commercial and public activity, it mounts pressure on existing infrastructure like water supply, traffic, sewerage, etc. The Draft Revised Master Plan 2015 has permitted, as a part of its zoning regulations, mixed use of residential areas for certain categories which are termed C2, I2, U3 and T2. Under C3 are all commercial/corporate offices, retail shopping complexes, restaurants, banquet halls and automobile garages apart from petty shops, nursing homes and kalyana mantapas. Under I2, IT/BPO activities, R&D labs and tiny industries are allowed. Under T2, transport garages and workshops are permitted while, in the U3 category, schools, hospitals, government buildings, auditoriums, dhobi ghat and broadcasting stations. Although the RMP 2015 makes it obligatory for all commercial activities listed under KSPCB to obtain necessary NOCs from the board, the cumulative impact of these commercial/public activities on environment is not considered at the time to giving the NOCs.

A PIL (3676/2008) was filed in the Karnataka High Court challenging the RMP 2015, prepared by the BDA that allowed commercial use of residential areas through mixed residential land use category. The HC issued an interim order on 21.1.2012 and another order on 12.11.2012, directing BBMP to not permit or grant any change of land use in all residential areas listed in RMP 2015. Subsequently, the division bench of the HC in February 2014 issued a direction to the state government to amend the land use change rules of the BDA's RMP to allow only ancillary use of residential properties instead of land use change. The BDA through an affidavit submitted to the HC agreed to amend the RMP. Till such time, the court directed that the interim order shall apply.

The *Karnataka Municipal Corporations Act 1976* gives a number of operational directives to the city corporations that have a positive bearing on the environment. However some of them seem to contradict the provisions of the RMP. For instance, Section 304 of the act states that sanction will not be given to buildings for entertainment within 200 m radius of residential areas, whereas the RMP may permit these buildings as part of the mixed land use regulations. There is also a provision in the act that it is not lawful for the owner of any dwelling to occupy, unless the corporation is convinced that sufficient water is available for domestic consumption within reasonable distance. In most cases where commercial activities are permitted, the requirement of water may be considerably high which, in turn, could lead to reduced availability of water for domestic consumption. Water may also be sourced from private tankers that draw water from a distant place depriving that place of the water availability as well. In such instances the corporation should not grant permissions on the basis of not being able to meet the provisions of this section. There are certain other provisions with good intentions but highly challenging to implement. For instance, Section 256 says that it is the responsibility of the owner of the premises to clean up any rubbish thrown nearby and not to accumulate such rubbish for more than 24 hrs. Similarly, no person is allowed to throw chemicals, liquids higher than 45 deg. Celsius into the drains. It could be very difficult to implement this since the origin of such material is difficult to trace.

Groundwater Law The *Karnataka Ground Water Regulation Act 2011* is another significant piece of legislation that aims to regulate the use of a very valuable natural resource. As per the act, areas where groundwater use and extraction need to be regulated have to be notified by the Ground Water Authority, and in such areas, prior permission is required to dig or use an existing borewell. Such permits can be granted only after due consideration is given to factors such as availability of groundwater, use of groundwater, competitive users and long-term effects of groundwater use in that area. In case anyone fails to get a permit, then BESCOM shall not supply power for borewell extraction as well as no institution can lend finance for the same purpose.

Bengaluru figures as one of the cities in Karnataka where all groundwater quality contaminants show higher than permissible standards which are fluoride, nitrates, iron and salinity. It is also one of the cities notified for GW development. Some reports portray a dismal picture of the city's GW condition. The saturation level for water table recharge has gone up to 141% in Bengaluru as against the desired level of 70% beating even the state average of 68%. The maximum exploitation of GW is taking place in the peripheral areas of the city which are also the areas which have the least amount of recharge. The GW levels are highest in HSR, Kallikere and Yelahanka, while the least levels are in Sadashivanagar and Pattangere.

Urban Transport Law *The Motor Vehicles Act 1988* mandates rules to control emission of air pollutants from motor vehicles, installation of catalytic convertors and fix standards for emissions in consultation with MoEF. The State Road Transport Authority has powers to enforce the provisions of the act in each state. Most motor vehicles that belong to the category of light vehicles have pre-installed catalytic convertors that follow Euro II norms and in some cases Euro III norms. All vehicles have to compulsorily get a Pollution Under Control (PUC) certificate as per the PCB norms. The RTA has powers to inspect and check the certificates. In case the owner possesses a certificate but the vehicle continues to emit smoke, the RTA can get the vehicle tested. In case it is found in excess of the limits, the vehicle owner can be given a time period to get the level of pollution within limit and report compliance but cannot be prosecuted (Sushil Kumar Dutta v. State of WB-2004).

What is of concern is the growing number of vehicles in Bengaluru. It has more than 30 lakh vehicles registered, of which 50% are two-wheelers. The daily pollution load of Bengaluru shows 207 MT of CO emitted every day followed by 117 MT of HC. The Nox and Sox levels are under control given that there is use of unleaded petrol and catalytic convertors as well as shifting HTV outside the city. According to studies, the increase in vehicular population has also led to CO2 emissions.[4] As per estimates, CO2 emissions from road transport have increased from 4.5 mt in 2000 to 11.2 mt in 2011 which is almost three times increased. The RTA says that there is no provision under law to restrict the number of vehicles on the road or the number of vehicles that can be purchased. The only way is to see how to reduce the fossil fuel consumption which will bring down the CO2 emissions.[5]

Disaster management law Karnataka Disaster Management Authority has been set up under the *National Disaster Management Act 2005*, to coordinate with various agencies at times of disaster and also focus on preparedness and early warning. In 2007, the Department of Science and Technology, GOK, has established the Karnataka Natural Disaster Monitoring Centre with the objective of providing scientific inputs on disaster management, hazard mapping, vulnerability assessment and early warning system for natural disasters. Subsequently, in 2008 a Disaster Management Cell for Bengaluru was set up with the deputy commissioner as the head and 17 agencies including BBMP, fire department, traffic police, etc. to coordinate during times of disaster. Noteworthy is the fact that the Environment Department is not included.

Most natural disasters in Bengaluru are related to fire, building collapse, tree fall during heavy rains, floods in low-lying areas and sometimes spread of diseases such as bird flu, dengue, etc. Often, these disasters are caused due to negligence on the part of government agencies to enforce compliance or take preventive measures. The Disaster Management Cell has not been able to do much to prevent the disasters. If the cell takes up periodic vulnerability assessment for old buildings, overgrown

[4] Karnataka Climate Action Plan, CSD, 2012.

[5] Seminar on Air Pollution by CSD in July 2013.

trees, fire-prone zones and so on, it will help the concerned agencies to act on time. There are also unnoticed natural disasters such as toxic pollutants in groundwater, storage of explosive fuels and substances, open drains and stagnant ponds that can have long-term disastrous effects. The cell should take note of these potential threats and communicate to the concerned agencies with a time-bound action plan.

4.4 Key Environmental Issues and Challenges

Environmental governance should also be analysed in the context of prevalent laws and institutions administering the environmental policies and programmes. Compliance, enforcement, violations and dispute settlement are the key components of environmental governance. It is proposed to examine these components in the context of urban environmental laws in Bengaluru.

Compliance It is hard to find an answer to a question on the extent of compliance to environmental laws in the city. There are no information resources on environmental compliance. For instance, if one wants to know the establishments in the city that comply with Solid Waste Management Rules, against those that do not comply, such information would be found wanting. The only agencies that could possibly have this information are KSPCB and BBMP. But neither has recorded data. KSPCB records air pollution data at hotspots in the city which show that air pollution levels are quite high. The source of such pollution is not clearly known and therefore compliance is evaded. Measuring environmental compliance requires proper record-keeping and monitoring system which at present is frugal in Begaluru. The KSPCB maintains the consent to establish records of applicants but how many are unregistered is not known and whether those registered establishments are following the law is also not clear. The BBMP too has no centralised data on solid waste and is difficult to gather data from each regional office. In conclusion it is difficult to pass a judgement on compliance without proper and reliable information.

Enforcement There is ample evidence to show that enforcement of environmental laws in the city is weak. One of the most significant cases is the groundwater pollution. Both the Lake Development Authority (LDAs) and the Central Ground Water Board have said that groundwater pollution is a serious problem especially caused by industry. Recent reports of the DMG (2011) also suggest the extent of groundwater pollution in the city. Despite the Ground Water Regulation Act 2011 being in force, there is no account of the actual number of borewells in the city and action taken to curb illegal extraction of groundwater.

The LDA has complained of the illegal encroachments on lakes in Bengaluru, silt being removed for brick making and dumping of sewage and solid waste in lakes. All these activities should have been prohibited by the enforcing authorities. The LDA on an average receives 10–12 complaints every month. Recently the residents of Madiwala Lake filed a complaint with KSPCB that BWSSB was illegally discharging raw sewage into the lake through storm water drains and the water had become so polluted that it was unfit even for animal consumption. KSPCB has issued a show-cause notice to BWSSB.

A recent study by TERI on Bengaluru's air pollution reveals that 42% of the pollution is caused by vehicles, while the growth of vehicles is about 10.5% in 2012–2013. The study also observes that the number of vehicle owners' penalised for pollution is very low. The Transport Department too has admitted that there is a shortage of vehicle inspectors.[6]

In so far as land use violations are concerned, there is no official statistics, but reports suggest that only 10–15% of the applications get 'change in land use certification'. The rest are not authorised.[7] The growing number of noise pollution complaints in the recent past shows that in many residential areas, commercial and industrial activities have been permitted. KSPCB has also accepted these findings in a recent High Court case.

The above instances show the lack of proper enforcement by the regulatory agencies.

4.5 Violations and Penalty

The National Green Tribunal (NGT) ordered status quo on constructions that covered drains in Delhi. As per EIA notification 2006, it is necessary that construction of buildings on drains has to obtain environmental clearance from the state-level Clearance Committee. Since most of the buildings had not obtained clearance, the NGT found them to be in violation of the law and so constructions was stopped.

A municipal solid waste management plant in Bareilly, UP, was asked to shut down and shift by the NGT based on complaints and protests from local residents. The tribunal observed that the plant was set up without the necessary environmental clearances from the state and also violated the Master Plan 2021 in which the site was earmarked for social and cultural activities. Further the tribunal in its ruling said that the larger interest of the public must prevail over the narrow end of collecting and disposing the waste.

As per Environment Protection Act (1986), the penalties specified in Section 15–17 of the act:

Section 15 deals with penalty for contravention of the act and rules making it a jailable offence for a term of 5–7 years.

Section 16 deals with offences by companies making the person at the time of offence responsible and deemed guilty.

After the Supreme Court order in 2003 on hazardous waste, stringent actions were prompted by various SPCBs particularly in the states of Gujarat, Maharashtra,

[6] *The Hindu* dated 10 May 2014.

[7] *Times of India* 12 June 2005.

Kerala and Delhi. The Gujarat Pollution Control Board had directed the closure of 957 industrial units for a period of 1 month and was reopened only after necessary authorisations were issued. In Kerala 195 industrial units, including 32 major industries, were issued specific direction for closure in August 2004. In Maharashtra 75 factories were directed to be shut down and could be reopened after they provided bank guarantees of 25,000 each and which would stand forfeited if the pH of the discharge is below 5. One company in Goa had to pay a hefty fine for discharging cosmetic contaminated waste.

The Supreme Court Monitoring Committee noted that after its directions to clear up illegal waste dumpsites, about Rs 100 millions were spent by industries in Gujarat, Maharashtra and WB for removal and safe disposal of hazardous waste.

When it comes to penalties and punishments, it has been limited and not effective. This is perhaps due to violation being rectified through non-legal measures or inadequate capacity to inspect and file proper reports, based on which penalties could have been imposed. The reason why violations are high is because punishment and penalties are few. Except for some state governments as mentioned earlier, which recently issued closure orders or levied fines for a few industries, in most cases, the violator is not penalised heavily. Obviously, the costs of abatement are more than the costs of violation. Violations occur maximum at the level of informal recyclers and small-scale hazardous waste-generating industries and commercial establishments. They are difficult to track, monitor and report. The existing system of environmental governance lacks the appropriate framework for proper monitoring and verification at grass root levels which is the reason large number of violators escape any punishments.

A stricter enforcement regime could have enhanced the effectiveness of the law. But the point to be made here is that the activities and tasks involved in implementing the provisions of the law are multifarious and complex. It also requires proper technical knowledge and competence by the persons implementing the law. At present this capacity is low in the SPCBs and CPCB. A decentralised framework for administering the Hazardous Waste Management Rules is necessary. If enforcement is to be strengthened, coordination mechanism at the state level are important to ensure that different agencies are in the know of the law and the steps being taken to control hazardous waste.

Stringent punishments and heavy fines could have been good deterrents for violators. Merely issuing show-cause notices and revoking closure orders based on 'cosmetic compliance' steps in industry or paltry sums of fines will never stop an industry from generating and disposing hazardous waste. In the absence of a 'carrot', at least the 'stick' would have to work.

Where the law is expected to be stringent, there should also be supportive system to help the industry comply with the law. When Hazardous Waste Management Rules were formulated in 1989, most of the generators were not even aware that the waste they produced was hazardous to the environment and that disposal in any available natural sink was the right thing to do. The law has since then evolved and has become more streamlined. But the generator still finds it difficult to follow because there is no means either financially or technically to comply with law, which to him is a costly affair. The government may have thrust the onus of compliance on the generator or the facility, but without providing the proper means to achieve the objectives, the law will continue to be violated.

Dispute Settlement The National Green Tribunal has been established under the *National Green Tribunal Act 2010* for effective and expeditious disposal of cases relating to environmental protection and conservation of forests and other natural resources including enforcement of any legal right relating to environment and giving relief and compensation for damages to persons and property. It is a specialised body equipped with the necessary expertise to handle environmental disputes involving multidisciplinary issues. The tribunal is mandated to make and endeavour for disposal of applications or appeals finally within 6 months of filing of the same.

So far NGT has given about 520 judgements since it has been set up, including those related to urban environment. The NGT has played a crucial role in the adjudication of environmental cases. Its judgements are based on three principles – polluters pay, precautionary principle and sustainable development. By and large the NGT has been commended for its role in protecting environmental degradation through speedy disposal of cases. Recently, the Chennai HC has said that the orders of NGT can be challenged in the higher courts. Those accused of polluting may now have the chance to appeal further.

Apart from the NGT, courts of subordinate judge can be approached for the grant of permanent injunction and compensation in cases of environmental degradation due to mismanagement of civic amenities and commercial establishments operating illegally in residential buildings causing environmental problems to other residents.

Special courts can also be constituted at state level to try cases under Air, Water, and Environment Protection Acts. Two such courts were set up in Haryana and one in UP. Regular writs and PILs can be filed in the High Courts against the administrative machinery for failing to act against violations under the environmental regulations.

Administrative bodies have provision for dispute settlement mechanism. Grievance Cell was constituted in the MoEF in 1991 to attend to complaints from public regarding environmental problems. The SPCB and the municipal/corporation commissioner have the authority to hear and act on complaints regarding pollution affecting civic amenities.

Citizens therefore have a number of avenues for dispute settlement, but most of them are not effective. This is more so in the case of administrative bodies/authorities who fail to act against the violators. It is also discouraging to note that the penalties handed down by these bodies including the courts are not effective deterrents. Hence pollution in cities continues to be unabated.

Emerging Challenges Three questions need to be addressed: Have urban environmental laws been successful in checking the rampant growth of the city? How can compliance to environmental laws be strengthened? How can factors such as technology, stakeholder participation, capacity building and financial mechanism play a role in effective environmental governance?

The Akrama Sakrama scheme recently approved by the government shows the extent of illegal buildings in the city (it is difficult to put a number as they run into thousands) and the callousness towards urban laws. Allowing violations in building norms or unauthorised land use change or illegal groundwater extraction has led to the unrestricted and unorganised growth of the city which in turn has drastically impacted our environment and resources. In 2011–2012 the number of complaints received by the KSPCB from different parts of the city is about 471, of which 313 have been attended to in the same year. The number of writ petitions pending in the High Court is 36 and in lower courts 80. These numbers indicate the growing environmental problems in the city and the lack of effective implementation of laws. If Akrama Sakrama were to be implemented, what impact would it have on the environment? Will KSPCB document those violations that create environmental hazards and provide a mechanism to restrict the damage in the future? At least for the present, there is no move to safeguard the city's environment in the wake of regularising the violations.

One of the main reasons for poor compliance is multiple laws and multiple agencies working at cross purposes. What was once the sole responsibility of the municipal corporation such as garbage disposal, water pollution, noise and so on has now been divided between the municipal corporation, KSPCB and CPCB. This has resulted in each agency passing the buck without being accountable, and this seems to have helped the violator to get away scot-free. Take four specific cases in which compliance has eluded the law.

Air Pollution It is known that about 40% of air pollution comes from vehicles. Most light motor vehicles are Euro II or III compliant, many autos have converted to LPG, heavy vehicles except BMTC buses are not allowed to ply in the city's main roads and PUC certificate is mandatory for all vehicles. Despite these measures, air pollution levels have remained high in all the major junctions of the city. The real cause is construction activity, street sweeping, disposal of debris and burning of litter which contribute to high SPM levels. Narrow and bad condition of roads adds to emissions from vehicular pollution that gets stagnant and enhances the levels of air pollution. The BBMP is primarily responsible for all of these functions that actually create high levels of air pollution. But BBMP is legally not responsible for such pollution, and therefore the blame often goes to the PCB.

Water Quality BWSSB ensures that quality of water supplied is tested and found to be of permissible standards. However, in areas where water is supplied by private tankers or borewells, the quality of water is found to be contaminated (ERC). Who is to test the quality of water in this case? The Ground Water Authority is the agency to test borewell water quality, and the BBMP Health Department is responsible for water quality if supplied through private sources. The KSPCB should also ensure that water is not polluted and has the authority to inspect and take action against any source of water pollution. In practice, none of these agencies test water quality provided through tankers or borewells, and therefore people living in areas out of BWSSB purview continue to use contaminated water.

Contamination of water bodies and sites – Among the most polluted places in a city are its lakes, tanks and open wells which become the obvious dumping yards for all types of pollutants. It is very difficult to trace the polluter since it could be an individual, commercial establishment or an industry. Who is expected to control such pollution? Again the responsibility is divided between the city corporation, BDA, LDA and KSPCB depending on who has control over the specific water body. The plight in case of contaminated sites is even worse. Many open sites and government land often becomes the sink for pollutants including some very carcinogenic and heavy metal substances. These contaminants over a period of time destroy groundwater sources and also render the land unusable for human activity. Most of the time, this is not taken cognisance, and the land is used for habitation or commercial purposes. The PCB is supposed to monitor the extent of pollution in these sites along with support from other departments. However, this has become a no man's island and continues to evade the law.

Another reason for poor compliance is that there are no proper prosecutions. The KSPCB is the agency which can file legal case against polluters and prosecute them. According to a newspaper report (DNA 12/4/2011), there is not a single person prosecuted by KSPCB so far though it has the powers to do so under the Water Act. The RFD of the KSPCB reveals that there is no mention of any prosecutions by the board, though they claim that action taken on non-compliance is 100%. In most cases, the board orders a temporary shutdown of operations till such time that the erring party takes action to become compliant. In 2012, the KSPCB ordered the shutdown of Ramky for improper waste management at Mavallipura and issued a show-cause notice to the BBMP commissioner for not taking action to prepare a plan for waste disposal and stop sending waste to the landfill. But here too the board stopped short of initiating any legal proceedings against the BBMP or Ramky.

There is a general lack of awareness about environmental laws by those violating them, including citizens, real estate developers, builders, architects, commercial establishments and SMEs. The PCB and the Environment Department have not taken sufficient steps to promote awareness of environmental laws that affect us in daily life. Karnataka is one of the lowest spenders among states on environment. The environmental budget for the state is 0.004% of the GDP (ESI, CFD, 2009). Apart from setting up of eco-clubs in schools, the department's expenditure does not show any specific awareness programmes on legal rights of citizens on environmental issues.

The state government and the PCB lack adequate trained manpower to enforce the laws in an effective manner. The DFEE has one senior technical director and two technical officers apart from the administrative staff headed by principal secretary and one undersecretary (RTI Manual). The KSPCB has about 130 technical officers of different ranks. Given the magnitude of laws to be administered, the number is inadequate. According to the board, it requires twice the number to ensure proper monitoring (Deccan Herald). Most of them are also not fully trained on subject matters pertaining to the law.

Effective environmental governance cannot come only by enforcement of laws but also the role of other factors such as technology, stakeholder participation, capacity building and a defined financial mechanism. Technology is a vital element for both compliance and effectiveness. More than legislation, the availability, accessibility and affordability of technologies are the key. Ozone-depleting substances (ODS) could be easily phased out, because alternatives were available and even affordable. There are not many incentives given for adoption of technology. Although broad mentions are made in the National Waste Policy on hazardous waste, climate action plan and so on, what concrete actions are being taken to make technologies available at affordable costs is still unclear. What is more surprising is that neither of the laws in question mentions the type or standards of technologies required. The USEPA has documented technically and commercially viable technologies for waste treatment that waste generators can adopt. The law also provides for a corrective action process wherein the violator has to adopt selected technologies for treatment.[8] Therefore, technology must become a part of the regulations, supported by a financial mechanism for technology transfer.

Stakeholder participation has been much advocated in clearance of environmental projects and is mandatory under the EIA. In most cases, this exercise is more cosmetic than realistic, and therefore many projects get environmental clearance. When it comes to urban environment, this process is virtually absent. It is one of the reasons why Bengaluru has been witnessing a lot of protests from residents and others in various parts of the city. For example, 2 years ago, residents of Malleshwaram strongly objected to felling of trees on Sampige Road which was to be done for road widening. The local residents were not consulted by the BBMP or the Forest Department before deciding on the work. Similarly, PCB had to order the closure of two landfill sites due to strong protests from the residents of Mavallipura and Mandur. The sites were allocated without any stakeholder consultation. The Bengaluru Governance Bill has provided for Neighbourhood Committee that will comprise of local resident representatives who would be involved in the planning process of the neighbourhood. However this bill is in the limbo and is yet to become an act.

There is a debate on the need for new environmental laws to address emerging urban issues like climate change and disaster management. Climate change is a complex phenomenon and multidimensional in its effects. Most of the GHG emissions that cause climate change originate from industries like thermal power plant, cement and steel industry which are not located in cities. Similarly methane emissions largely come from agricultural waste which is confined to agricultural and peri-urban areas. The major contributors of emissions in the city are vehicles, solid waste disposal, felling of trees and consumption of fossil fuel for generating power by DG sets. Laws such as the Air Act, Environment Protection Act and Energy Conservation Act already exist to deal with these issues. Besides each sector has its own set of policies and laws that can mitigate climate impacts.[9] There have been good initiatives taken by various agencies like KSRTC, BESCOM, etc. to reduce stress on resources which helps mitigate climate change. The Karnataka Climate Action Plan (2012) has given in detail the policies of the government of Karnataka that already exist to deal with climate change. The real problem in cities lies with

[8] www.in.gov/idem

[9] Karnataka Climate Action Plan, EMPRI, 2012.

huge influx of migrants, uncontrolled expansion and dwindling natural resources. Unless there is strict enforcement of urban development laws, the emerging environmental issues cannot be fully redressed only through environmental laws. There should be greater involvement of environmental protection agencies at the time of drawing up the Master Plan for the city.

4.6 Discussion and Reflections

1. The current legal framework of urban environmental law is 'heavily loaded' and 'restrictive' in nature, devoid of any incentives to promote sustainable development. For example, laws under the KTP Act and KM Act place a lot of responsibility on the landowners and builders to follow a slew of regulations most of which they are not aware. In case of Municipal Solid Waste Rules, they overburden the generator with a lot of obligations which may act as deterrent to compliance. Therefore, implementation of such laws needs greater inspection, monitoring and penalty. The current establishment is grossly inadequate to meet the requirements of effective implementation. It is only wise to look at steps to deregulate a complex and overburdened urban environmental legal framework. As a first step towards effective urban environmental governance, list the laws that are redundant or unnecessary or overlap with other regulations and find ways of eliminating them. One such example could be to remove some of the existing powers of the corporation commissioner-related Section 192–2 where the commissioner should be convinced that sufficient water for domestic consumption is available before occupying a dwelling, Section 208 prevention of waste or misuse of water and Section 334 abatement of nuisance from dust, smoke, etc. in the neighbourhood. These laws overburden the commissioner and often are not enforced. In some instances, they may be used to harass residents and establishment owners. Such laws must be vested with a separate Urban Environment Compliance Unit created under the state's Environment Department which can exclusively facilitate compliance of environmental laws in tandem with other laws relevant to a city.

   The other big challenge is to allow development and commercial activities without harming the environment and health of the citizens. This calls for urban planning based on the principles of sustainable development. When Master Plans are being drawn, EIAs for different zones based on certain environmental and social parameters can be done rather than doing an EIA at the time of sanctioning specific building plans. This exercise can give an idea of the extent of environmental damage likely from cumulative activities in that area and not on an isolated basis.

2. Environmental law must be viewed differently from the regular civil law since environmental problems are more often between the state and citizen rather than between citizens (except in some nuisance cases which are dealt under IPC). More importantly, in environmental cases, often the origin of the problem is not known, and therefore the perpetrator is assumed to be the state for negligence

and inaction. In cases where the source is identified, the state initiates legal action against the polluter, usually after a complaint is filed. Solutions to environmental problems can be found with a mix of technology intervention, community action, evidence-based approach and strict enforcement.

3. To make implementation more effective, it is important to influence the target behaviour. The lawmakers should not confine the law to 'command-and-control' methods but enlarge the scope of environmental legislation to promote positive compliance. It means that, when laws are formulated, it is not sufficient to consider only issues of non-compliance but those that promote compliance. This is where factors (non-legal) such as economic, technological and social should be brought under the ambit of law to promote compliance and make implementation more effective.
4. Science policy interface is another key element to good environmental governance. Policy cannot be based only on scientific facts, but other factors like political issues and local conditions need to be taken into consideration. Environmental policies should specifically address development planning and poverty reduction in urban areas. Local data should be collected by the government so that preventive measures can be taken to avert any environmental disasters. For instance, if proper information on low-lying areas and the persons living there can be updated regularly, it could be useful to take timely measures during floods.

   Laying of roads, footpaths and drains should be done according to scientific inputs and standards. However, in cities like Bengaluru, these are done based on political considerations, resource constraints and lack of understanding the local topography. It is important that such basic infrastructure must be built only on scientific grounds for which there must be a strong policy in place that does not compromise on established scientific practices.

5. Environment is essentially a state subject, but ironically most of the environmental laws are nationally enacted like the Air, Water, and Environment Protection Act. However, these acts have enough powers delegated to states to enact stricter laws and take action against violators. The state governments have rarely made laws relating to the protection of their local environment. For instance, are there any ecologically sensitive areas declared in Bengaluru by the state, or why can't Bengaluru be made a plastic-free city? It is important to note that environmental issues differ from city to city and could be highly local. But current laws governing environment are more nationally oriented which mandatorily put more focus on areas not environmentally sensitive for a particular city or where even a law may not be required. This argument is not to dilute the importance of national laws but to reduce the number of such laws so that the burden of implementation becomes lesser for the local governments and, at the same time, they use their resources to handle more sensitive problems that plague them locally. For example, MSW Rules can be left to each city/state government to decide depending on the extent and type of the problem in each place. National laws on environment can be confined to larger issues that affect us at the national and international levels such as ozone layer protection, climate change, etc.

6. Financial mechanism is required to complement the regulatory framework to make compliance more effective. In addition to strict regulation, financial intervention is also needed to facilitate compliance. For instance, informal waste recyclers can be provided financial assistance to bring them under an organised system so that they can comply with the regulations without affecting their earnings. Those companies that will be covered under the amended Companies Act for CSR should compulsorily set up a fund for recycling water in the city. This could enable a large section of population to access water throughout the year. Financing technology is a very important aspect of ensuring compliance. Incentives must be given to technology manufacturers to promote clean technologies that will help reduce consumption of resources and make environment-friendly products more affordable. Fiscal incentives can enhance the volume of clean technology sales, which in turn can bring greater revenues to government through taxes on finished products. The more challenging issue is to liberalise the procedure for availing the benefits of fiscal or nonfiscal incentives.

Chapter 5
Bengaluru Metropolitan City: An Environmental Overview

Abstract The city of Bengaluru is one of the fastest-growing cities of the twenty-first century. From the past decade, the city has geographically grown enormously as a result of urbanisation and pursuing economic growth. Concomitantly economic setting of the city has undergone tremendous transformation which is characterised by the concentration of IT/BT industries largely located in the periphery of the city. Realistically, the city does not register growth with efficient access to infrastructure and basic amenities. As a result, there is a visible stress on local governance to promote quality of life in terms of protection and conservation of city's ecology and environment. This chapter examines how the city has grown geographically over the years as a result of rapid urbanisation. The chapter summarises how the growth patterns of both spatial and demographic have carried with them enormous environmental problems inevitably leading to the verge of environmental crisis-like situation for the city.

Keywords Bengaluru · City · BBMP · Slums · Environmental Stress · Local Institutions

5.1 Introduction

This chapter briefly traces the growth of the city both in geography and demographic as a result of urbanisation. The chapter summarises how the growth patterns of both spatial and demographic have carried with them enormous environmental problems inevitably leading to the verge of environmental crisis-like situation for the city. The city of Bengaluru[1] has witnessed tremendous growth both spatially[2] and in population. It is referred to as a 'million plus' city in India, and one of the

[1] The Greater Bengaluru City agglomeration is area of 741km^2 which includes 7 city municipal councils (CMCs) and 2 town municipal councils (TMCs) and 111 villages. The city of Bengaluru was reconstituted with gazette notification issued by the Urban Development Department in 2006 for constituting Greater Bruhat Bengaluru Mahanagara Palike merging the existing city with 8 urban local bodies (ULBs) and 111 villages of the Bengaluru Urban District (Sudhira et al. 2007: 379).

[2] The spatial area of Bengaluru in 1949 was 69 kms. In 2001, the area increased to 151 kms^2 and in 2009 further increased to 741 kms^2 (Parisara 2010: 3).

K. V. Raju et al., *Urban Environmental Governance in India*, The Urban Book Series, https://doi.org/10.1007/978-3-319-73468-2_5

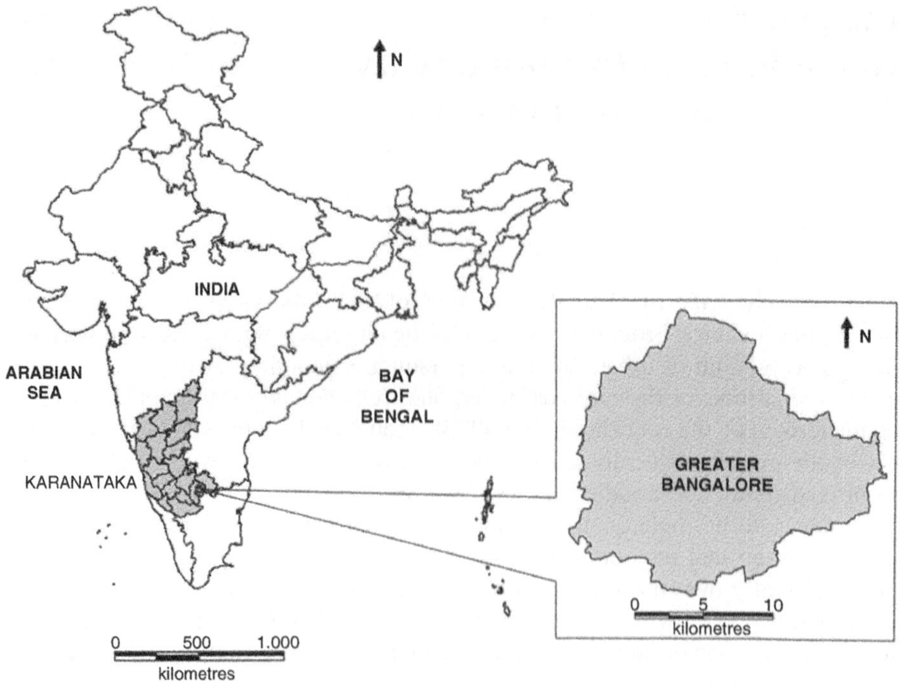

Source: Ramachandra and Kumar (n.d)

Fig. 5.1 Greater Bengaluru

fastest-growing cities of the twenty-first century. The city has grown spatially more than ten times since 1949 (Sudhira n.d.: 119)[3]; as a result tremendous physical growth has been recorded. The city is the fifth largest urban agglomeration[4] in 2001 (Narayana 2008: 5).[5] The city has been reconstituted administratively during 2006–2007, as Greater Bengaluru,[6] incorporating 100 wards. Now Greater Bengaluru constitutes 232 wards[7] (CDP 2006: 13) (Fig. 5.1).

Bengaluru City is the most populous city in Karnataka. Between 1871 and 2001, the population was 5.7 million as per within the area of 560 km^2, gradually increas-

[3] Sudhira, H. S. (2008). The Emergence of World City Bangalore.Conference Paper - 9th World Congress of Metropolis.http://eprints.iisc.ernet.in/24656/1/Chapter_4,_Vol_4,_Metropolis_Congress_2008.pdf.

[4] Urban agglomeration refers to the area formally administered by Bengaluru City Corporation (BCC) and eight councils (Sudhira 2008).

[5] Narayana (2008). *Globalisation and Urban Growth*: Evidence from Bangalore (India), CIRJE, Tokyo: University of Tokyo and Bangalore: ISEC.

[6] The city of Greater Bengaluru comprises of Bengaluru Mahanagara Palike (BMP) with an area of 226 sq. kms and 7 CMCs (city municipal corporation) with an area of 300.9 sq. kms and 111 villages (CDP 2006).

[7] Which include city corporation limits, 7 CMCs, 1 TMC and 111 villages of Greater Bengaluru.

ing to 7 million in 2007; similarly, as per the 2011 census, the city population[8] has increased to more than 9.6 million.[9] The city has become the hub of migrant population,[10] constituting 2 million during 2001 census (Sudhira n.d.: 124). The conurbation area of the city has increased from 365.65 sq. kms in 1971 to 445.91 sq. kms during 1981 to 531 sq. kms during 2001 (Narayana 2008: 6) and to 800 sq. kms by 2009 (CDP 2009:11).[11] Economic setting of the city has undergone tremendous transformation, characterised by IT[12]/BT sectors, garment sector and automobile, machine tool, defence, space and aviation industries. In addition, there is a huge concentration of small- and medium-based industries/enterprises (SMEs)[13] largely located in the periphery of the city. Economic growth rate of Bengaluru is highest in post-liberalisation, particularly, in terms of employment generation,[14] hosting labour force, manufacturing sector, business clusters, real estate and ownership of dwellings and activities such as trade, hotels and restaurants. Bengaluru has emerged as the fourth largest hub of technological cluster[15] in the world.

5.1.1 Geography and Environment: An Overview

The city of Bengaluru is situated in the south-east part of Karnataka. The city is situated at an elevation of 920 m above mean sea level. It is positioned at 12.97°N, 77.56°E, and covers an area of 2190 sq. kms (CDP 2006: 12). The city is divided into Bengaluru Urban and Bengaluru Rural Districts. The Bengaluru Urban District is divided into three taluks: Bengaluru North, Bengaluru South and Anekal. The city's land use pattern has been divided into five zones, namely, (i) the core city consisting of business areas; (ii) the pericentral area holds, planned residential area and surrounding core area; (iii) the recent extensions of the city (from the past 4 to 5 years) consisting of outer ring road; (iv) the new layouts in the peripheries of the city, vacant land and agriculture land; and finally (v) the green belt and agriculture area in the city outskirts including small villages (CDP 2009: 14–15).

---

[8] The city's population is expected to reach 108 lakhs 2021 (CDP 2006: 53).

[9] The decadal population growth rate of Bengaluru City is 37.7 per cent which makes the city one of the fastest-growing metropolis in India (CDP 2009).

[10] Bengaluru City is third highest number of migrants comprising 6.2 per cent of the total population for 2001 (Narayana 2008: 7), which increased to 45 per cent during 2001.

[11] CDP (2009). *Urban Infrastructure and Governance*, Revised City Development Plan: JNNRUM.

[12] Bengaluru's IT sector contribution alone amounts to US $12.2 billion (Rs. 54, 000 crores) to India. The city is divided into three IT clusters, namely, (i) Software Technology Parks of India (STPI), (ii) International Tech Park Ltd. (ITP) and (iii) Electronic City (CDP 2009).

[13] There were about 4000 SMEs in Peenya Industrial Region in Bengaluru, one of the largest industrial areas in South Asia (ibid 124).

[14] For instance, 66 per cent of employment was generated by service sector during 2004–2005 in Bengaluru (Narayana 2008: 9).

[15] Hence, Bengaluru is referred to as 'Silicon Valley', 'IT hub of Asia' and 'IT capital of India'.

The topography of the city is flat except in the middle running NNE-SSW. The city is located over ridges delineated by four watersheds, viz. Hebbal, Koramangala, Challaghatta and Vrishabhavathi. The undulating terrain of the region has accommodated a large number of water bodies that include lakes and tanks. During 1961, the number of lakes accounted to 262, and due to rapid urbanisation and increase in density of population, most of the lakes are fast disappearing. In 2010 (Nagendra 2010),[16] it was recorded that 210 lakes existed within the administrative boundary of Greater Bengaluru City, but by 2011, lakes were reduced to 189 under various institutional jurisdictions (Parisara 2010: 5). Meanwhile, official figures on number of lakes and tanks differ from 117 to 81, and satellite image for 2003 provides a picture showing only 33 visible lakes, out of which 18 lakes are clearly delineated and 15 show only faint existence (CES n.d.). The largest tanks are Madivala and Sankey tanks, and the largest lakes are Ulsoor, Yediyur and Hebbal.

The geographical area of Bengaluru Urban District constitutes 217,410 ha, out of which the forest area is 4198 ha including notified and other government lands under the Forest Departments. Around 1.97 per cent of the forest area comes under the Bengaluru Urban District spread over four revenue taluks and five forest ranges covering seven reserve forests (GoK 2009: 12).

The salubrious climate[17] in the city has earned its tags such as 'Pensioners Paradise' and 'Garden City[18]'. The city receives adequate rainfall[19] from both north-east and south-west monsoon. Natural green space like Lalbagh and Cubbon Park adds to the green landscape[20] of the city. The study by CES at Tata Institute of Science, Bengaluru, shows that about 40 species of mammals, more than 340 bird species, 38 varieties of reptiles, 16 amphibians, 41 types of fishes and 160 species of butterflies are recorded within the radius of 40 kms of Bengaluru City (CES n.d.). The green diversity of the city is under attack, and the city is gradually losing its green cover due to large-scale infrastructure projects (metro project, Bengaluru Corridor, road widening, construction of flyovers, etc.), real estate activities and encroachments. Similarly, increase in number of emanating vehicles has resulted in air pollution subsequently increasing the city temperature and inducing urban heat islands.

[16] Nagendra (2010)

[17] The city residents enjoy very pleasant climate throughout the year, temperature ranging from 33° C to 16° C with an average of 24° C.

[18] Bengaluru Urban District has naturally grown sandalwood trees.

[19] The mean annual total rainfall is about 880 mm for about 60 rainy days a year (CES n.d.). Please refer to article titled 'Geography and Environment of Bangalore', accessed in www.ces.iisc.er.net.in

[20] The percentage of green space constituting parks, open space, lakes and tanks is 2.5 to 3 per cent which is reduced from 4.4 per cent in 2002 (Sudhira n.d.). Please refer to his presentation titled 'Urbanisation in Bangalore', accessed in http://wgbis.ces.iisc.ernet.in/energy/lake2006/programme/programme/proceedings/Presentations/Lake%202006%20-%20Presentations/30%20Dec%202006/Session%20XI/Sudhira/Urbanisation%20in%20Bangalore.pdf

5.1.2 *Key Urban Issues*

The urbanised[21] area of Bengaluru City is 512 sq. kms (up to 2006) constituting 39 per cent of the total area. Realistically, the actual city does not concomitantly grow with access to infrastructure and basic amenities. As a result, there is visible stress on local governance to provide necessary basic needs and promote quality of life in terms of promoting both urban environment and ecology of the city. Some of the key urban issues of the city include:

(i) Urbanisation[22] in terms of (a) population growth and (b) spatial growth
(ii) Migration and urbanisation
(iii) Creation of urban slums (urban poor vs urban poverty)
(iv) Multiple stakeholders/complex institutions
(v) Volatile land use pattern

(a) **Population Growth in the City**

The city of Bengaluru is growing in unprecedented size with population and density. Rapid urbanisation is result of (i) rural-urban migration and (ii) reclassification of 'rural' to 'urban' which happening in most of the Asian cities (UN-HABITAT 2010: 7). In addition, factors such as concentration of service sectors like IT/BT sector and garment sectors are attributed for economic growth and followed by demographic changes with accompanied surge in urban population of the city. The city records highest growth rate in population. Such unmanageable population growth is beset with 'negative externalities' such as poor environmental and ecological quality in the city. The growth pattern deemed to be unsustainable and put heavy stress on delivery of urban services such as infrastructure and basic needs. The decadal growth rate of the city between 1991 and 2001 was as high as 61.4 per cent (see Table 5.1) which is attributed to the spatial expansion of the city (Ravindra 2005 in Sivanna 2014: 7) (Fig. 5.2).

(b)**Spatial Growth of the City**

Reclassification of 'rural' to 'urban' or expansion of the city into rural-periphery or absorption of population/settlements from urban periphery often referred to as 'in situ urbanisation' is one of the contributory factors in spatial expansion (often referred to as urban growth) in the cities of India (UN-HABITAT 2010/11: 48). The city of Bengaluru is spreading in all possible directions indiscriminately leading to urban sprawl.[23] Table 5.2 shows spatial expansion of the city over the years.

[21] The process of urbanisation is characterised by the proportional change in urban to total population (Sudhira n.d). Please refer to his presentation titled 'Urbanisation in Bangalore', accessed in http://wgbis.ces.iisc.ernet.in/energy/lake2006/programme/programme/proceedings/Presentations/Lake-Presentations/Dec2006/SessionXI/Sudhira/Urbanisation in Bangalore.pdf

[22] Urbanisation in Asia occurs due to transition from low-productivity agriculture to high-intensity productivity and services (UN-HABITAT 2010/11).

[23] Urban sprawl is generally uncontrolled expansion of urban area. It is defined as the increase in land development/use in suburban areas outside their respective urban centres (as defined by Cornell University's webpage).

Table 5.1 Population growth in city corporation and metropolitan areas of Bengaluru

Year	Population of BCC	Population of BMA	Population growth rate in BMA (%)	Population growth rate in BCC
1951	778,977	786,343	–	–
1961	1,093,798	1,206,961	53.5	40.4
1971	1,540,741	1,664,208	37.9	40.9
1981	2,476,355	2,921,751	75.6	60.7
1991	2,660,088	4,130,288	41.4	7.4
2001	4,292,223	5,686,844	37.7	61.4
2011	5,458,977	7,509,599	32.1	27.2

Source: Ravindra (2005: 5) in Sivanna (2014)

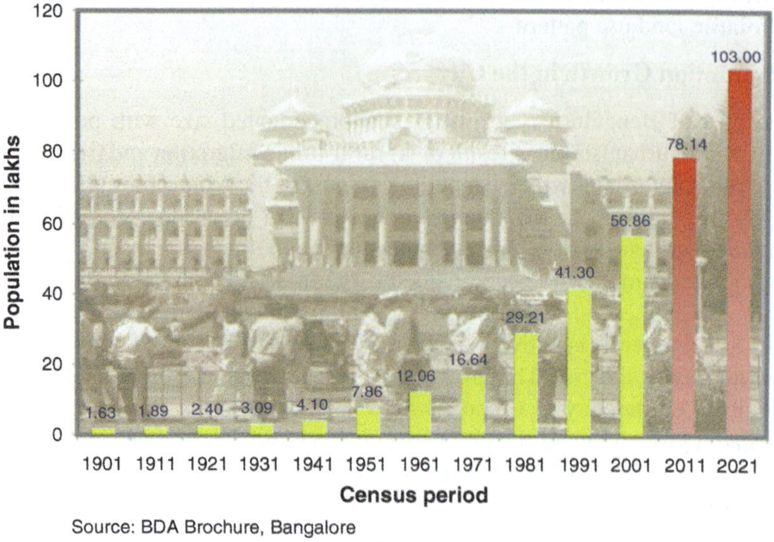

Fig. 5.2 Population growth of Bengaluru City over the years

Table 5.2 Spatial expansion of the. Bengaluru City Corporation

Year	Population (millions)	Density (people/km2)	Built-up area %
1971	1.7	9465	20
1981	2.9	7990	26
1991	4.1	9997	39
2001	5.7	11,545	69
2011	8.5	12,142	Na

Source: Mehta et al. (2013).

The spatial growth of the city is determined by both concentration of (i) industries and (ii) population growth. As a part of spatial/territorial restructuring, today the city's urbanisation stretches in all the directions like east and north-east (concentration of service sector like Whitefield and ITPL), west (Dasarahalli, Tumkur and Magadi Road), south (Electronic City and Bommasandra) and north (Devanahalli and international airport). Owing to the urban agglomeration economies, the city has witnessed growth of urban corridors (such as IT Corridor, Bengaluru-Mysore Corridor, Bengaluru-Tumkur Corridor, etc.). These areas signify intense economic activities which stand out as one of the key features of the agglomeration. The challenge is to sustain the environment and city's ecology from damage (Fig. 5.3).

Box 5.1: The Bengaluru: 'Beacon' of IT Revolution

Over the past several years, Bengaluru, a city located right in the middle of southern India, has been put on the global map by its achievements in the information technology sector (IT). According to Software Technology Parks of India, Bengaluru's information technology exports have risen from about US $1 billion in 2001 to more than US $10 billion in 2006. The city has also benefited from the employment spin-offs of the information technology industry: according to India's National Association of Software Services Companies (NASSCOM), one IT job results in four indirect jobs. This is the success of Bengaluru, which has triggered a software revolution in Indian cities. The IT industry tends to flourish where technology professionals are available. Bengaluru-based Infosys Limited, India's second largest IT firm, concentrates most of its staff, or nearly so, in a single location. Interestingly, these professionals are not located in Bengaluru but some 750 km up north in Pune (south-east of Mumbai). From an international perspective, the rapid rise of Bengaluru has forced major software companies to consider alternative locations outside the developed world. Cities in India are now competing against each other in order to attract major software companies. Apart from Pune, cities like Bhubaneswar, Chennai, Gurgaon, Hyderabad and Jaipur are all following Bengaluru's example of excellence in the software business.

Source: UN-HABITAT 2010: 94.

The observed spatial/territorial growth, as well as the projections, indicates that the development process has spread to the newly acquired territories and surrounding regions. Post-2006 with the declaration of Bruhat Bengaluru, much of the spatial growth has occurred in the surrounding municipal areas. These areas fall within town, municipal and village panchayat jurisdiction. With the expansion of the city, some of the key changes that occurred include real estate development, acquisition and transformation of agriculture land to nonagriculture and industrial purpose, shortage of urban services, encroachment and so on.

Administrative Areas within the BDA

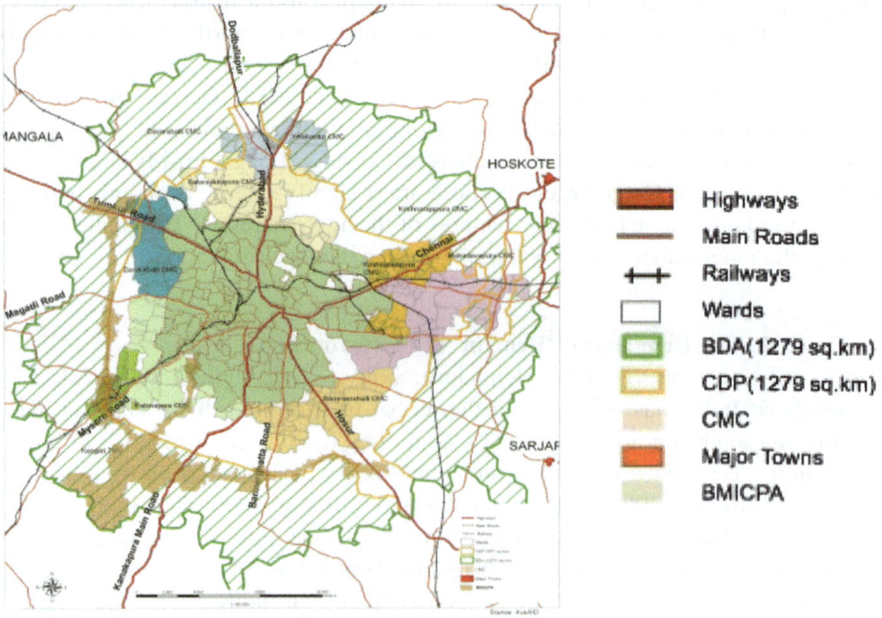

Source: BDA Brochure, Bangalore

Fig. 5.3 Administrative areas under BDA

(c) Migration and Urbanisation

Migration is the strategy adopted by the rural population to improve the family livelihood and get benefitted by the services available in the urban regions (UN-HABITAT 2010/11: 49). Particularly, the process of 'urbanisation' seems to be the 'pull' factor for urban migration in Karnataka (Roychowdhury et al. 2012: 36). The trend of urban migration in the state of Karnataka[24] has increased past 2001 to more than 34 per cent (ibid 13). Similarly, the trend of migration to the city is one of the primary contributors for the urbanisation process and demographic changes. Particularly rural-urban migration is most prevalent followed by urban-urban due to the emergence of IT/BT corridors. Therefore, the spin-off from the concentration of service sector and industries in the city further attracts the migrant population. Overall in-migration to the city constitutes 45 per cent to the total population (CDP 2009: 14). As per census of India (2001), out of 5.7 million population of the city agglomeration, 2 million are migrants, out of which 1.2 million are from within Karnataka state and 0.8 million are from the outside state. The city of Bengaluru attracts more than 36 per cent of in-migrants as compared to the state average (Roychowdhury et al. 2012: 21).

[24] Incidentally, the state of Karnataka is experiencing a rapid economic growth during post-1990s (Roychowdhury et al. 2012: 13).

One of the driving factors for rural-urban migration into the city is absence of daily-wage opportunities in rural areas and access to informal jobs[25] in construction sectors, service industries, or garment industries located in the city. Most of the in-state migrants are from rural Karnataka, in search of wage employment in the city. These in-migrants are mostly employed in garment,[26] construction and other day-to-day service sectors which is referred to as 'informal[27]' employment[28](ibid.: 27). Reversing or reducing the rural-urban migration has been one of the policy goals seriously pursued by the city governments in India to regulate spatial distribution.

Such city-led economic growth does not increase concomitant urban resources to access and consequently bring in significant changes in land use pattern. Dense concentration of the population such as Bengaluru City by in-migrants and floating population does indirectly link the city's global/economic activities with rural through the emergence of conurbation. All these factors have, in fact, contributed to the urban demographic growth in terms of size and expansion. Therefore, India is going to add 226 million people to its cities in next two decades, with its rate of urbanisation reaching 39.7 per cent by 2030 (UN-HABITAT 2010/11: 42). Therefore, the city of Bengaluru seems to offer opportunities for employment and access to social services like health, education, and income mobility to a certain extent. But rapid growth of population, lack of access to urban basic services and subsequent accompanying demographic changes have led to crisis in urban environmental governance. Therefore, city's inability to provide necessary infrastructure and urban basic services including land in response to expanding population has direct bearing on local environment sensitivities and ecology.

(d) **Emergence of Slums**

In South Asia alone, the number of urban poor[29] has increased to 107–125 million between 1993 and 2002 (UN-HABITAT 2010/11: 14). Emergence of slums[30] is identified as 'physical manifestation of glaring demand for labour in Asia cities and inadequate access and supply of affordable housing and infrastructure for decent living' (ibid 14). In-migration could be one of the major contributing factors for emergence of urban slums.[31] The rural population, particularly the landless labour, migrate to city in search of employment opportunities and settle in shanty or temporary dwellings (Picture 5.1).

---

[25] The Peenya Industrial Region holds more than 2500 small-scale units; the city is the second largest industrial estate, situating around 3000 manufacturing units (Roychowdhury 2003: 5282).

[26] Officially there are 788 garment-manufacturing units in Karnataka, of which 729 are in Bengaluru alone (Roychowdhury 2005: 2251). Garment industries are primarily located in two industrial belts: Mysore Road, Yeshvanthapura and Peenya II Stage (Roychowdhury et al. 2012: 44).

[27] Informal sector is referred to the employment opportunities in unregulated sector like garment, domestic service, street-vending, etc.

[28] More than 28 per cent of migrants migrated to the Bengaluru Urban District for employment opportunities (Roychowdhury et al. 2012: 33).

[29] The region of Asia still hosts half of the world slum population over 505.5 million slum dwellers (UN-HABITAT 2010/11: 17).

[30] Over 900 million of world's poor still live in Asia (UN-HABITAT 2010/11: 108).

[31] A slum household consists of group of individuals living under the same roof in urban area lacking amenities like (i) house, (ii) sufficient living area, (iii) access to improved water and sanitation and (iv) secure land tenure status (UN-HABITAT 2010/11: 181).

Picture 5.1 Temporary settlers in Bengaluru in urban periphery

The city of Bengaluru has experienced exponential growth of slums in the past decade of post-1990s, from 444 slums in 1991, with the population of 1.12 million, to 763 slums in 1998–1999 with a population of 2.2 million; about 20 per cent of the city population constitutes urban slums (Schenk 2001 and Roychowdhury 2003: 5281). As per 2001 census, the city had 733 slums. The CDP (2009: 20–21) figures show 640 slums in Greater Bengaluru City comprising core city, periphery and villages. Unofficial figures estimate that the city hosts more than 1500–2000 slums in both core city and periphery areas. At present, there are 450–500 slums[32] coming under direct management of BBMP, while official figures from Karnataka Slum Development Board indicate 587 slums in Bengaluru City. There is no proper consolidated data on slums for the city of Bengaluru. Under JNNURM-BSUP programme, BBMP has rehabilitated 75,000 families (housing programme). For rehabilitating the slums, BBMP is tied up with KSCB. Approvals and funds are transferred to KSCB for rehabilitation. Recently, one lakh houses are planned to be constructed for slum rehabilitation by BBMP.

The living conditions of the urban slum dwellers are very poor. They are practically deprived of access to basic services like water, sanitation-latrines, garage/waste collection, drainage and so on in the city. Lack of secure land tenure status further aggravates the vulnerabilities of the urban poor either to occupy vacant land or encroach the ecologically sensitive/agriculture land. They are further excluded

[32] Interview with commissioner, BBMP, Bengaluru, dated 24 May 2014.

Table 5.3 Growth of slums in BMR

Year	No. of slums	Population
1990s	444	1.12 million
1998–1999	763	2.2 million
2001 (census)	733	
2006	542	
2009	640	703,037

Source: BBMP (2009), Roychowdhury (2003), and Sudhira et al. (2007)

Source: Sudhira et al. (2007)

Picture 5.2 Slum in Bengaluru City

through 'gentrification[33]' process by evictions/demolition. Thus, the city has many unplanned settlements which pose serious threat to the environment and ecology (Table 5.3 and Picture 5.2).

(e) **Multiple Stakeholders**

The urban/city/local governance is concerned with the dynamics of interface between urban administration, local elected representatives, bureaucracy, private sector, NGOs, community and local residents.

---

[33] Gentrification is the process of renewal and rebuilding accompanying the influx of middle-class or affluent people into deteriorating areas that often displaces poor residents (www.merriam-webster.com).

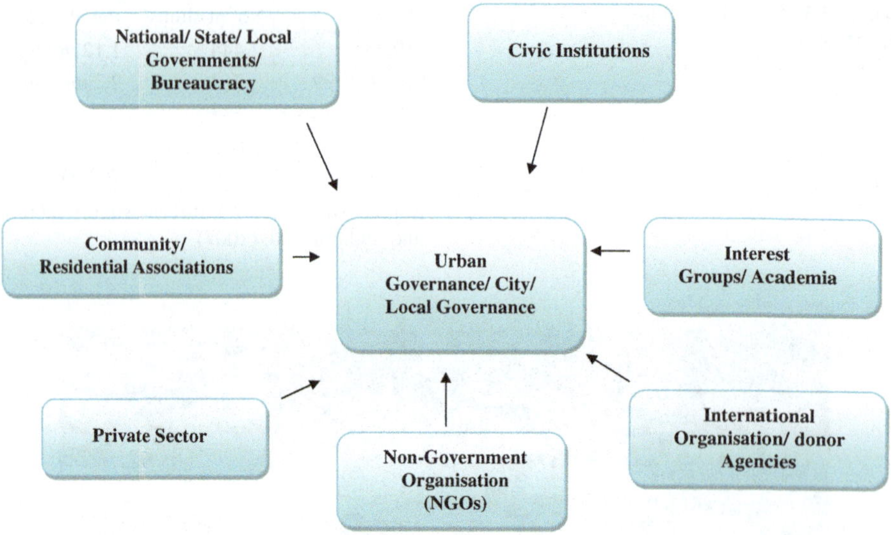

These multiple stakeholders[34] present governance structures and processes framing multiple laws, regulations, plans, standards and financial schemes for managing urban affairs and environment. Therefore, effective urban governance structure and ensemble of institutions are crucial for promoting sustainable cities. Appropriate institutional mechanisms are critical for promoting environmental governance. Details on the role of institutions and governance structures and its impact on environmental governance are presented in Chap. 7.

(f) **Volatile/Shifting Land Use Pattern**

Geographical expansion of the city, followed by demographic density towards corridors and periphery, has in fact changed the land use pattern. Population density (in population and migration) and creation of urban slums have put additional pressure on the access to urban land market for housing. With the emergence of IT/BT and industrial clusters in the city, demand for land has increased multifold (CDP 2009; Sudhira et al. (2007): 380). The entire city land pattern has been divided into five zones, namely, (i) east and north-east (concentration of service sector like Whitefield and ITPL), (ii) west (Dasarahalli, Tumkur and Magadi Road), (iii) south (Electronic City and Bommasandra), (iv) north (Devanahalli and international airport) and finally (v) agriculture land and green belt region quickly urbanising. With undue stress on urban economic growth, land developers and real estate use vacant land for commercial purpose; thereby there is rapid decline of public land.

[34] Please visit Chap. 3 for more details on institutions and governance structures in India and Bengaluru.

5.1.3 *Critical Environmental and Ecological Pressures/Stress*

The city of Bengaluru has doubled its population twice in the period of 8 years and grew in its area from 226 to 800 sq. kms within 10 years (Ranganathan 2013).[35] The city is, moreover, caught between industrial development and intense population growth exacerbated by the deficit of human and financial resources, increase in slums and thereby growth in urban poor population. Consequently, this has resulted in environmental and ecology degradation, loss of city's biodiversity, increase in poverty and growing rich-poor divide/inequality. One of the greatest challenges facing the local government is to promote environmental and ecological sustainability.

5.1.3.1 Defining Key Environmental Challenges

Some of the key environmental and ecological pressures faced in the city include the following: (i) emphasis on the pace of economic growth and development is much greater than the concomitant capacity of the local government to tackle challenges related to poverty, environmental pollution and access to basic needs; (ii) many labour-intensive and service-oriented industries have in fact attracted FDIs and rural population in search of better employment opportunities but often with heavy environmental and ecological cost; (iii) city's industrialisation-stimulated spatial and demographic growth has led to pre-urban/suburban expansion with dense population growth putting pressure on land and resulting to depletion and contamination of water bodies affecting city ecology; (iv) today city's 'ecological footprint[36]' is in excess of population, unable to sustain them, a phenomenon that impacts the environment; (v) the undulating terrain and topography of the city, combined with high population density, lack of effective urban planning and undertaking of large-scale infrastructure projects (like metro and corridors and so on), makes it vulnerable to climate change and its impacts (natural disasters like urban floods and urban heat islands).

5.1.3.2 Environmental Impact in the City

The city's rapid economic growth and urbanisation process combined with inconsistent planning and implementation of environmental standards have negative impact on urban environment governance.

The city is facing various kinds of pollution (i) air, (ii) water, (iii) sound/noise, (iv) visual and (vi) vehicular impacting on the quality of life. Due to rapid growth, the city got crowded and congested with people and vehicles. The city's vehicle

---

[35] Ranganathan (2013). Sewerage is a much bigger problem than water supply in Bangalore, the alternative, 25 March 2013. Please access www.thealternative.in.environment

[36] The 'ecological footprint' is an average measure of the amount of land required to sustain an individual (UN-BABITAT 2010/11: 17).

Table 5.4 Environmental service in the city of Bengaluru

Environmental services	Quality
Air	Air quality has been fast deteriorating in the city of Bengaluru
Water	CPHEEO standards are followed to ascertain water quality. However, tests like turbidity, trihalomethane (THM), aluminium and pesticide residues are yet to be conducted
Waste water	Treatment capacity of 93 per cent of estimated waste water generated is currently available
Solid waste	One hundred per cent of door-to-door waste collection is available in the core city. Thirty-four per cent of waste water is treated. No disposal facility. Treatment and disposal facilities of 2000 TPD capacity is available.

Source: BDA (2007)

pollution is greater than any city in India. It has lost much of its green canopy/space including water bodies like lakes and tanks (Sudhira et al. 2007: 383). The city in 1961 recorded 262 lakes; official figures at present vary from 117 to 81 lakes. But satellite figures show 33 lakes from which 18 are not visible (ibid 382). The changing land use pattern has pushed the migrant population to settle down in temporary shanties like slums under very poor and hygienic conditions without access to basic amenities (ibid 387). Inefficient land use pattern of the city for commercial, industrial and large-scale infrastructure projects like road widening and construction of flyover/underpass has severely eroded the environment and increased pollution. In addition, poor quality of air has caused respiratory illness affecting the health of the city's population.

The problem of catering drinking water supply[37] to the growing population particularly urban periphery region and newly added 110 villages constitutes a daunting problem to the governing authorities (like BWSSB and BBMP). Recently, under Karnataka High Court instruction, BWSSB has decided to supply water through tankers (with capacity of 600 litres with 30 tankers). Similarly, BBMP has decided to supply water through 599 borewells to 110 villages (Deccanchronicle.com, September 2013) (Table 5.4).[38]

The concentration of the poor in informal settlements such as slums under unhygienic conditions has resulted in the spread of communicable and non-communicable diseases, and unhealthy living conditions are aggravated by the inadequate access to water supply, toilets and garbage collection and waste disposal methods in the city (Tables 5.5 and 5.6).

[37] The city's total requirement of drinking water supply per day is 1150 million litres (mld), but the BWSSB is able to draw only 835 mld. So the city falls short of 250 mld of water (The Hindu, 23 May 2013). Please access article titled 'Water Supply resumes in Bangalore', www.thehindu.com

[38] Please access article titled 'Bangalore Water Supply and Sewerage Board to Supply Water to 110 villages', www.deccanchronicle.com. Accessed on 18 Nov 2013.

Table 5.5 Status of infrastructure in the city of Bengaluru

Sector	Parameters	Service delivery levels
Water	Coverage	BMP 100 per cent, CMCs and TMC 20 per cent
	Quantum of water availability	995 mld
	Average per capita water supply	73 lpcd
	Frequency of water supply	3–5 hours alternate days
Sanitation and sewerage	Coverage	225 sq. km area/40 per cent of total area
	Disposal of sewerage	408 mld
	Present operating capacity	306 mld (3/4 of the capacity)
	Waste water generation	721 mld
Municipal solid waste management (SWM)	Coverage	100 per cent BMP area
	Waste generated	3395 TPD
	Waste collected	2715 TPD
	Collection efficiency	80 per cent
	Segregation	10 per cent
	Treatment and disposal	Treatment facilities for 1000 TPD/landfill facilities being constructed
Roads	Quality	80 per cent tarred
	Length of the roads in BMP area	3500 km
	Length of the arterial	250 km
	Length of NH and SH 100 km	100 km
	Length of roads in ULBs	2400 km
	No. of street lights	2.5 lakh
Transport	No. of registered vehicles	23 lakh
	No. of buses (BMTC)	3300
	Daily passenger trips	32 lakh
	Congestion exceeds	Exceeds 1, in 52 corridors/links
	Noise decibels	Above 80 in most areas (beyond permissible levels)
	Average speed of vehicles	12–18 kmph
	No. of accidents	7575 (in 2005) 3654 (up to 30–06-06)

(continued)

Table 5.5 (continued)

Sector	Parameters	Service delivery levels
Parks	Coverage area	14 per cent
	Four important parks	Lalbagh, Cubbon Park, Bannerghatta National Park, Dhanvantari Vana
	Small parks	365
	Well developed	55
	Partially developed	105
	Not developed	180
Lakes	Coverage area	3 per cent of total CDP area

Source: CDP (2009: 21)

5.1.4 *Institutional Settings*

Effective urban environmental governance can be achieved through compliance and enforcement of laws, regulations, norms, practices and standards. Moreover, environmental governance can be strong with local collaboration and cooperation of different stakeholders. Undoubtedly, due to unprecedented scale and pace of urbanisation, institutional synergy will promote effective urban environmental planning and management.

The sprawling city of Bengaluru is governed by a plurality of bodies, and institutions as a result suffer from territorial, institutional and administrative fragmentation while carrying out planning and implementation of environmental governance. In the event of increasing emphasis on 'good urban governance', the Master Plan 2015 for Bengaluru City was published in 2007. It is a revised version of Comprehensive Development Plan 1995. The Master Plan was developed by Bangalore Development Authority (BDA) as per Karnataka Town and Country Planning Act 1961. The Revised Master Plan 2015 covers Bengaluru City, a local planning area of 1307 sq. kms consisting of 387 villages, 7 city municipal councils and 1 town municipal council (Mahendra et al. 2010: 7). The Master Plan aimed at balanced and equitable urban growth and promotes efficiency in spatial distribution and infrastructure development. To promote sustainability of urban services, the Master Plan has incorporated various provisions for environmentally sustainable activities like rainwater harvesting, solar heating and lighting and tree planting.[39] The BWSSB (Bengaluru Water Supply and Sewerage Board) has made rainwater harvesting mandatory for all the plots which are more than 240 sq. metres and residential buildings. The Master Plan laid special emphasis on protection and conservation of water bodies and green lounge like lakes, tanks, national parks and forests including protection of natural heritage and historical monuments of the city (ibid 12–13). The slum population constitute 20 per cent of the total population of the city. The Master Plan has integrated urban poverty reduction strategies under JNNURM into local development planning.

[39] See Chap. 6 on sustainable practices in the city.

Table 5.6 Shortfall in delivery of urban services in Bengaluru

Delivery of urban services	Requirement	Shortfall
Water supply	1150 mld, but now the city is drawing only 835 mld. Seventy-nine per cent (900 mld) of drinking water is met by surface water Water requirement per day is 150 lpcd (litres per capacity per day)	250 mld Twenty-one per cent of drinking water is met by groundwater (250 mld) Forty-five per cent (43 per cent loss/400 million litres per day) of the surface water is lost due to unaccounted-for water in slums of the city Existing supply is only 110–120 lpcd Shortfall in 2007 was 349 million litres per day Demand by 2011 = 1567 mld Present supply = 900 mld Deficit = 676 mld The city has nearly 312,000 borewells which draw about 300 MLD. As a result water table is fast dwindling RWH has covered only 40 per cent properties out of 18 lakh properties in the city (dna, 23 June 2013)
Solid waste	Garbage/waste door-to-door collection is available only in core city area	Waste collection is not implemented in the periphery and 110 villages of BBMP area.
Waste water	Generation of both sold and liquid waste constitutes 1000 mld (Ranganathan, 2013) 14 secondary treatment plants 4 tertiary treatment plants 920 mld generated and 720 mld treated	Only 30 per sewerage is treated, and most of the waste water is polluting the lakes (Vrishabhavathi, Koramangala and Hebbal) and tanks inside the city. Most of the lakes and tanks have become cesspools, causing immense health hazard to the residents (Balasubramanain 2013) The periphery and 110 villages newly added region to Bengaluru City corporation limits do not have underground drainage system (UGD) provided by BWSSB Waste water and leaking underground water pipes contaminate the groundwater table
Lakes	Initially 927 lakes	Now reduced to less than 200 lakes in the city

Source: Compiled from various sources

5.1.5 *Environmental Governance and Organisations in Bengaluru*

To address Bengaluru City's growth, development and planning, various local self-governments are constituted from time to time. The presence of multiple agencies carries with it various systemic issues such as multiple institutional jurisdictions, planning for service delivery, organisational development and accountability issues and so on impacting on the environmental governance.

Table 5.7 Institutions in Bengaluru

Organisations/Agencies	Area covered/authorities
Bruhat Bengaluru Mahanagara Palike (BBMP)	City Corporation
	Bommanahalli CMC
	Byatarayanapura CMC
	Dasarahalli CMC
	KR Puram CMC
	Mahadevapura CMC
	RR Nagar CMC
	Yelahanka CMC
	Kengeri TMC
	110 villages (VP)
Bengaluru statutory authorities	Bangalore development authority (BDA)
	Bangalore metropolitan regional development authority (BMRDA)
	Bangalore water supply and sewerage board (BWSSB)
	Bangalore metropolitan transport corporation (BMTC)
	Bangalore metropolitan land transport authority (BMLTA)
	Lake development authority (LDA)
	Karnataka slum clearance board (KSCB)
	Karnataka urban infrastructure development and finance corporation (KUIDFC)
	Bangalore international airport area planning authority (BIAAPA)
Ward committees	30 ward committees
Semiformal organisation	Bangalore agenda task force (BATF)
	ABIDe
	City infrastructure review committees
	JNNURM consultations
Elected representatives	MLAs, MPs and Ward Councillors
NGOs and RWAs	APSA, HASIRU-USIRU, etc.
Institutional stakeholders	Trade, industrial associations
Plan instruments	City development plan, master plan, comprehensive development plan, etc.

Source: BDA (2007)

As per the City Development Plan (2009) and Revised Master Plan (2015) (BDA), these institutions can be classified as (i) urban local bodies (ULBs), (ii) statutory authorities and (iii) government departments. As far as these organisations are concerned, three significant factors determine their impact on city's environment and ecology; they include (i) role and functions of institutions involved in the service delivery, (ii) the planning instruments to regulate city's growth scenario and facilitate service delivery and (iii) the approach to long-term city development (BDA 2007; BBMP 2009).

This section presents various institutions and organisations (see Table 5.7) affecting city's environmental governance. The section proceeds to discuss institutional and legal framework in the city's environmental governance and the functional areas and overlaps.

Interplay of various regulatory authorities and developmental agencies with their respective laws and overlapping jurisdiction has in fact increased the complexity of environmental governance (Table 5.8).

5.1.5.1 Bruhat Bengaluru Mahanagara Palike (BBMP)

The traditional form of local government of the city constitutes Bruhat Bengaluru Mahanagara Palike (BBMP)[40] or city corporation. During 1949, both the city municipality and cantonment municipality were amalgamated to form the corporation for the city of Bengaluru. The erstwhile BMP[41] consisting of 100 wards was merged with 7 city municipal councils (CMCs), 1 town municipal council (TMC) and 110 villages to form the Bruhat Bengaluru Mahanagara Palike[42] in 2007. At present, the city corporation area is divided into 198 wards also referred to as 'Greater Bengaluru'. BBMP is responsible for the delivery of the following services along with other organisations and departments in the city: (i) water supply, (ii) sewerage, (iii) storm water drainage, (iv) solid waste disposal (collection and removal of garbage), (v) municipal transport (including flyovers), (vi) street lighting, (vii) health immunisation services, (ix) removal of accumulated water on streets and public places due to rain and other causes, (x) horticulture and (xi) public health.

Best Practices of BBMP, Bengaluru

- Two units are handling collection and recycling of hotel waste in the city: one in Magadi Road and another one in Doddaballapura. They convert and produce cylinder gas up to 10 tonnes.
- Coconut shells are collected to make cakes. At present two firms are sanctioned. But they are not functional.
- Vendors are tied up with BBMP for collection of municipal waste from bulk producers particularly from apartments.
- A firm is picking up used thermocoal for recycling and producing thermocoal.
- 'Adopt-Park' scheme was introduced to maintain urban parks with local communities
- 'Swachh Bangalore Campaign' was introduced by BBMP for 'door-to-door garbage collection'. 'Suchi Mitras' or citizen volunteers were encouraged to assist BBMP to keep the city clean.

Source: BBMP, Bangalore

---

[40] The Karnataka Municipal Corporation Act of 1976 mentions the functions of BBMP (Sivanna 2014: 8).

[41] The government passed Karnataka Municipal Corporation Act of 1976 to govern all municipal corporations in the state (Sivanna 2014: 8).

[42] Government of Karnataka (GoK) on 2 November 2006 was officially notified for the creation of Greater Bengaluru (Bruhat Bengaluru Mahanagara Palike). The process was completed by the months of April–May 2007.

Table 5.8 Functional responsibilities of various agencies in the delivery of urban services

Delivery agencies	Planning and design/construction/operation/maintenance
Water supply	BWSSB, BBMP
Sewerage	BWSSB, BBMP
Storm water drainage	BBMP
Solid waste disposal	BBMP
Municipal roads	BDA, PWD, BBMP
Street lights	BBMP

Source: Revised City Development Plan (2015), JNNRUM, Bengaluru: BBMP
Note: Bangalore Water Supply and Sewerage Board (BWSSB), Bruhat Bengaluru Mahanagara Palike (BBMP), Bangalore Development Authority (BDA) and Public Works Department (PWD)

(a) **Department of Horticulture Under BBMP**

True to the city's name as 'Garden City', a large number of urban parks and tree management are taken care of by the Department of Horticulture under BBMP. The main function of the Department of Horticulture is to develop and maintain parks, by planting roadside saplings, and boulevards. The BBMP maintains more than 435 large and small parks around 200 open spaces and the greenery along road margins and medians. At present there are about 811 horticulture workers (Sivanna 2014: 13).

(b) **Jawaharlal Nehru National Urban Renewal Mission (JNNURM)**[43]

One of the flagship programmes of the government of India for introducing and implementation of urban sector reforms for the development of metropolitan cities is the JNNURM. The city of Bengaluru has been one of the recipients of JNNURM funds for the improvement of the city's infrastructure and additional component for the urban poor. The city of Bengaluru has been selected under A category cities, and the central government provides 35 per cent, while the government of Karnataka shares 15 per cent and the concerned institution 50 per cent of the cost for all projects undertaken. Under this scheme, City Development Plan (CDP) 2006 (later Revised City Development Plan 2009) has been prepared with an estimated budget of 22,536 crores which is to be implemented within seven years from 2007. The main thrust of the project is divided into two components (i) urban governance and infrastructure (storm water drains, flyovers, etc.) and (ii) basic services for the urban poor.

Bangalore Development Authority (BDA)

Bangalore Development Authority is a premier urban planning and development authority which was constituted in 1976.[44] BDA combined the both planning and development functions of erstwhile City Improvement Trust Board (CITB) and City

[43] Please refer to Sivanna N. (2014). Urban Governance and Organisational Restructuring: The Case of Bruhat Bangalore Mahanagara Palike (BBMP), Monograph series 31, Bangalore: Institute for Social and Economic Change.

[44] Under a separate state legislature act, viz. the BDA Act of 1976.

Planning Authority. Under Karnataka Town and Country Planning Act, 1961, the BDA is designated as the planning authority.

Since the inception of BDA, the city has witnessed tremendous growth from 531 sq. kms in 2001 to 800 sq. kms by 2012. Similarly, BDA has a jurisdiction of 1219 sq. kms including the area under the Bruhat Bengaluru Mahanagara Palike (BBMP). Besides city planning and land regulation, BDA is also responsible for providing key developmental and infrastructure facilities like roads and other transport-related infrastructures for the entire Bengaluru Metropolitan Area (BMA) and integrates with urban environmental governance. Metropolitan area of BDA covers the following taluks as per 1995 Revised CDP; they include (i) Bengaluru North, (ii) Bengaluru South, (iii) Bengaluru East (now called), (iv) Anekal (part), (v) Hoskote (part), (vi) Devanahalli (part), (vii) Magadi (part) and (vii) Nelamangala (part). As a planning authority, BDA has to prepare the Comprehensive Development Plan (CDP) for Bangalore Metropolitan Area. The total jurisdiction of BDA as per Revised CDP is 1279 sq. kms. Out of this, the green belt area covers 682 sq. kms. The conurbation area (urbanisable area) is 597 sq. kms including spotted development. Out of this conurbation area, about 225 sq. kms is the Bengaluru Mahanagara Palike area.

BDA Key Functions Planning Functions

- Preparation of development plan for Bengaluru
- Preparation of scheme plans
- Approval of development plans for group housing and layouts
- Approval of building plans
- Other statutory functions under KTCP Act

Development Functions

- Planning and implementation of schemes to provide for residential sites, commercial sites, industrial sites, civic amenity sites, parks and playgrounds
- Construction of commercial complexes and construction of houses for economically weaker sections, low-income group, middle-income group and high-income group
- Development of major infrastructure facilities

Source: http://www.bdabangalore.org/functions.html

5.1.5.2 Bangalore Water Supply and Sewerage Board (BWSSB)

Bangalore Water Supply and Sewerage Board (BWSSB) is solely responsible for providing drinking water supply to the looming population of Bengaluru City. BWSSB[45] was constituted in 1964 delivering both water supply and sanitation services. The main functions of the BWSSB are as follows:

[45] BWSSB was constituted under the Karnataka State Legislature vide notification No.PLM/15/MNY/64 dated 30 September 1964, and BWSSB came into existence from 2 October 1964.

- Providing water supply and making arrangements for the sewerage and disposal of sewage in the existing and developing new regions of Bengaluru Metropolitan Area
- Investigating adequacy of water supply for domestic purpose in Bengaluru Metropolitan Area
- Preparation and implementation of plans and schemes for supply of water for domestic purposes within the Bengaluru Metropolitan Area to the required standards
- Preparation and implementation of plans and schemes for proper sewerage and disposal of sewage of the Bengaluru Metropolitan Area
- Levy and collection of water charges on 'no loss no profit basis'

As the city of Bengaluru expanded from mere 100 wards to 195 wards with the merged 7 CMCs, 1 TMC and 110 villages by 2006, consisting more than 9 million people by 2013, the per capita supply of drinking water[46] slides, thus increasing the gap between demand and supply.

Sustainable Initiatives Undertaken by BWSSB

- Cauvery Water Supply Scheme stage IV phase II for additional 500 mld water at a cost of Rs. 3384 crores for providing drinking water for newly merged 7 CMCs, 1 TMC and 110 villages.
- Providing UGD facilities to erstwhile 7CMCs and 1 TMC at a cost of Rs. 1300.00 crores. Works are under progress, and 30% of works are completed.
- Water supply infrastructure has been provided to the area under erstwhile 7CMCs and 1 TMC at a cost of 537 crores.
- Six sewerage treatment plants treating approximately up to 418 million litres of waste water. This includes 60 mld + 10 mld tertiary treatment capacity.
- Restoration of 280 kms of dilapidated water supply pipeline at a cost of Rs. 77 crores.
- Restoration of 301 kms of dilapidated waste water pipeline at a cost of Rs. 48 crores.
- Renovation of 70 kms of waste water sub-mains at a cost of Rs. 370 crores.
- Scheme for zero discharge of wastewater into the Hebbal Valley storm water drains at a cost of Rs. 45 crores.
- Waste water source line of 1200 mm dia from Krishnappa Garden, Madiwala, to Central Silk Board at a cost of Rs. 18.00 crores.
- Under 'social mobilisation' component during October 2010, free installation of water meters to 22,000 households in 350 slums at a cost of Rs. 2.20 crores was initiated.
- Procuring 30 jetting machines for cleaning UGD.

(continued)

[46] Complete details on demand-supply gap of drinking water supply are presented in Chap. 5.

- Installation of 218 electronic bulk flow water meters at a cost of Rs. 14 crores for plugging water leakage.
- Construction of ten water reservoirs at various places at a cost of Rs. 46 crores.
- Construction of six reservoirs at a cost of Rs. 140 crores under Cauverystage IV phase II scheme.
- Amendments to the rules to make rainwater harvesting mandatory.
- Setting up of a rainwater harvesting theme park in Jayanagar at a cost of Rs. 2 crores.
- Free rainwater harvesting training to 1200 plumbers and others. Setting up of a help desk for providing information to public (23348848–49) regarding rainwater harvesting.
- The rainwater harvesting system has been installed in 23,000 buildings in Bengaluru City.
- Pilot project on leakage detection adopting international technologies at the cost of Rs. 48 crores with JIBC funds. The project will be extended to entire Bengaluru at the total cost of Rs. 400 crores.
- It is proposed to replace old corroded pipe lines in 5 water supply division of the BWSBB.

Source: http://bwssb.org/achivements/

5.1.5.3 Bangalore Metropolitan Transport Corporation (BMTC)

BMTC is responsible for providing public transport serving Bengaluru urban, sub-urban and rural areas. The BMTC constitutes one of the largest public transport operators in India, consisting of mammoth fleet of 6500 buses daily plying more than 5 million passengers accounting to 42 per cent of all motorised trips in the city (Steer 2014). Some of the key functions include the following:

- Provide people-centred (quality, efficient, integrated and safe) services
- Provide commuter-responsive service planning and promotion
- Optimise resources and build capacity
- Adopt environment-friendly and sustainable practices
- Strengthen commuter feedback mechanism
- Modernise and maintain zero breakdown fleet
- Evolve effective mechanism to monitor service performance
- Conduct safety training, performance audits and awareness for stakeholders
- Increase commercial revenue through monetizing land, buildings and buses
- Increase efficiency in operations and administration
- Ensure interagency coordination and multimodal integration
- Formulate and enforce police measures for sustainability of the service provision
- Implement intelligent transport system to improve the quality of service
- Extend travel concession to the weaker sections of the society
- Act as an agent for cultural synthesis and national integration
- Promote research on urban transport

The city's looming population consisting of more than 9 million has put enormous pressure on public transport services. The creation of Greater Bengaluru has increased registered vehicles from 6.28 lakhs in 1990 to 25.27 lakhs by 2006 (Abide 2010: 35). The annual growth rate of vehicle population between 1995 and 2000 constitutes 14 per cent in Bengaluru which is higher than other metropolitan cities such as New Delhi, Chennai, Hyderabad, Kolkata and Mumbai (Abide 2010: 40). The city of Bengaluru is among the 14 cities which are among the 'high' bracket for high particulate pollution. The situation, therefore, has considerably strained urban mobility in terms of access to pedestrian walking paths, traffic congestion, travel time and poor air quality.

Sustainable Initiatives by BMTC

1. Introduction of novel campaign called 'Bus Day' a traffic demand management strategy to encourage citizens to travel by bus services, at least a month, to ease the congestion in the city and subsequently quantify its benefits. Long-term benefits include decrease in journey time, overall increase in productivity and improved air quality in the city.
2. Similarly, under JNNURM scheme, 'eco-friendly' BMTC daily service buses are introduced.
3. BMTC on 16 September 2013 launched the BIG Bus Network with an initial 62 buses on Hosur Road. This is expected to reduce the commuting time by 56 per cent (Steer 2014).
4. Creation of suburban 'feeder' routes, connecting peripheral destination to the BIG Bus Network, is expected to impact an estimated 2.5 million passengers by 2016 (Steer 2014).

Source: BMTC, Central Office, Bangalore.

Urban Development Department (UDD)

Urban Development Department, a nodal agency is responsible for planning, coordinating and supervising orderly development within Bengaluru Metropolitan Region (BMRA) with the help of Bangalore Metropolitan Region Development Authority (BMRDA) and under Bangalore Metropolitan Region Development Authority Act of 1985. Some of the key functions of the UDD include:

- To plan, regulate, control, monitor and facilitate urban development for creation of major infrastructure facilities, development of residential layouts and construction of houses for the underprivileged and rejuvenation of lakes of Bengaluru City with the help of Bangalore Development Authority (BDA) under BDA Act of 1976.
- To create and maintain infrastructure for water supply and sewerage facilities in the Bengaluru Metropolitan Area to meet the growing demands till 2015 with the help of BWSSB under BWSSB Act of 1964.

- Under Karnataka Municipal Corporation Act of 1976 and Karnataka Municipalities Act of 1964, UDD will facilitate, supervise and control municipal governance in 8 municipal corporations, 44 city municipal councils, 94 town municipal councils, 68 town panchayats and 6 notified area committees.
- Municipal governance of Bruhat Bengaluru Mahanagara Palike (BBMP) is facilitated by the state government and other BBMP; municipal governance of CMCs, TMCs, town panchayats and notified area committees is facilitated through the directorate of municipal administration and the state government.

Environmental Functions of UDD

Basically, UDD is concerned with the critical areas of (i) municipal governance; (ii) creation, improvement and maintenance of infrastructure and civic amenities and delivery of services with the help of urban local bodies and statutory authorities; (iii) under Karnataka Urban Water Supply and Drainage Act of 1973, facilitating and financing of the creation of infrastructure for delivery of water supply and sanitation in the urban areas outside Bengaluru Metropolitan Area (excluding BBMP, CMCs, TMCs and town panchayats) through Karnataka Urban Water Supply and Drainage Board (KUWSDB); (iv) within the investment programmes of urban local bodies (ULBs) and statutory bodies through Karnataka Urban Infrastructure Development Corporation (KUIDFC), facilitating of improved and sustainable infrastructure services; (v) under Karnataka Town and Country Planning Act of 1961 and Karnataka Urban Development Authorities Act of 1987, facilitating of technical assistance for the preparation of Master Plans and development schemes/layouts by urban development authorities and planning authorities; (vi) preparation of city mobility plans for urban areas in the state and providing of technical guidance and capacity building including promotional campaigns for sustained mobility through comprehensive traffic and transport studies; and (vii) finally, enabling of improved quality of life of the cities and creation of economically vibrant, efficient and sustainable basic infrastructure. UDD will facilitate policy and legislative framework to equip governing bodies with required capacities, human resources and funds.

As per Municipalities Act of 1964, UDD enforces building by-laws for city's environment and resource management. They include the following:

1. City Municipalities (Model) Building Bye-laws, 1979
2. Municipalities (Advertisements) (Model) Bye-laws, 1966
3. Municipalities (Levy of Fees on Buses Using Bus-Stand) (Model) (Bye-laws)
4. Municipalities (Licencing of Hand Carts and Hand-barrows) Model Bye-laws, 1966
5. Municipalities (Licencing of Premises Used as Factory) Model Bye-laws, 1967

(continued)

6. Municipalities (Licencing of Surveyors and Plumbers) Model Bye-laws, 1966
7. Municipalities (Model) Bye-laws regulating the conditions on which permission may be given for the temporary occupation of or for the erection of temporary structures on public streets or projections over public streets, 1966
8. Municipalities (Protection, Regulation of Management and Use of Public Parks, Gardens and Open Spaces Belonging to the Municipal Council) Model Bye-laws, 1966
9. Municipalities (Registration of Births and Deaths and Collection of Statistics) Model Bye-laws
10. Municipalities (Regulation and Inspection of Lodging and Boarding Houses) (Model) Bye-Laws, 1966
11. Municipalities (Water Supply) Model Bye-laws, 1966
12. The Karnataka Town Municipalities (Building) Model Bye-Laws, 1981

Source: http://uddkar.gov.in/municipality_byelaws

Lake Development Authority (LDA)

Lake Development Authority (LDA) was constituted in 2002,[47] a registered body under the Karnataka Societies Registration Act of 1960. The jurisdiction of LDA extends over the metropolitan area of Bengaluru, BMRDA area, including greenbelt areas of Bengaluru. Meanwhile, in 2003, the jurisdiction of LDA[48] was further extended over the lakes in the city municipal corporation of the state which form the main source of drinking water. LDA is the regulatory, planning and policymaking body with nodal functions for protection, conservation, reclamation, restoration, regeneration and integrated development of lakes in the jurisdiction of the authority. It is the nodal agency for the National Plan for Conservation of Aquatic Eco-systems (NPCA) (merging existing National Lake Conservation Plan and National Wetlands Conservation Programme) and took up 16 lakes in Karnataka under NLCP, with 5 of them in Bengaluru City and 6 projects in Karnataka under the National Wetlands Conservation Programme.

The LDA is an autonomous, regulatory, planning and policymaking body with nodal functions for protection, conservation, reclamation, restoration, regeneration and integrated development of lakes whether natural or man-made in the jurisdiction of the authority. It works for the regeneration and conservation of lakes in the designated area of operation.

[47] LDA vide GO No. FEE/12/ENG/2002 Bengaluru, dated 10 July 2002.

[48] As per the government of Karnataka (GoK) vide GO No. FEE 12 ENG 2002, dated 30 June 2003. (http://www.karnataka.gov.in/ldakarnataka/Pages/Home.aspx).

The aim of the LDA is to protect and preserve the urban wetland which includes conservation of water bodies like lakes and tanks. The main functions of the LDA constitute the following:

- To establish a strong system of well-linked lakes and tanks free from organic and chemical pollutants
- To intensify official concern and motivate community vigilance to the extent where pollution and encroachment of lakelands would become impossible
- Restoring lakes and facilitating restoration of depleting groundwater table
- Diverting/treating sewage to generate alternative sources of raw water and prevent contamination of underground aquifers from waste water
- Environmental impact assessment studies
- Environmental planning and GIS mapping of lakes and its surrounding areas
- Improving and creating habitat for water birds and wild plants. Reducing sullage and non-point water impacts
- Improving urban sanitation and health conditions especially of the weaker section living close to the lakes
- Impounding run-off water to ensure recharge of groundwater aquifers and revival of bore wells
- Monitoring and management of water quality and lake ecology
- Utilising the lake for the purpose of education and tourism
- Community participation and public awareness programmes for lake conservation

Source: LDA, Bengaluru.

Notwithstanding its lofty mission declared by the LDA, but in practice LDA lacks organisational and statutory powers to deal with the complex issues involving the lake conservation in the state, particularly in the city of Bengaluru. Besides, LDA grossly lacks staff and funds to undertake its functions. The need for the preservation and conservation of urban ecosystem particularly lakes is more pronounced in the context of rapid urbanisation, and to deal with various kinds of offences related to lakes, the government of Karnataka is contemplating to legislate Karnataka Lake Development Authority Bill of 2014. If the bill is passed and enforced, LDA can be expected to play a critical role in the conservation of lakes in and around the Bengaluru City as well as in other urban areas of the state.

5.1.5.4 Bangalore Electricity Supply Company Limited (BESCOM)

BESCOM is an outcome of comprehensive power sector reforms embarked by government of Karnataka in the year 1999. As a first step, Karnataka Electricity Reforms Act was passed by the Karnataka Legislature; as part of corporatisation, the

Table 5.9 Profile of BESCOM

Details		BESCOM profile
Area covered		41,092 sq. kms
Districts		8
Population		20.7 million
Distribution transformers as on 31 march 2012		161,905
Length of HT line ------- " ------- (Prov)		75,074 circuit kms
Length of LT line ------- " ------- (Prov)		150,105 circuit kms
No. of employees	Sanctioned	16,902
	Working	10,580
Turnover during 2011–2012 (Prov)		Gross block, Rs. 4122.08 Crs Net block, Rs. 2842.69 Crs
Turnover during 2011–2012 (Prov)		Rs. 9406.53 Crs
EHV stations		401
Total no. of consumers (Prov)		79.65 lakh as on 31.03.2012

Source: http://bescom.org/en/about-us-2/

erstwhile Karnataka Electricity Board ceased to exist, and Karnataka Power Transmission Corporation Limited (KPTCL) was constituted from 1 August 1999. Subsequently, KPTCL was separated and remained as transmission company. This was followed by constitution of Karnataka Electricity Regulatory Commission (KERC) in November 1999. In the next phase of the reform process, the transmission and distribution business managed by KPTCL were unbundled in June 2002. Four new distribution companies were formed to distribute power in Karnataka.

Therefore, by 2002, as a part of the reforms, the distribution sector was further divided into four companies, viz. Bangalore Electricity Supply Company Limited (BESCOM), Hubli Electricity Supply Company Limited (HESCOM), Mangalore Electricity Supply Company Limited (MESCOM) and Gulbarga Electricity Supply Company Limited (GESCOM). These companies came into existence from 1 June 2002. Bangalore Electricity Supply Company Limited (BESCOM) has taken over the responsibility from KPTCL for the distribution of electricity in eight districts and commenced its operations from 1 June 2002. Bangalore Electricity Supply Company came into existence post-reform process in 2002. The main objective of the BESCOM[49] is to develop infrastructure, commensurate with the growth continuously ensuring reliable and quality power supply. BESCOM strives to use best technology and best practices in power sector (Table 5.9).

The BESCOM is responsible for distribution of power to eight districts in Karnataka which include Bengaluru Urban/Rural Districts, Kolar, Tumkur, Ramanagara, Chikkaballapur, Chitradurga and Davangere. BESCOM covers an area of 41,092 sq kms with a population of over 207 lakhs. The company has three operating zones – Bengaluru Metropolitan Area Zone, Bengaluru Rural Area Zone and Chitradurga Zone.

[49] Please check Table 4.9 in the Annexure for category of BESCOM consumers.

The dominant function of BESCOM is to retail supply of electricity in the designated area. Power is purchased from generating companies like Karnataka Power Corporation Limited, central generating companies and independent power producers (IPPs) at the agreed rates and sold to different categories of consumers determined by Karnataka Regulatory Commission. A chain of offices have been established to enable to attend to the needs of consumers in terms of servicing of installations, collection of revenue and maintenance of lines, attending of consumer complaints, augmentation of infrastructure, etc.

The main duty of BESCOM is selling power to consumers. In this process, supplemental duties which are incidental to main function are the following.

Duties of BESCOM

1. Selling power to consumers at the rates for and by KERC.
2. Supply at specified voltage and frequency.
3. Maintenance of lines and equipments to ensure smooth and quality power supply.
4. Augmentation of infrastructure to meet the demand.
5. Ensuring safety of public and animal life by taking suitable actions to minimise risk of accidents.
6. Perspective planning of activities in relation to demand and supply of power.
7. Detailed duties and responsibilities are also available in KERC website.

Source: http://bescom.org/en/about-us-2/

Sustainable Initiatives by BESCOM[50]

Several initiatives have been taken up by BESCOM for conservation and saving of energy. Specifically to promote energy conservation, BESCOM has constituted Demand Management Cell (DSM) for promoting solar energy and initiating education and awareness programmes on energy conservation among the consumers. DSM is a mechanism to influence customer's '*CAPABILITY*' and '*WILLINGNESS*' to reduce electricity consumption among the consumers. Demand-side management (DSM) programmes consist of planning, implementing and monitoring activities of electric utility that are designed to encourage consumers to modify their level and pattern of electricity usage.

DSM provides enhanced reliability to the energy system by reducing overall demand through energy efficiency and by reducing peak demand through dispatchable programmes. For promotion of various energy-efficient/DSM methods in the state, the following steps were taken:

[50] Details on sustainable initiatives by BESCOM are collected from DSM, Corporate Office, BESCOM, on 25 April 2014.

- National Energy Conservation Day observed on 14 December (GOI has enacted the Energy Conservation Act of 2001).
- Earth Hour observed by all the employees of BESCOM in offices and their residences. Public were also requested to participate (Earth Hour was observed on 23 March 2013 from 8.30 PM to 9.30 PM by switching off all the unnecessary lights).
- Under 'Belaku Yojana' BESCOM is contributing indirectly to saving energy.

Under DSM, many energy saving initiatives have been undertaken they include the following:

(i) Awareness programmes among school students.
(ii) Constituting Demand Management Cell (DSM) for promoting solar energy and conservation among the consumers.
(iii) Under WENASA programme, farmers are motivated to replace inefficient pumpsets with the assistance of BESCOM.
(iv) 'Belaku Yojana' BESCOM is contributing indirectly to saving energy.
(v) Installing solar rooftop is another major programme initiated by BESCOM for energy conservation.
(vi) 'Niranthara Jyothi' is one of the initiatives for farmers at rural areas.
(vii) A new scheme called 'Adopt Transformer'.
(viii) Use of solar power pumpset has been promoted among agriculture farmers under 'Ganga Kalyana' scheme.

Source: BESCOM, Corporate Office, Bengaluru

(i) **BESCOM Efficient Lighting Programme**

A pilot project was undertaken during 2005–2006 for the first time in India under USAID in Bengaluru City. Under this scheme, consumers were encouraged to purchase CFLs directly from suppliers or from BESCOM on a monthly instalment basis paid through electricity bills. Bachat Lamp Yojana scheme, a scheme to promote replacement of incandescent lamps with CFLs was implemented during 2011.

(ii) **Installation of Solar Water Heater**

The use of solar water heaters is now made mandatory for all nondomestic and domestic customers in buildings with a built-up area of 600 sq. ft. and above constructed on site area of 1200 sq. ft. and above falling within the limits of municipalities, city corporations and BDA sectors. So far, between April 2012 and March 2014, 169,150 solar water heaters are fixed by BESCOM.

(iii) **Niranthara Jyothi: In Rural Areas**

Niranthara Jyothi project is a prestigious scheme of the government of Karnataka where segregation of nonagricultural loads and agricultural loads in rural areas was undertaken to provide 24-hr three-phase power supply to nonagricultural loads like domestic, commercial, water supply, street light, rural industries, milk dairies, etc.

This will enable BESCOM to give 24×7 supply to nonagricultural loads in rural areas and also the following advantages such as (a) reduction in distribution transformer failures, (b) increase in metred sales, (c) reduction in technical losses, (d) improved load management and (e) improved standard of living in rural areas.

'Niranthara Jyothi' is one of the sustainable initiatives taken up by BESCOM for farmers at rural areas. At present, both phase I and phase II are completed. Niranthara Jyothi project is a prestigious scheme of the government of Karnataka, to provide 24-hr three-phase power supply to nonagricultural loads like domestic, commercial, water supply, street light, rural industries, milk dairies, etc., in rural areas by segregating the agricultural loads, which is a boon to the rural economy. This will enable BESCOM to give 24×7 supply to nonagricultural loads in rural areas. The aim is (i) reduction in transformer failure, (ii) increase in metred sales, (iii) reduction in technical losses and (iv) improved load management.

- DDIS (Diesel Direct Injection System) – running 'energy-saving tips' in kiosks in public places
- DEEP (Distribution Energy Efficiency Programme) – in rural areas, to reduce power cuts and voltage problems

(iv) **WENEXA**

BESCOM launched Distribution Reform, Upgrades and Management (DRUM) project and Water and Energy Nexus Activity phase II, both conducted under the aegis of India's Ministry of Power (MoP) and the United States Agency for International Development (USAID).

WENEXA project has been taken up on a pilot basis in Doddaballapura subdivision for improving the energy efficiency of pumpsets through an ESCO model under USAID-assisted programme by replacing old inefficient pumpsets by new high energy-efficient pumpsets. By April 2014, 277 of irrigation pumpsets have been replaced by high energy-efficient pumpsets. Following the successful implementation of WENEXA pilot project, BESCOM has decided to extend the implementation of WENEXA project by selecting one of the most backward taluks in each of the six districts of Karnataka for replacement of low energy-efficient pumpsets with high energy-efficient ones.

(v) **Solar Rooftop: Grid Connection**

For installing solar rooftop, BESCOM has initiated Solar Rooftop Policy. Installing solar rooftop is another major programme initiated by BESCOM for energy conservation. Two tenders are called, i.e. one tender for selected 15 to 20 agencies for installing solar rooftop and another for selecting agencies for third inspection for ensuring safety measures and monitoring quality. Agencies are selected based on 2 years of experience in solar installation. Solar rooftop power

Table 5.10 No. of solar heating systems. Installed by BESCOM

Year of installation	No. of solar water heating systems installed
2008–2009	6187
2009–2010	47,218
2010–2011	52,491
2011–2012	72,084
2012–2013	95,819
2013–2014	41,832

Source: BESCOM, Corporate Office, Bengaluru
From: April 2013 to October 2013

generators connected to the grid can sell excess power generated to BESCOM at the rate of Rs. 9.56 per unit. A proposal has been sent to KERC to make use of solar panels compulsory for multi-storied buildings. All the new buildings to be constructed in the government/government-aided sectors will incorporate energy-efficient building design concepts including renewable energy technologies with effect from 1 year from the date of issue of this order (Table 5.10).

(vi) **Street Lights: Switch Timers**

BESCOM has installed switch timers on street lights in the city of Bengaluru. By 2013–2014, 12,241 switch timers are installed in the city. This was initiated mainly to ensure that the street lights are switched on only during the nights and avoid manual switching off thereby contributing to saving energy.

(vii) **Vidyuth Jagruthi Yojana: Save Energy Campaign**

Under Vidyuth Jagruthi Yojana, an awareness programme has been introduced for higher primary and high school children on contract period of 2 years. This programme is expected to bring awareness among school students who will communicate their learnings to their families. In this project, electricity consumption data of selected schools and from school students will be collected for the last 2 years, and those in charge will prepare a working model, study materials, booklets, exhibits, drawings, posters, PPT models, awards and prizes for those school students on energy conservation measures.

In addition, near railway stations, BESCOM captions/hints appear on TV panels to create awareness. A booklet with information on electricity generation, transmission and distribution, regulators and energy conservation methods was distributed among the children. An objective and pick and speak test were conducted after a week and cash prizes awarded for the top three students in both tests.

Under Vidyuth Jagruthi Yojana, a tender amounting to Rs. 1.5 crores (approximately) has been called for design and implementation of conducting energy efficiency measures and demand-side management awareness activities in selected 100 schools in Bengaluru Urban/Rural Districts and to sensitise approximately 50,000 students and to bring down energy consumption on contract period of 2 years.

(viii) **'Adopt Transformer': By BESCOM**

'Adopt Transformer' is new scheme which is at planning stage to encourage individual or community to adopt a transformer for its monitoring as well as maintenance.

(ix) **'Ganga Kalyana' Scheme**

The use of solar power pumpset has been promoted among agriculture farmers under 'Ganga Kalyana' scheme.

(x) **Initiatives Towards Renewable Energy**

BESCOM is also contributing towards renewable energy; particularly a proposal was given for installing 'wind energy'. The state of Karnataka is contributing 25 per cent in the area of renewable energy through wind power, while 30 per cent is through generating hydroelectric power.

(xi) **Regularisation of Unauthorised IP Sets**

Unauthorised IP sets are existing in the state of Karnataka, due to which transformers are overloaded which leads to failure of transformers and hence occurrence of commercial and distribution loss. To reduce transformer overloading and losses, the state government has introduced a scheme called Regularisation of Unauthorised IP sets in the year 2005 and 2011. These unauthorised IP sets are regularised after the payment of regularisation fee of Rs.10,000 – with other deposits and necessary infrastructure provided on seniority basis. With BESCOM, 85,241 have registered by 2012, and 42,223 have been regularised after the payment of regularisation fee and other deposits.

(xii) **BESCOM Efficient Irrigation Pumpset Programme**

A pilot project was conducted during 2005–2006 for the first time in the country. Innovative BESCOM Efficient Irrigation Pump Sets Program (BEIPP) scheme was launched, supported by the International Copper Promotion Council (India). A total number of 1074 submersible pumpsets were replaced by energy-efficient IP sets under the scheme. Manufacturers offered a warranty of 18 months for pump and offered attractive rates for energy-efficient IP sets.

(xiii) **Lifestyle Change Approach to Best Energy Conservation Practices: A Pilot by CSD**

Centre for Sustainable Development were engaged for bringing in awareness among residential/commercial customers for adoption of energy-efficient methods in their daily lifestyle as a pilot project in Brigade Road, M.G. Road, Residency Road and Halasur. CSD have submitted the final report, and the results are encouraging which indicate a saving of an average 8–10% of energy by all participating customers in this scheme. This pilot is the first of its kind in India conducted by a public utility using an NGO. BESCOM has undertaken replacement of existing copper chokes in fluorescent tube light fittings of BESCOM buildings by energy-efficient electronic ballast through M/s KEONICS, a state government undertaking. This resulted in saving of energy to the tune of 0.35 million units per year considering average usage of TL fittings of 10 hours per day for 25 days in a month.

BESCOM is insisting on BBMP and Bengaluru Metropolitan Area Zone to make fixing automatic timer switches to street light circuits mandatory. Studies show that the usage of electronic timer switches saves energy up to 30%. And also it ensures timely switching on of streetlight circuits at 6 PM and switching off at 6 am exactly.

5.1.5.5 Central Ground Water Board (CGWB) and Karnataka Ground Water Board (KGWB)

Central Ground Water Board (CGWB) was constituted in 1970 and is a subordinate office under the Ministry of Water Resources, government of India. It is a national apex agency with the responsibility of providing scientific inputs for management, exploration, monitoring, assessment, augmentation and regulation of groundwater resources of the country. CGWB is a multidisciplinary scientific organisation consisting of hydrogeologists, geophysicists, chemists, hydrologists, hydrometeorologists and engineers and has its headquarters at BHUJAL BHAWAN, NH 4, Faridabad, Haryana. The board has 18 regional offices, each headed by a regional director, supported by 17 engineering divisions and 11 state unit offices for undertaking various field activities. Various activities related to regulation of groundwater development in the country are being looked after by the Central Ground Water Authority (CGWA),[51] constituted under the Environmental (Protection) Act of 1986.

Major activities being taken up by Central Ground Water Board (CGWB) include macro–/micro-level groundwater management studies, exploratory drilling programme, monitoring of groundwater levels and water quality through a network of groundwater observation wells comprising both large-diameter open wells and purpose-built bore–/tube wells (piezometers) and implementation of demonstrative schemes for artificial recharge and rainwater harvesting for recharge augmentation. Geophysical studies, remote sensing and GIS studies and groundwater modelling studies are taken up to supplement these activities. The board also takes up special studies on various aspects of groundwater sector such as groundwater depletion, sea water ingress, groundwater contamination, conjunctive use of surface and groundwater, water balance and so on. It also organises various capacity building activities for its own personnel as well as central/state government organisations engaged in various activities in groundwater sector as well as mass awareness campaigns on the importance of water conservation and judicious groundwater management.

[51] Central Ground Water Authority (CGWA) was constituted under subsection (3) of Section 3 of the Environment (Protection) Act of 1986 for the purposes of regulation and control of groundwater development and management in the country.

Various initiatives have been taken up by Central Ground Water Control Board, Karnataka, to conserve and manage groundwater resources. Some of the significant measures include the following:

(i) Firstly, various levels of training have been imparted to NGOs, academicians and government officials on methods of artificial recharge and groundwater management.
(ii) Secondly, at the community levels, awareness is promoted using local language involving teachers, farmers, members of panchayats, etc.
(iii) Thirdly, national-level painting competition will be organised among the youth.
(iv) Fourthly, training will be given on aquifer mapping to identify how deep the groundwater is, how much water has been used and how to manage it.
(v) Fifth, tier 2 programme will be conducted in a year involving officials from departments of watershed, groundwater and panchayats.
(vi) Awareness programmes are organised at educational institutions such as Kendriya Vidyalaya conducting drawing competitions, on themes of conservation models, and exhibits.
(vii) Celebration of 'World Water Day' and 'Rainwater Harvesting' day.

Source: Central Ground Water Board, Bangalore.

Karnataka State Pollution Control Board (KSPCB)

The Karnataka State Pollution Control Board (KSPCB)[52] for prevention and control of water pollution was constituted by the government of Karnataka in 1974 in pursuance of the Water (Prevention and Control of Pollution) Act of 1974. The Water Act will provide for the prevention and control of water pollution and maintaining or restoring of wholesomeness of water. After the enactment of the Air (Prevention and Control of Pollution) Act of 1981, the enforcing responsibility was entrusted to the KSPCB. As such, the board was later renamed as the Karnataka State Pollution Control Board in 1985. The Air (Prevention and Control of Pollution) Act of 1981 is an enactment to provide for prevention, control and abatement of air pollution.

The KSPCB is also enforcing the following acts and rules with respect to environment protection and management of the city:

1. The Water (Prevention and Control of Pollution) Cess Act, 1977, and as amended from time to time
2. The Water (Prevention and Control of Pollution) Cess Rules, 1978

(continued)

[52] The important functions of KSPCB are presented in the Annexure.

The following rules and notifications framed under Environment (Protection) Act, 1986:

- Hazardous Waste (Management, Handling and Transboundary Movement) Rules, 2008
- Environmental Impact Assessment Notification, 2006
- Biomedical Waste (Management and Handling) Rules, 1998, and Amendment Rules 2000
- Plastic Waste (Management and Handling) Rules, 2011
- The Noise Pollution (Regulation & Control) Rules, 2000.
- Municipal Solid Wastes (Management & Handling) Rules, 2000.
- E-waste (Management and Handling) Rules 2011
- Batteries (Management and Handling) Rules, 2001
- The Public Liability Insurance Act, 1991.

Source: http://kspcb.kar.nic.in/about.html

5.1.5.6 The Karnataka Renewable Energy Development Ltd. (KREDL)

The KREDL is an organisation under the purview of the Energy Department, government of Karnataka. The main objective of KREDL is to promote renewable energy, i.e. nonenergy conventional energy sources, and support all the efforts towards energy conservation in the state. The KREDL works through various agencies which includes government, private, NGOs and accredited energy auditors. The organisation supports all forms of nonconventional energy sources such as wind, small hydro, biomass, solar and energy recovery from waste through private investment. The company advises the government of Karnataka on policies for ensuring systematic growth of projects for harnessing renewable energy sources and catalogue 'best practices' on renewable energy sources in the state.

5.2 Impact of Shortcomings in the Present Institutions on Environmental Governance in Bengaluru

Many shortcomings are clearly visible and practically hinder the implementation of effective environmental governance in the city. Some of the key shortcomings are listed below:

1. Over the years, many institutions and governance structures have emerged that directly or indirectly impact on environmental governance. Particularly, post-urbanisation, new environmental laws, rules and regulations such as Hazardous

Waste Rules, Bio-Waste Act (amended in 2000), E-waste Rules of 2011, Plastic Recycling Rules of 2011, Ground Water Regulation Act of 2011, etc. and constitutions of specific instruments such as the Lake Development Authority (LDA) have certainly resulted in proliferation of institutions for environment governance in the city. Creation and implementation differ from sector to sector which inevitably increase the fragmentation of most often contradictory institutions. Fragmentation always leads to inefficiencies and lack of synergy often creating overlaps and duplication of rules and regulations complicating coherence.
2. One of the issues seriously affecting environment and ecology of the city is sector-wise departments and concerned organisations which are not equipped sufficiently in terms of sustainable institutional vision or long-term sustainable plans. Therefore, concerned organisations lack commitment towards promoting sustainability for their sectors. Similarly, concerned local bodies, development authorities and specific agencies lack adequate stable and predictable resources. The fact that most of the departments are understaffed and underfinanced, e.g. Lake Development Authority (LDA), creates imbalance in the network of institutions.
3. Most of the key institutional rules and regulations are violated. Organisations lack monitoring agencies to elicit compliance and accountability.
4. Finally, the relationship and complementarities of environment with economic and social aspects within the sustainable development are yet to be clarified and operationalised.

5.3 Conclusion

The city of Bengaluru is growing in leaps and bounds. The city is reeling under the increasing impact of urbanisation process. The greatest challenge so far has been imbalance created due to the growth of population and the capacity of local bodies to cater the basic services. The local bodies and authorities are overburdened with population seriously suffering from lack of sufficient resources and infrastructure. Promoting urban environmental governance, therefore, is the major challenge for the present as well as for the future. The city of Bengaluru is no wonder a part of an urban crisis-like scenario triggered by economic growth impacts. Though the city is offering many opportunities and jobs, concomitantly this has increased the social inequality and poverty. Benefits are not equally shared resulting in the lopsided development. Local environmental conditions are therefore critically linked to well-being and development. The city is witnessing quantifiable impact on environment and ecology such as loss of green space, destruction of local ecosystems, floods and pollution. Chapter 6, therefore, clearly presents the current environmental resource trends and use in the city of Bengaluru.

Annexure

UDD Complies with both State and Central Legislations Impacting Environmental Governance

Bangalore City Planning Zonal Regulations (Amendment and Validation Act, 1996 (2 of 1996)

Bangalore Development Authority Act (1976) (12 of 1976), amended by Acts 8 of 1981, 37 of 1982, 17 of 1984, 34 of 1984, 11 of 1988, 18 of 1991, 6 of 1993, 17 of 1994, 26 of 1995, 1 of 2000 and 22 of 2000, 19 of 2002 and 19 of 2005

Bangalore Metropolitan Region Development Authority Act, 1985 (39 of 1985), amended by Act 8 of 2005

Bangalore Water Supply and Sewerage Act, 1964 (36 of 1964), amended by Acts 6 of 1966, 10 of 1966 and 18 of 1984

City of Bangalore Improvement Act (Karnataka Act 19 of 2002) (Mysore Act V of 1945)

Improvement of Boards Act, 1976 (11 of 1976), amended by Acts 68 of 1976, 15 of 1984, 34 of 1984, 13 of 1985 and 40 of 1986 and 12 of 2001

Local Authorities (Official Language) Act, 1981 (30 of 1981)

Municipal Corporations Act, 1976, amended by Acts 24 of 1978, 11 of 1979, 21 of 1979, 28 of 1978, 40 of 981, 8 of 1982, 13 of 1983, 34 of 1984, 21 of 1986, 32 of 1986, 20 of 1987, 2 of 1990, 19 of 1991, 22 of 1991, 32 of 1991, 35 of 1994, 14 of 1995, 25 of 1995, 24 of 1998, 27 of 98, 9 of 2001, 31 of 2001, 5 of 2003, 8 of 2003, 32 of 2003 (this amendment act further amended by Act No. 17 of 2004) 39 of 2003, 5 of 2005, 1 of 2007, 14 of 2007 and 17 of 2005, 1 of 2007, 14 of 2007 and 17 of 2007

Karnataka Municipal Corporations and Certain Other Law (Amendment) Act, 2011 (Karnataka Act No. 24 of 2011)

Municipalities Act, 1964 (22 of 1964), amended by Acts of 34 of 1966, 2 of 1976, 39 of 1976, 83 of 1976, 13 of 1979, 21 of 1979, 22 of 1981, 26 of 1982, 28 of 1982, 12 of 1983, 2 of 1984, 33 of 1984, 34 of 1984, 33 of 1986, 20 of 1987, 2 of 1990, 22 of 1991, 36 of 1994, 20 of 1995, 24 of 1995, 24 of 1998, 22 of 2000, 28 of 2001, 31 of 2003 (This Amendment Act further amended by Act No. 17 of 2004) 40 of 2003, 5 of 2005, 1 of 2007)

Municipal Taxation (Karnataka Extension) Act, 1973 (Karnataka 13 of 1974), Central Act of 1881

Open places (Prevention of Disfigurement) Act, 1981 (35 of 1982) Amended by Act No. 15 of 1984.

Parks, Playfields and Open Spaces (Preservation and Regulation) Act, 1985 (16 of 1985) Amended by 42 of 2003.

Regularisation of unauthorised constructions in Urban Areas Act, 1991 (29 of 1991) Amended by Acts 9 of 1994, 29 of 1994, 27 of 1995 and 17 of 2007.

(continued)

Town and Country Planning Act, 1961 (11 of 1963) Amended by Acts 14 of 1964, 2 of 1998, 12 of 1976, 39 of 1985, 34 of 1987, 2 of 1991, 17 of 1991, 8 of 1994, 18 of 2003, 23 of 2004, 1 of 2005, 1 of 2007 and 2 of 2007.

Urban Development Authorities Act, 1987 (34 of 1987) Amended by Acts of 17 of 1991, 14 of 1992 and 12 of 1996.

Urban Water Supply and Drainage Board Act, 1973 (25 of 1974) Amended by Acts 7 of 1976, 20 of 1977, 45 of 1981 and 19 of 1993.

Employment of Manual Scavengers and Construction of Dry latrines (Prohibition) Act, 1993 (Table 5.11).

Table 5.11 Category of BESCOM consumers

Category of consumers	Consumers up to 31 March 2012
Domestic	61.48 lakhs
Commercial	07.35 lakhs
Irrigation pumpsets	06.78 lakhs
Industrial	01.53 lakhs
Water supply	00.43 lakhs
Public lighting	00.50 lakhs
Others	01.50 lakhs
Total	79.57 lakhs
Industrial, commercial, lift irrigation and residential	00.08 lakhs
Grand total	79.65 lakhs

Source: http://bescom.org/en/about-us-2/

The important functions of the Karnataka State Pollution Control Board (KSPCB) under Water (Prevention and Control of Pollution) Act, 1974, and Air (Prevention and Control of Pollution) Act, 1981 (Section 17), are:

(a) To plan a comprehensive programme for the prevention, control or abatement of pollution of streams and wells in the state and to secure the execution thereof
(b) To advise the state government on any matter concerning the prevention, control or abatement of water pollution and air pollution
(c) To collect and disseminate information relating to water pollution and air pollution and the prevention, control or abatement thereof
(d) To encourage, conduct and participate in investigations and research relating to problems of water pollution and prevention, control or abatement of water pollution
(e) To inspect sewage or trade effluents, works and plants for the treatment of sewage and trade effluents and to review plans, specifications or other data

(continued)

relating to plants set-up for the treatment of water, works for the purification thereof and the system for the disposal of sewage or trade effluents or in connection with the grant of any consent as required by this Act

(f) Laydown, modify or annul effluent standards for the sewage and trade effluents and for the quality of receiving waters (not being water in an interstate stream) resulting from the discharge of effluents and to classify waters of the state

(g) To evolve economical and reliable methods of treatment of sewage and trade effluents, having regard to the peculiar conditions of soils, climate and water resources of different regions and more especially the prevailing flow characteristics of water in streams and wells which render it impossible to attain even the minimum degree of dilution

(h) To evolve methods of utilisation of sewage and suitable trade effluents in agriculture

(i) To evolve efficient methods of disposal of sewage and trade effluents on land, as are necessary on account of the predominant conditions of scant stream flows that do not provide for major part of the year the minimum degree of dilution

(j) To laydown standards of effluents of sewage and trade effluents to be discharged into any particular stream and the tolerance limits of pollution permissible in the water of the stream, after the discharge of such effluents

(k) To make, vary or revoke any order

  (i) For the prevention, control or abatement of discharges of waste into streams or wells

  (ii) Requiring any person concerned to construct new systems for the disposal of sewage and trade effluents or to modify, alter or extend any such existing system or to adopt such remedial measures as are necessary to prevent control or abate water pollution

(l) To laydown effluent standards to be complied with by persons while causing discharge of sewage or sullage or both and to laydown, modify or annul effluent standards for the sewage and trade effluents;

(m) To advise the State Government with respect to the location of any industry the carrying on of which is likely to pollute a stream or well.

(n) To plan a comprehensive programme for the prevention, control or abatement of air pollution and to secure the execution thereof;

(o) To inspect, at all reasonable times, any control equipment, industrial plant or manufacturing process and to give, by order, such directions to such persons as it may consider necessary to take steps for the prevention, control or abatement of air pollution;

(p) To inspect air pollution control areas at such intervals as it may think necessary, assess the quality of air therein and take steps for the prevention, control or abatement of air pollution in such areas;

(continued)

(q) To laydown, in consultation with the Central Board and having regard to the standards for the quality of air laid down by the Central Board, standards for emission of air pollutants into the atmosphere from industrial plants and automobiles or for the discharge of any air pollutant into the atmosphere from any other source whatsoever not being a ship or an aircraft; Provided that different standards for emission may be laid down under this clause for different industrial plants having regard to the quality and composition of emission of air pollutants into the atmosphere from such industrial plants;
(r) To advise the State Government with respect to the suitability of any premises or location for carrying on any industry which is likely to cause air pollution;
(s) To perform such other functions as may be prescribed or may, from time to time, be entrusted to it by the Central Board or the State Government.

Source: http://kspcb.kar.nic.in/about.html

References

Abide. (2010). Plan Bengaluru: bringing back a Bengaluru of Kempe Gowda's dreams. Bengaluru: Agenda for Bengaturu Infrastructure and Development Task force.

Balasubramanain, V. (2013). Bangalore water crisis, the big hoodwink, March 15, DNA. www.dnaindia.com

BBMP. (2009). Baseline survey report. BBMP: Bruhat Bengaluru Mahanagarapalike.

BDA. (2007). Revised master plan – 2015, Bangalore, Volume 2, BDA: Bengaluru.

CDP. (2009). Urban Infrastructure and Governance, Revised City Development Plan: JNNRUM. Bengaluru: BBMP.

CES. (n.d.). Geography and environment of Bangalore. www.ces.iisc.er.net.in

Comprehensive Development Plan (CDP). (2006). Bangalore Development Authority.

GoK. (2009). *State of environment report, Bangalore 2008*. Bengaluru: Department of Forest, Ecology and Environment and Government of Karnataka.

Karnataka Slum Clearance Board. (2017). http://housing.kar.nic.in/DOH_KSCB_page.htm

Mahendra, B., Harikrishnan, K., & Krishne, G. (2010). Urban governance and master plan of Bangalore City. *India Journal, 7*(2), 1–18.

Mehta, V. K., Rimi, G., Eric, K. B., Sekhar, M., & Deepak, M. (2013). Social ecology of domestic water use in Bangalore. *Economic and Political Weekly, XLVIII*(15), 40–50.

Nagendra, H. (2010). Maps, lakes and citizens. www.india-seminar.com. Document accessed on 18 Sept. 2013.

Narayana, M. R. (2008). *Globalisation and urban growth: Evidence from Bangalore (India), CIRJE*. Tokyo: University of Tokyo and Bangalore: ISEC.

Ranganathan, S.S. (2013). Sewerage is much bigger problem than water supply in Bangalore, thealternative, March 25, 2013. www.thealternative.in.environment

Ravindra, A. (2005). *Bangalore Multiplicity*. Bengaluru: Gangarams Publications Pvt Ltd.

Roychowdhury, S. (2003). Old classes and new spaces: Urban poverty, unorganised labour and new unions. *Economic and Political Weekly, 13*, 5277–5284.

Roychowdhury, S. (2005). Labour activism and women in the unorganised sector garment export industry in Bangalore. *Economic and Political Weekly, 40*(22/23), 2250–2255.

Roychowdhury, S., Sivanna, N., & Vani, B. P. (2012). *Migration, informal work and welfare: A policy perspective on Karnataka Cities, Project Report*. Bangalore: ISEC.
Schenk. (2001). *Living in India's slums: A case study of Bangalore, IDPAD*. New Delhi: Manohar.
Sivanna, N. (2014). Urban Governance and Organisational Restructuring: The Case of Bruhat Bangalore Mahanagara Palike (BBMP), Monograph series 31, Bangalore: Institute for Social and Economic Change.
Steer, A. (2014). 7 environment and development stories to watch in 2014. March 24, Huffingtonpost. https://www.huffingtonpost.com/dr-andrew-steer/7-environment-stories-2014_b_4644525.html
Sudhira, H. S. (2008). The emergence of world city Bangalore. Conference Paper – 9th World Congress of Metropolis. http://eprints.iisc.ernet.in/24656/1/Chapter_4,_Vol_4,_Metropolis_Congress_2008.pdf
Sudhira, H. S., Ramachandra, T. V., & Balasubramanya, M. H. (2007). City Profile: Bangalore, Cities, 24(5): 379–390.
Parisara. (2010). *Conservation of water bodies in Bangalore*. Bangalore, ENVIS: Environment Management & Policy Research Institute (EMPRI).
UN-HABITAT. (2010). *State of the world cities 2010/11*. Japan: UN-HABITAT and ESCAP.

Chapter 6
Trends and Status of Environmental Resource Use in Bengaluru Metropolitan Area (BMA)

Abstract The city of Bengaluru has outgrown in terms of the proportion of population and geography with profound impacts on both the physical and natural resource use. A combination of factors such as urbanisation, migration, urban poverty, complex institutions and so on have a tremendous impact on how the city is coping up with varied urban-related and environmental challenges. Urban environmental quality is one of the significant factors associated with the ability of the city to promote environmental governance. Therefore, this chapter presents an analysis of the trends and status of environmental resource use across different sectors such as water supply, sanitation, solid waste, urban ecology, etc., in the Bengaluru Metropolitan Area (BMA). The analysis is based on the methodology developed to assess the current environmental status of the city. The purpose is to illustrate an overview of better or worse conditions across different urban sectors that affect environment and ecology of the city.

Keywords Environmental resource · Environmental quality · BMA · Bengaluru · Water · Sanitation · Urban lakes · Solid waste · Urban climate change · Urban floods

6.1 Introduction

The city of Bengaluru (urban) division is located on the Deccan Plateau between 12°14 and 13°30 north and 77°3 and 77°59 east. The topography is almost flat with moderate slopes, while the southern part is undulated and hilly. The elevation varies from 835 to 953 m. The urban division has plenty of water tanks spread across the city region (BBMP 2006: 12; GoK 2009: 11). The temperature of the city ranges from 35 °C during the day to 21 °C at night during winter; similarly, during hot months of April–May, temperatures vary from 38 to 39 °C (BBMP 2006: 12; GoK 2009: 11). The Greater Bengaluru area is reported to have increased by 632% between 1973 and 2009 with 76% and 79% of vegetation and water bodies (Ramchandra and Kumar 2010).

K. V. Raju et al., *Urban Environmental Governance in India*, The Urban Book Series, https://doi.org/10.1007/978-3-319-73468-2_6

The population of Bengaluru[1] City has increased considerably since the 1980s, from 12 lakh in 1970 to 56.86 lakh as per census 2001 and to 65 lakh by 2008 (GoK 2003: 1). The CDP (City Development Plan) has included population in the peripheral villages which is estimated to reach 61.70 lakhs (BBMP 2006: 14). The city's urbanisation can be identified with a demographic explosion in terms of population growth, and poverty induced various kinds of migration (Ramachandra et al. 2012: 229), namely, (i) migration within Karnataka, (ii) from other states of the country, and (iii), to a lesser extent, from outside the country *(ibid)*. The city of Bengaluru is a hub of multifarious growth of industries like IT/BT, manufacturing and services sectors and hospitality industries. A large number of multinational companies have earned the city its name as 'Silicon Valley' and 'International City'. Improved quality of life, better educational and employment opportunities and access to infrastructure, health, transport and communication largely attract migrants from both within and outside the state (GoK 2009: 1).

The burgeoning population growth coupled with a rapid urbanisation influenced by migration, industrialisation and technology has resulted in a tremendous strain on access to and quality of land and local renewable and nonrenewable resources, threatening the preservation and protection of biodiversity and ecology of the surroundings. Urban environmental problems stem from weak institutional capacity to respond to inadequate planning, management and governance and mismatch of demand-supply between growing population and access to basic amenities such as water supply, toilets, drainage facilities, electricity, transport, road network and waste management. Concomitantly, environmental problems trigger varied and widespread pollution like air, noise, water and land, seriously impacting the general health and thereby the quality of life in the city (Ramanchandra et al. 2012: 329). From the past 5 years, the emergence of new kinds of disease[2] in the city is inevitably linked to the exposure to environmental hazards (Fig. 6.1).

The diseases range are water-borne, dengue virus, influenza and acute respiratory infections like bronchial asthma, allergy, fluorosis, etc. Lack of water, sanitation and waste management (safe disposal) has been one of the significant reasons for the outbreak of gastroenteritis, viral hepatitis and typhoid (GoK 2003: 9).

This chapter captures the current trends and status of environmental resource use in the Bengaluru Metropolitan Area (BMA). The aim here is to assess trends and status of environmental resource use across different sectors such as water supply, sanitation, solid waste, urban ecology, etc. in the Bengaluru Metropolitan Area (BMA). The chapter, therefore, provides an overview of a micro-level assessment of

[1] Between the months of July–August and October, the mean annual rainfall is around 875 mm spread over 50 days in a year. The city also receives cyclone-induced rainfall between the months of November and December (GoK 2009: 1).

[2] Outdoor pollution due to an increase in vehicular traffic, number of vehicles and industrial pollution has infected the city with pollen, dust mites, fungi and irritants like tobacco and cleaning agents leading to various kinds of respiratory illness like asthma, bronchitis wheezing, irritation to eyes, etc.

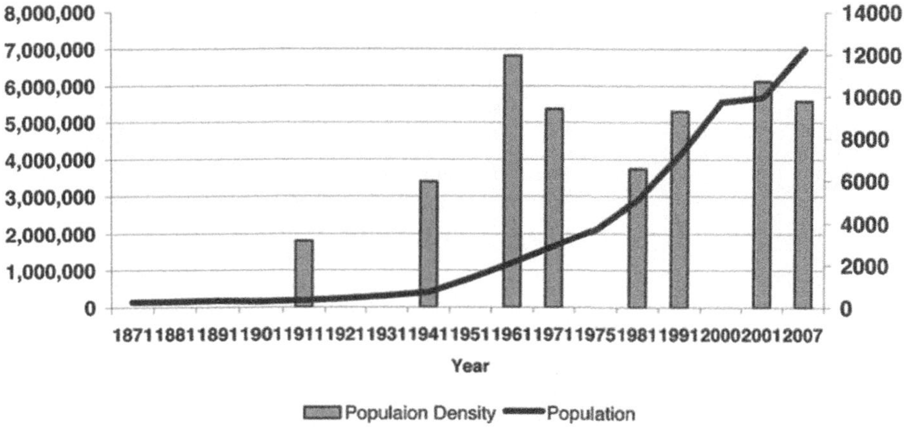

Source: Ramachandra and Mujumdar (2009: 11)

Fig. 6.1 Population growth and density

several issues across the urban sectors[3] that affect the environment and ecology of the city. The analysis is based on the methodology developed to assess the current environmental profile of the city as presented in the next section.

6.2 Methodology

The Bengaluru Metropolitan Area (BMA) has been chosen for the study. The Bengaluru Metropolitan Area (BMA) includes Bengaluru urban and rural districts falling under the jurisdictions of Bengaluru Development Authority (BDA), Bengaluru International Airport Planning Authority (BIAPA), Bengaluru Mysore Infrastructure Corridor Area Planning Authority (BMICAPA), Magadi Planning Authority, Nelamangala Planning Authority, Kanakapura Planning Authority, Ramanagara-Channapatna Urban Development Authority and Bengaluru Metropolitan Region Development Authority.[4]

BBMP has been formed by merging eight urban local bodies along with the erstwhile Bengaluru Mahanagara Palike (BMP). BMP had three administrative divisions called zones. At present, BBMP has been divided into eight administrative zones consisting of 198 wards (Table 6.1).

[3] Water supply, sanitation (toilets, sewerage and UGD), solid waste management, land use change, urban transport, energy and urban wetlands (lakes and tanks).

[4] *The Economic Times* (2006). Jurisdictions of authorities defined, 28th January. Visit the link (http://articles.economictimes.indiatimes.com/2006-01-28/news/27447719_1_jurisdiction-villages-layouts).

Table 6.1 BBMP zones

BBMP zones	Extent in kms
BMP zones (south, east and west)	240.40
Rajarajeshwari Nagar	144.75
Dasarahalli	89.24
Yelahanka	113.85
Mahadevapura	172.00
Bommanahalli	109.59

Source: CSD (2012)

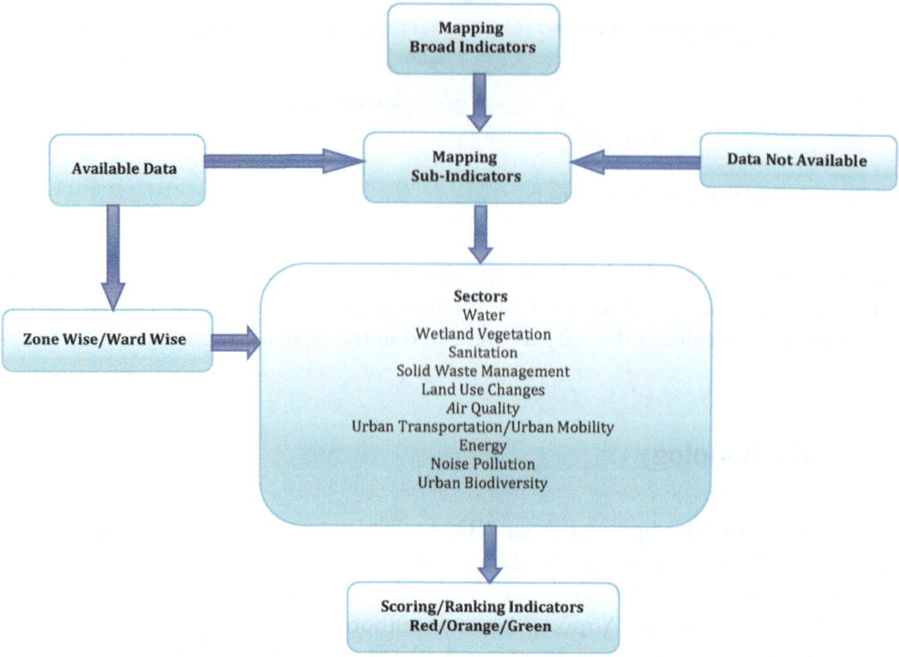

Fig. 6.2 Methodology for ranking economically sensitive zones in BMA (Bengaluru Metropolitan Area)

6.2.1 Methods Used for Analysis

A predictive analysis has been used for the identification of the current state of environmental resource use and trends and their impact on the environment of Bengaluru Metropolitan Area (BMA). Such an analysis helps identify the facts for a better understanding of the performance and outcomes and also the potential risks and opportunities involved in service delivery across sectors such as water supply, sanitation, waste management, groundwater management and urban parks to name few. For analysis, sector-wise secondary data has been collected, collated and analysed (Fig. 6.2).

The methodology includes the following stages:

1. In the first stage, environmental indicators and sub-indicators are mapped across sectors to measure the performance and the consequent impact on the environment. It is necessary to develop a set of indicators clearly exploring the approach and outcomes from an institutional dimension. Basically, indicators are diagnostic tools and instruments to assess the multi-faceted issues, concerns and shortfalls in the context of policy interventions. It includes mechanisms for evaluating governance bottlenecks, i.e. to identify under what circumstances environmental policies and planning are determined and to help ensure the steps for improving environmental governance in the short, medium and long runs. Thus, indicators help evaluate and assess the state of environment for establishing a casual linkage between the current status and its outcomes. A detailed list of indicators is presented in the Annexure.
2. At the second stage, a score board is developed to measure the indicators across sectors. Subsequently, scores are used for ranking and identification of economically sensitive zones (ESZs) within BMA limits.
3. In the third stage, the identified ESZs are categorised as (i) mild, (ii) moderate and (iii) acute/stressful, based on the score board. To each categorisation, a separate colour is assigned for, i.e. orange and green (for mild green, moderate orange and for acute red).
4. Finally, data sets are collected from secondary sources regarding different sectors for arranging and analysing each zone based on indicators and sub-indicators. Ranking has been done based on the weighted averages.
5. The chosen sectors include (i) drinking water, (ii) sanitation (toilets, sewerage and storm water drains), (iii) groundwater status and exploitation, (iv) urban water bodies, (v) solid waste management (SWM), (vi) urban land use, (vii) urban open space and parks, (viii) urban transport and (ix) energy. In addition data has been collected on (i) the status of pollution and (ii) climate change.

6.3 Environmental Resource Use in BMA: Trends and Status

The question of service delivery and the status of environmental resource use have become very critical due to spatial growth and an undue emphasis on the dynamics of a rising economy. The annual spatial expansion of the city's built-up area over the past 30 years constitutes 5.4% (BDA 2007: 17). The general observation is that the quality of urban life does not correspond to the spatial growth. There is an ever-growing gap between the needs of people and delivery of services. The following section briefly evaluates the levels of urban services offered to the city dwellers in the Bengaluru Metropolitan Area (BMA) and the subsequent impact on the environment.

6.3.1 Drinking Water

Urbanisation is one of the defining features of Asian cities (UN-HABITAT 2010). In Asia, 7% of the urbanites do not have access to safe drinking water. Similarly, in India, 7% of the entire population lacks access to safe and sufficient drinking water (World Bank 2006). Access to basic services like safe drinking water supply, toilets, sanitation, sewerage connection and waste collection is very critical to public health safety particularly in the cities of Asia. In fact ensuring access to safe drinking water supply is one of the Millennium Development Goals (MDGs). The universal coverage of access to drinking water in India is very slow as compared to Bangladesh, Indonesia, Myanmar, and Nepal (in fact has declined by 3–12%) (*ibid:* 143). This trend is attributed to an increase in the number of slums/informal settlements in the context of twin problems: (i) affordability and (ii) lack of formal tenure in the metropolitan cities of India (*ibid*: 144).

The city of Bengaluru in Asia has transformed itself into a bustling metropolis as the result of a combination of factors such as (i) migration and (ii) rapid industrialisation, particularly a very high concentration of high-end services. The transition of the city from the erstwhile BMP (Bengaluru Mahanagara Palike) to Bruhat Bengaluru Mahanagara Palike (BBMP) consisting of Bengaluru Urban and Rural areas has given rise to a number of problems including the loss of both the agricultural lands and water bodies. In addition, the concentration of commercial and industrial establishments around the city has put an immense stress on access to quality services. The heterogeneous composition of the city has impacted the regular and reliable water supply (BBMP 2006: 53; Smitha 2006: 394). If the population growth[5] continues at the present rate, the city shall be home to a population of 108 lakh by 2021 (*ibid*), thus severely straining the local governance in terms of providing basic services.

The Bengaluru Water Supply and Sewerage Board (BWSSB) is primarily responsible for the provision of safe drinking water and sewerage facilities to the population of the city. Initially, the service was restricted to the erstwhile BMP area (consisting of 100 wards), but after the expansion of the city, following the inclusion of 7 CMCs, 1 TMC and 110 villages,[6] the service coverage has increased. Table 6.2 presents the status of water supply in Bengaluru up to 2006.

River Arakavathy and TG Halli reservoir, located at 18 km and 4.28 km off from Bengaluru, were the traditional surface water sources. Later, due to an increase in demand, River Cauvery was tapped for providing drinking water to the city's growing population. Table 6.3 presents the water supply coverage in Bengaluru Metropolitan Area (BMA).

[5] The city of Bengaluru registered a population growth of 4.10% in 2001 and 4.75% in 2011 and is expected to register a population growth of 5.52% by 2021 (BBMP 2006: 57).

[6] The new amalgamation constitutes Bengaluru Metropolitan Region (BMR). The local governance was reconstituted as Bruhat Bengaluru Mahanagara Palike (BBMP) in 2007.

Table 6.2 Status of water supply in Bengaluru up to 2006

Parameter	Amount	Comments
Water availability		2009–2010 estimates
Installed capacity	Cauvery – 810 mld Arkavathi – 184 mld	
Daily increase	900 mld	
Daily demand	1125 mld	
Estimated groundwater extraction	200 mld	Decreasing levels of groundwater (especially post monsoon)
Source of water supply	98 km from Bengaluru	
Water supply coverage and metering		
Domestic consumption	52	
Nondomestic consumption/ others	9	
Estimated UFW (unaccounted for water)	39	20,000–30,000 unauthorised connections
Availability	90–110 lpcd	Complaints of a nonuniform supply and low-residual pressure in outlying areas
Wastage of water	47–50%	
Area coverage	100% in core city Between 10% and 60% in noncore city	
Sewerage		
Tertiary capacity	72 mld	At V-valley and Yelahanka
Proposed additional capacity	339 mld	From the major and minor STPs
Estimated usage of treated water in industries	4 mld	
Estimated capacity utilisation	75%	
Area covered through sewer system	40%	About 245 sq. km
Consumer redress		System is in delay

Source: BBMP (2006: 66)

6.3.1.1 Access to Water Supply

The city of Bengaluru has witnessed a phenomenal population growth rate of 4% over the past half a decade. Additionally, the population spread across 772 km^2 is expected to increase by one core in 2016 (Balasubramanian 2013). Access to water supply[7] is determined by (i) quantity of supply and (ii) no. of connections. According to a CSD (2010) study, the city is home to a population of 7 million and is going to

[7] River Cauvery's supply constitutes 1400 MLD which equals to 18 TMC, serving the entire urban population of Bengaluru (Balasubramanian (2013).

Table 6.3 Water supply coverage in Bengaluru Metropolitan Area (BMA)

Urban local bodies	Water supply coverage (% area)
Erstwhile BMP	100
Erstwhile CMC areas	–
Yelahanka	60
Rajarajeshwari Nagar	25
Mahadevapura	20
KR Puram	20
Bommanahalli	Yet to be covered
Dasarahalli	10
Byatarayanapura	10
Erstwhile TMC	60
Kengeri	–

Source: BBMP (2006: 67)

Table 6.4 Different sources of water and percentage share of supply

Source	For drinking purposes (%)	For other purposes (%)
BWSSB/Cauvery supply	40	40
Tanker/private sources	40	30
Bore well	20	20

Source: CSD (2010: 10)

reach between 85 lakh and 95 lakh by 2020[8]; the supply of water to the city is 840 mld, while the demand is 1050 mld at the supply rate of 140 lpcd. Correspondingly, the demand and supply gap[9] amounts to 210 mld and 560 mld.

Surface water is one of the major sources of drinking water to the city's population. The city receives water from two rivers, namely, Arkavathi and Cauvery. Providing water from River Cauvery was commissioned from Stage I 1974, 135 mld; Stage II 1982, 135 mld; Stage III 1993, 270 mld; Stage IV Phase I 2002, 270 mld; and Stage IV Phase II 2011, 500 mld to meet the growing demands of the city's population (GoK 2009: 83). Further the inability on the part of BWSSB to meet the growing demands has necessitated the tapping/extracting of groundwater resources for drinking purpose. Table 6.4 presents different sources of water for Bengaluru City.

During 1991 and 2003, 'metered domestic connections' increased from 213,000 to 400,000, an increase from 25% to 35%, while newly merged areas of CMCs and TMCs account for less than 10% of individual connections (BDA 2007: 47) (Picture 6.1).

[8] Hedge and Chandra (2012): Resource availability for water supply to Bengaluru City, Karnataka, Current Science, 102(8): 1102–1104

[9] Access http://www.rainwaterharvesting.org/Crisis/Urbanwater-scenario.htm for details on demand and supply gap presented in Fig. 6.2.

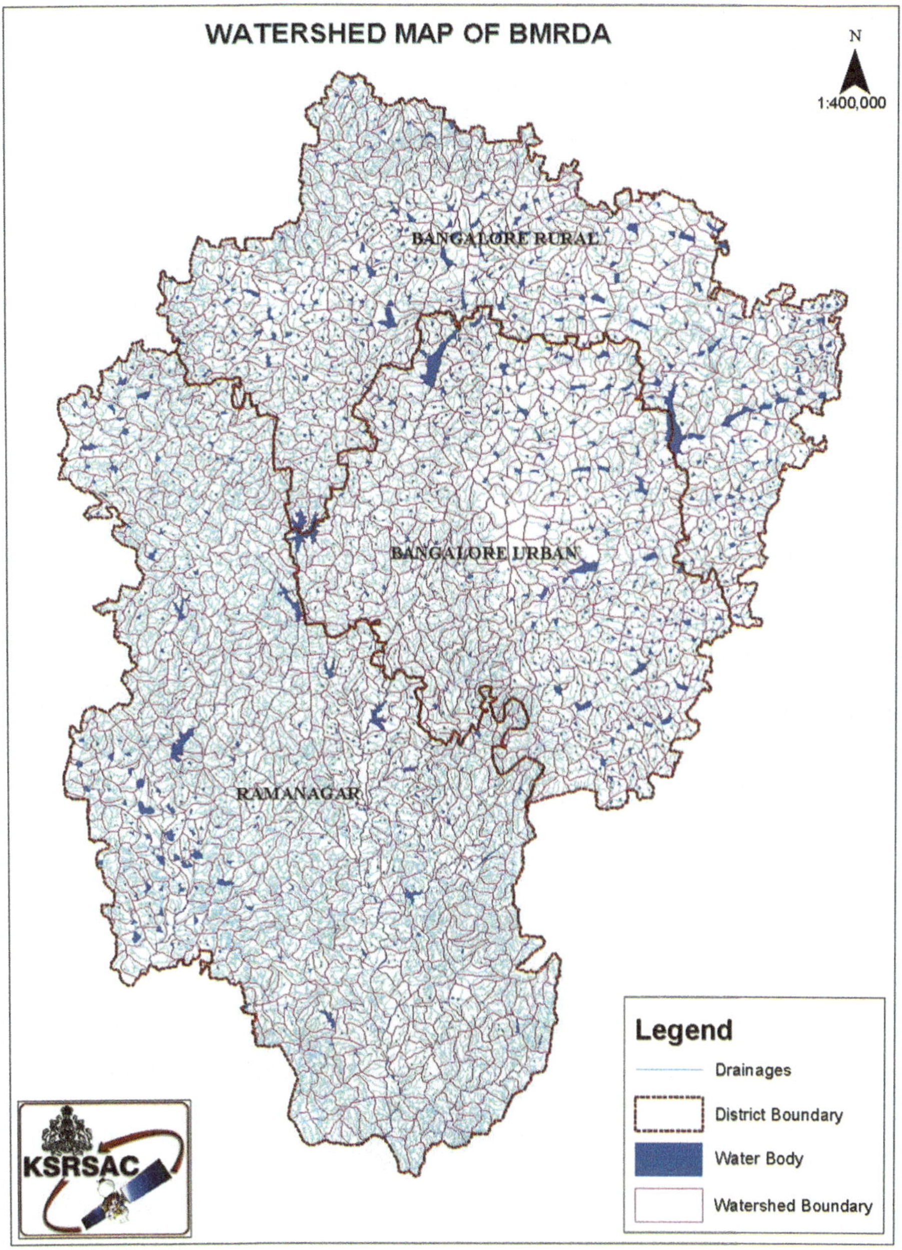

Source: GoK (2009: 86)

Picture 6.1 Groundwater potential of BMR

6.3.1.2 Coverage of Water Supply

Although, at present, the city of Bengaluru spreads over 786 sq.km of conurbation area of CDP (City Development Plan), water supply covers for about 226 sq.km of BMP area and 334.9 sq.km of the newly merged areas[10] – 7 CMCs, 1 TMC and 110 villages and also industrial areas. But the water supply distribution system is available only for 226 sq.km of BMP area, while the remaining peripheral region does not have developed water supply system (CSD 2008: 26).

According to the Environment Report Card of Bengaluru 2012 brought out by CSD (Centre for Sustainable Development), while the west zone receives the lowest water supply (expressing dissatisfaction at the service provision), nearly 31% of their water requirements are met by multiple sources like groundwater, private tankers and package water, whereas the south zone is highly satisfied with water supply, a well-fed zone in the city. However the highest dissatisfaction levels are found in the east zone with a severe shortage of water supply, and its demand for water is met through groundwater (25%) followed by private tankers and package water. It is interesting to note that Revised Master Plan 2015 prepared by BDA claims 30% of groundwater being tapped to fulfil the city's water requirements through about 5850 borewells and 15,180 public taps (BDA 2007: 47) (Fig. 6.3).

BWSSB has initiated Cauvery Phase Stage IV and Phase II for supplying drinking water to the areas of Greater Bengaluru (Ray 2012b). The Cauvery IV, Phase II project was initiated in 2005 with the aid from Japan International Cooperation Agency (JIBC). Under this project, an additional 500 mld and a total of 1400 mld of water are expected to be supplied to the city (The Hindu 2012a). The percentage of water supply coverage grossly varies from the erstwhile BMP areas to newly added

Fig. 6.3 Satisfaction levels of quantity of water supplied by major sources

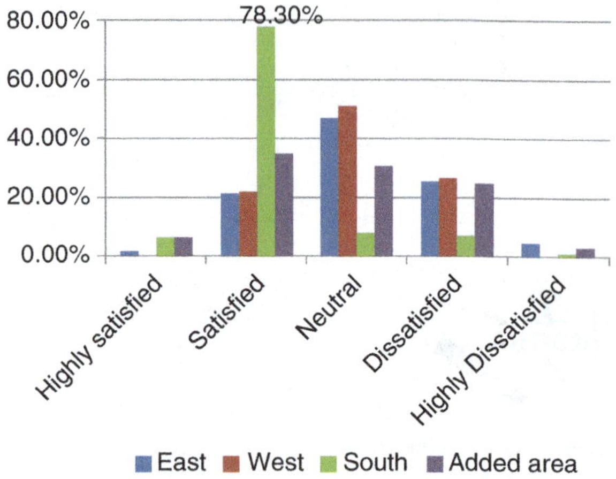

Source: CSD (2012: 25)

[10] The merged areas comprise 54 wards with a population of more than 2 lakh (The Hindu 2012).

Table 6.5 Water supply coverage in ULBs (urban local bodies)

Urban local bodies	% of coverage
Erstwhile BMP	100
Erstwhile CMC	No data
Yelahanka	60
Rajarajeshwari Nagar	25
Mahadevapura	20
KR Puram	20
Bommanahalli	Yet to be covered
Dasarahalli	10
Byatarayanapura	10
Erstwhile TMC	60

Source: BBMP (2009)

Table 6.6 Status of water supply and groundwater resources in towns of BMR

City municipal council and town municipal councils	No of borewells	Total supply (mld)	No. of days of supply in a week	No. of hours of supply in a day	Per capita supply (lpcd)
Nelamangala town panchayat (TP)	37	0.04	2	1	1.58
Doddaballapura CMC	48	3.62	2	1	50.55
Devanahalli TMC	–	1.46	4	1	62.38
Hoskote TMC	21	NA	NA	NA	NA
Magadi TMC	–	2.05	6	8	81.90
Ramanagara CMC	206	8	3	1	100.76
Channapatna CMC	NA	NA	NA	NA	NA
Kanakapura TMC	135	.066	NA	NA	14.02
Anekal TMC	NA	.03	1	12	.90
Kengeri TMC (now part of BBMP)	NA	.56	7	9	13.19
Vijayapura TMC	43	NA	2	0.05	NA

Source: BMRDA 2009: 306

areas such as 7 CMCs and 1 TMC. Particularly in CMCs, there are about 80,000[11] unauthorised connections. Total households are 1.6 lakhs; water pipes have reached near the premises of 1.3 lakh households while only 65,000 households have meters in CMCs (Table 6.5).

The status of water supply and groundwater in the towns of BMR region is presented below (see Table 6.6).

Although the water supply coverage is 100% in BMP area, with the formation of BBMP area, BWSSB is unable to provide sufficient drinking water to the CMCs and TMCs. Tables 6.5 and 6.6 clearly illustrate the mismatch between the levels of services offered by BWSSB.

[11] Details collected during an interview with the Chairman and Executive Engineers of BWSSB, on April 9, 2014.

Table 6.7 Availability of water based on population capacity for BMR

Water source	Population catered to considering 300 lpcd (in millions)
Surface water from Cauvery and Arkavathi (1535 MLD)	5.12
Groundwater through borewells	5.34
Rainwater harvesting – Considering 20% of rainfall = 2487 MLD	8.21
Reuse of water 400 MLD by 2021 600 MLD by 2031 700 MLD by 2041 800 MLD by 2051	2.33
Total water available catered to the population	21

Source: BMRDA (2009: 308)

The major source of water in CMCs and TMC is groundwater, leading to a further depletion of groundwater table. Although drinking water is supplied to some newly added areas, it is insufficient as it is supplied only for 1 day per week and that too for 1 h only. Both northern and north-east parts of the BMR depend on groundwater which is exploited to the maximum extent of 85%, resulting in the deterioration of groundwater source (BMRDA 2009: 158). As Table 6.7 shows, nearly 5.12% of the population is supported by Arkavathi river, while 5.34 million population living in the periphery is covered by groundwater source (borewells) (Picture 6.2).

6.3.1.3 Adequacy of Water

Domestic connections[12] constituting 90% have increased to 6,60,355[13] in 2013 as against nondomestic connections consisting of only 5%, but this does not correspond to the growing population of the city. Therefore, 8% of the city's population lacks access to safe drinking water (GoK 2009: 87). While the actual supply[14] of BWSSB water is 900 MLD, the demand for water comes to 1125 million litres per day. Thus there is a shortage of 21 MLD (CSD 2012: 16). With the expansion of the city beyond city corporation boundaries (former BMP area), an additional 500 MLD is drawn from Cauvery Phase II, Stage IV (Citizen Matters 2013) to quench the thirst of a looming population of expanded boundaries.

The demand and supply gap over the years clearly shows the gross supply and net supply from the 1990s to 2006 (see Fig. 6.4). Similarly, the Table 6.8 clearly shows that as the demand for water increases, there is a persistent shortfall in the supply

---

[12] The drinking water pipeline network of 7500 km carries water to 6.1 lakh households (The Hindu 2012).

[13] Citizen Matter (2013): Where does Bengaluru get its water from? Citizen Matter Desk. Access the following link: http://bangalore.citizenmatters.in/articles/5306-understanding-bengalurus-water-sources-cauvery-groundwater-borewells-rainwater).

[14] The actual quantity of water supply has considerably declined from 145 lpcd in 1995 to 75 lpcd in 2007 (GoK 2009: 87).

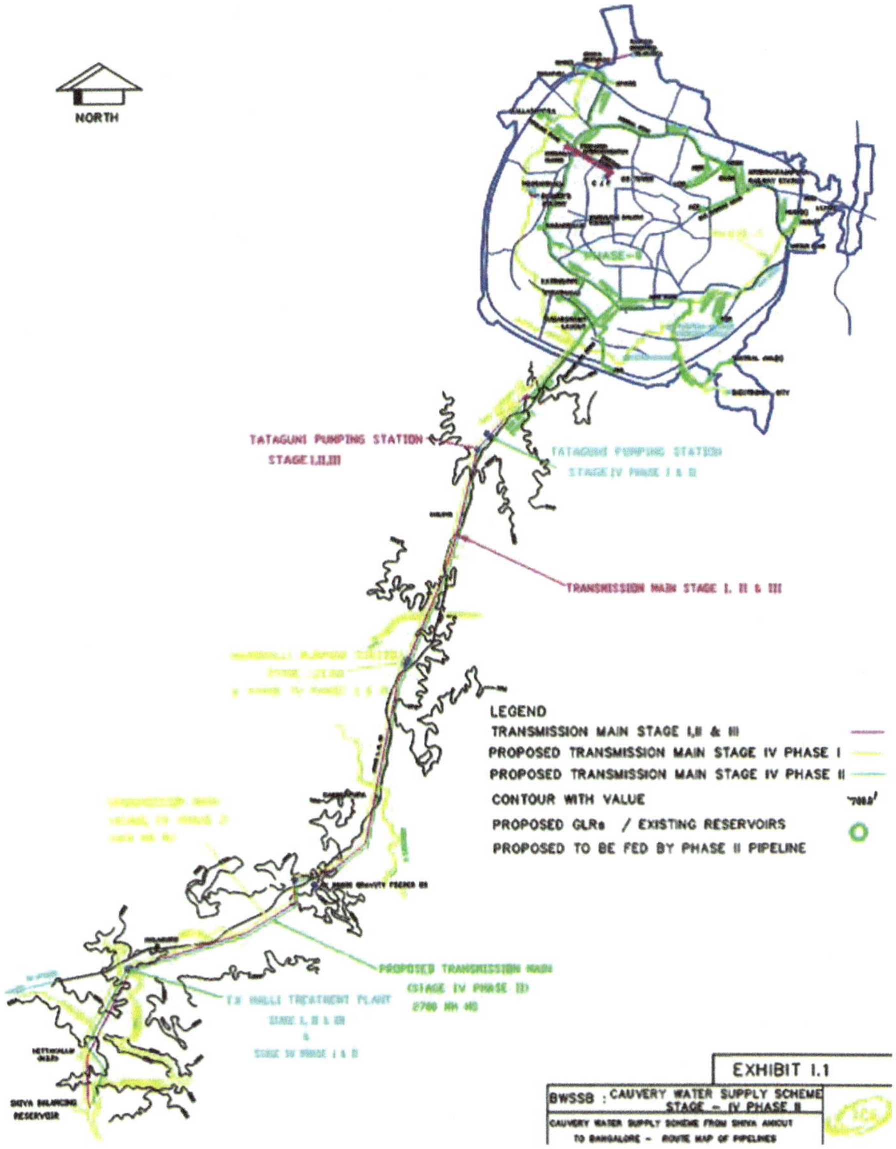

Source: GoK(2009)

Picture 6.2 Drinking water supply sources and routes to Bengaluru

over the coming years. There is a need for a strong policy intervention as the situation is very critical which can further worsen in the coming years. At present, the demand is 1600 MLD[15] for BBMP area. A DPR projection is done up to 2050: an additional 10 TMC of water is required for servicing the additional areas of the city.

[15] Details collected during an interview with the Chairman and Executive Engineers, BWSSB, on 9th April, 2014.

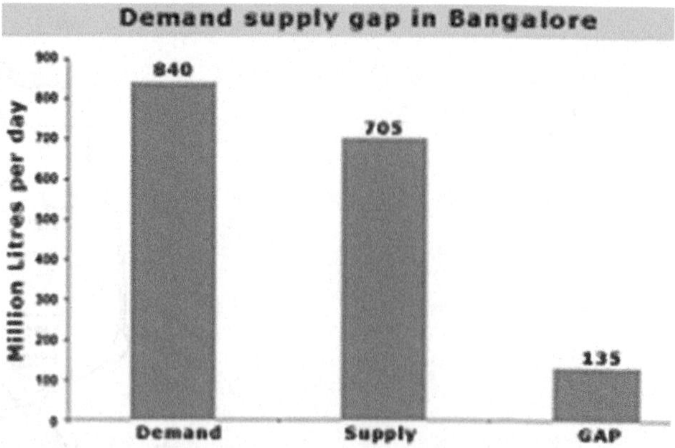

Source: Bangalore Water Supply and Sewerage Board

Fig. 6.4 Demand and supply gap in Bengaluru

Table 6.8 Water demand and supply scenario for Bengaluru

Year	Population (lakhs)	Water demand (MLD)	Present supply (MLD)	Shortfall (MLD)	
				MLD	**TMC**
2011	84.99	1683	960	723	9.33
2021	105.81	2095	1460	635	8.4
2031	142.96	2831	1460	1371	17.69
2041	170.85	3383	1460	1923	24.81
2051	205.61	4071	1460	2611	33.69

Source: Sawkar (2012b)

6.3.1.4 Frequency of Water Supply

Usually the frequency of water supply varies across zones and wards of BBMP areas. While in the core city of Bengaluru, the frequency of supply is alternate days, it is only for 2 days in a week in CMCs (see Table 6.9).

Key Issues in the Supply of Drinking Water

1. The city of Bengaluru does not have access to perennial river source. Therefore, it has to draw water from reservoirs and River Cauvery from a distance of 98–100 km away from the city (Suresh 2001: 98; BBMP 2006: 68; GoK 2009: 81).
2. Fast vanishing water bodies/wetlands like tanks and lakes from the city. This is due to an unabated expansion of the city, occupying land for residential/commercial and industrial purposes.

Table 6.9 Frequency and coverage of water supply

Drinking water supply	2012
Coverage area	100%
Frequency of water supply	Varies from 2 to 4 h on alternate day
Metering	100% in BMP area
Unaccounted for water	48%
Sanitation coverage	75%
Consumer redressal forum	Redressed within 2 days

Source: BBMP (2009: 80)

3. Due to a further expansion of the city,[16] the extraction of groundwater resources has become inevitable to meet the increasing demand for water (Anand et al. 2006: 330; BBMP 2006: 67).[17] Borewells are the main source of water[18] for CMCs and TMCs and 110 villages. It is estimated that 282 mld of groundwater is extracted from more than 7000 borewells. Further, an estimated 120 mld of water is drawn from nearly 80,000 private borewells in the city. Such extraction accounts for 28% of the total drinking water supply for 2001–2002 (GoK 2009: 85).
4. In addition, the prevalence of an informal association between suppliers[19] and the private water market[20] plays a significant role in catering to the increasing demand for water, particularly in the context of an ever-growing population of the city. Owing to the lack of effective management and legislation, the private market is resorting to an unabated abstraction of groundwater resources (Suresh 2001: 100). Nearly 20% of the city's population constitutes slums/informal settlements. About 30% of the households in slums do not have piped connections due to the lack of a tenure status. Consequently, they rely on leakages, public stand posts, unaccounted for water and other informal vendors.
5. The minimum water requirement of the city's population of 6.5 million in 2006 (at 200 lpcd) was around 1136 mld, but the actual supply of water was only between 90 and 100 lpcd after excluding an unaccounted for water of 533 mld (i.e. 40–50% of loss) (Anand et al. 2006: 331). The overall leakage and wastage of drinking water constitute about 48% (Smitha 2006: 400; GoK 2009; Vyas 2014). Some of the key issues and their environmental impact have been summed up in the Table 6.10.

[16] The geographical area of the city has expanded to 2190 of sq.kms following the inclusion of 3 taluks, i.e. Bengaluru north, Bengaluru south and Anekal, with a population of 6.5 million and a total of 1285 villages (Anand et al. 2006: 330).

[17] Visit the BWSSB's webpage for more information (http://bwssb.org/water-supply-sources-shortage/).

[18] In 2006, 3568 borewells supplied 56.74 mld of water to households located in 7 CMCs, 1 TMC and 110 villages of Bengaluru Metropolitan region (Anand et al. 2006: 331).

[19] The volume of water supplied during off season (January–February) is estimated at 95.76 mld, and during summer the supply increases to 227.16 mld (Anand et al. 2006: 335).

[20] There are absolutely no official records for keeping track of such informal and private water markets in Bengaluru City.

Table 6.10 Key issues in the supply of water and environmental impact in Bengaluru

Water supply	Issues	Environmental impact
Inadequacy of resources for augmenting future growth	Limited availability of water from Cauvery (after utilisation of 600 cusecs). No water available from adjoining river sources such as Hemavathy, Netravathi, etc. Changing land use pattern	Scarce water resources and resource limitation Encroachment
Access to safe drinking water to meet national standards of 150 lpcd	Relatively high UFW Nonavailability of distribution systems in the former CMCs/TMC and newly added BMP wards; Arkavathi source gradually depleting 20% of urban slums do not have access to water and sanitation Only 3–4 h of water supply for alternate days 29% of informal settlements depend on public fountains	Gradual groundwater depletion Shortage of supply
Groundwater	Indiscriminate withdrawal and sub-standard water quality 40% of groundwater usage The bottled water market is growing at 55% annually	Emergence of water markets Depletion of groundwater table beyond its carrying capacity Deterioration in both Groundwater quality and quantity and high presence of both organic and inorganic substances
Water quality issues (in distribution network and raw water)	Cross connections/back-siphon in distribution resulting in water-borne diseases. Raw water deterioration due to pesticides and chemical pollutants discharged by industries and sewage from upstream	Public health issues Health impacts due to water-borne diseases and water that is unfit for drinking. Diseases such as typhoid and cholera are highly prevalent in the city due to contamination of drinking water sources About 1812 cases of water-borne diseases were reported in 2007 59% of costs result from health impacts of water pollution
Uneven distribution and Intermittent supply	Parts of the city receive a higher quantum of water and for a longer duration when compared to certain other areas which receive a lesser quantum that too and only for a short duration; erratic growth assets needing rehabilitation; some areas get water only for 3–5 h on alternate days	Water pollution Shortage of water coverage

(continued)

Table 6.10 (continued)

Water supply	Issues	Environmental impact
High UFW	Absence of reliable source production (bulk metering) on all major water sources Nearly 30–40% of water is accounted for waste. Absence of metering on public fountain consumption amounts to a loss of water Poor accuracy and serviceability of consumer metering High non-physical loss due to consumption from Unauthorised connections and inaccurate/inoperable Meters Unauthorised connections are estimated to number about 30,000	Decrease in per capita availability of water Groundwater exploitation
Water wastage	Wastage of water due to Frequently leaking pipes by DA pipes Division-wise loss of water due to leakages, theft or spillage; absence of metering of public taps and fountains and old and corroded pipes; and unauthorised connections Loss and depletion of ground-level reservoirs The study reveals that 5% of leakage is due to faulty meters NRW from 47% to 16%	Decline in availability of fresh water for human consumption

Source: BBMP (2006: 69); GoK (2009)

6.3.1.5 Unaccounted for Water (UFW)

Another major problem facing the city is UFW[21] (unaccounted for water) at alarming levels, constituting 48% to 50% of the total supply (GoK 2009: 99; Smitha 2006: 8) and considered to be the fourth highest among urban utilities in the country (GoK 2009: 88). This actually means a loss of 50 crore to BWSSB every year (Vyas 2014). A study conducted by ISEC points out that the total UFW is as high as 509 MLD (million litres daily) in Bengaluru which could otherwise have met the water demand of 150 lpcd (litres per capita per day) for the city (Vyas 2014). The supply leakage works out to 500 MLD, 150 MLD supplied to industries; the balance 750 MLD (constituting 75 litres per capita per day) serves approximately 8.5 million people of the city (Fig. 6.5).

[21] Unaccounted for water refers to the difference of water between the actual water produced minus water consumed and water in store.

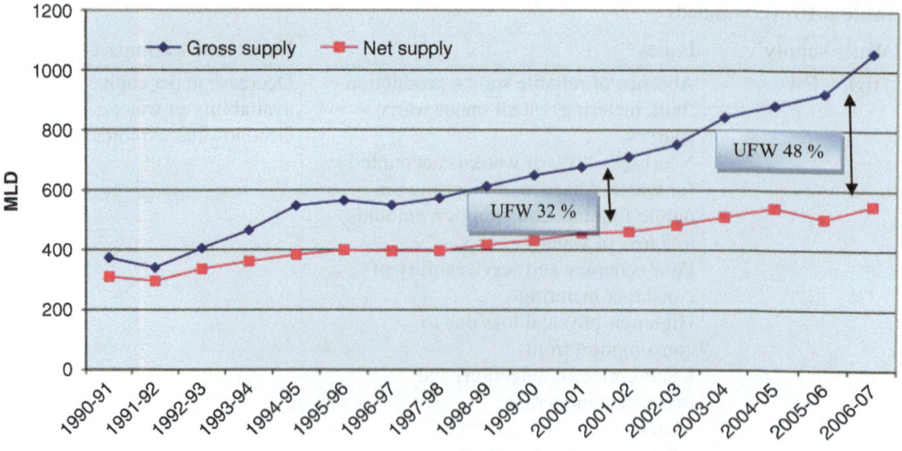

Source: GoK (2009)

Fig. 6.5 Trends in water supply and UFW in Bengaluru

One of the major stumbling blocks in terms of reaching 150 lpcd is UFW which is, at present, higher than the domestic consumption of 36% (GoK 2009: 103). In addition to the groundwater supply, people of Bengaluru City rely on tankers for supply of water. There are 100–200 tanker companies with approximately 3000 tankers in the city (Dayalapwar n.d.: 6). The number of public fountains rose from 32,999 in 1990–1991 to 54,912 in 2006–2007 (GoK 2009: 89). The water supply of 1125 MLD is unaccounted for, especially when the city is facing a severe shortage of 222 MLD (Yasmeen 2013) (Picture 6.3)

Under PPP (public-private partnership), a massive project was initiated by BWSSB to control distribution losses with the assistance of JIBC (Japanese International Bank Corporation). The project was launched in June 2003 on a pilot basis in five service stations at a cost of Rs. 480 million, funded by JIBC (www.bwssb.org). The pilot project is implemented in the south division covering BTM layout, Jayanagar, Chamrajpet, Cottonpet and Padmanabhanagar with an estimated cost of Rs. 2–3 crore annually. The project is expected to cover approximately 35,000 connections out of 650,000 in the core city area. The amount of water expected to be saved by plugging leakages works out to 16% (Balasubramaniyan 2013: 4; Vyas 2014). Similarly, under Phase II, BWSSB is implementing a Rs. 153 crore project (started in 2012) covering 52 km. The project has been contracted out to L&T and is scheduled to be completed by 2015. The 52 km area is going to be divided into 'district metered areas' (DMA) consisting of 1000–3000 connections (Subramanyan 2013).

Picture 6.3 Unaccounted for water through leakage in the city

Source: Vyas (2014)

6.3.2 Water Conservation: Rainwater Harvesting (RWH)

Rainwater harvesting is one of the significant sources of water for domestic and industrial usage. Increasing population and spatial growth have led to an inevitable stress on access to water in Bengaluru. The condition has further deteriorated with groundwater depletion and pollution of water bodies like lakes and tanks. The city of Bengaluru is the first in the state of Karnataka to adopt and implement RWH in the by-laws. The Bengaluru Mahanagara Palike building by-laws make RWH mandatory for both domestic users and industries (Smitha 2006; Umamani and Manasi 2013; Ravi 2005). Rainwater[22] supply of nearly 3000 million litres per day to the city (Vishwanath 2007) enhances water tables and prevents soil erosion and flooding especially in urban areas (GoK 2009: 108). It is estimated that an average consumption of 100 lpcd rainwater is sufficient for 1954 days (*ibid*: 109). RWH has been made mandatory for the construction of new buildings[23] in the Bengaluru Metropolitan Region (BMR). Since 1974, River Cauvery has been the main source of drinking water drawn from 98 km away and pumped from nearly 1500 feet below the city altitude. The spatial expansion of the city after the incorporation of 7 CMCs, 1 TMC and 110 villages into the metropolitan region and the

[22] By February 2010, the total number of RWHs in the BMR came to 10,072 (www.rainwaterclub.org).

[23] As per the law, 'every building with a plinth area of exceeding 100 sq. mtrs and built on a site measuring not less than 200 sq. mtrs shall have one or more rainwater harvesting structures having a minimum total capacity as detailed in the Schedule' (Ravi 2005).

deterioration of groundwater has added to the water woes of the city. The city of Bengaluru (1250 sq. km) received 187,500 million litres of rain for 2013. At 1200 million litres per day, this is equivalent to 156 days of water (www.rainwaterclub.org). Therefore, the solution lies in effectively promoting RWH and rooftop harvesting. Mr. S. Vishwanath (Rainwater Club,[24] Bengaluru) is one of the activists extensively engaged in promoting RWH in the city. He points that as RWH is not suitable for all types of locations, promoting rooftop harvesting is one of the best options for water conservation. Both rooftop and RWH can yield approximately 20 litres of water which is sufficient for an entire family. As various reports and documents suggest, with the water crisis further deepening, the best alternative seems to be the promotion of RWH. Though installing RWH has been promoted vigorously by BWSSB, yet the ground reality shows that households are yet to fully comply. The study by Manasi and Umamani (2013) reveals that 94% of the households in the city have installed RWH out of compulsion and 82% (control group) were not willing to respond fearing violation of non-adaptation. While 81% did not follow proper technical procedures.

6.3.3 Management of Open Wells

Open wells in Bengaluru constitute one of the rich sources of water, particularly for domestic consumption. In some of the old city areas like Malleshwaram, Jayanagar, Basavanagudi and VV Puram, households have large wells (Chandrashekar 2008). The city of Bengaluru has over 2 lakh unregistered borewells, while the BWSSB has about 12,000 borewells. The reckless exploitation of groundwater resources is one of the reasons for a decline in the yield of community open wells system in the city (Rao n.d).

6.3.4 Groundwater Utilisation/Use of Alternative Water Sources

Due to rapid urbanisation, there is a heavy reliance on groundwater resources for quenching the thirst of Bengaluru City citizens even as domestic, commercial and industrial usage of groundwater has increased substantially in the city. To compensate for the supply deficiency, BWSSB relies heavily on groundwater as an alternative supply source. A study conducted by CSD (2008) points out that 20% of water supply is supplemented by private borewells. The total supply from private borewells for 2008 amounts to 170 MLD and in 2012 to 200 MLD (CSD 2012: 16). There are about 4455 operating borewells of which 3295 are equipped with hand pumps while the remaining 1159 with electricity-run submersible pumpsets. According to a study conducted by Hedge and Chandra (2012), as against the

[24] Visit www.rainwaterclub.org for details on RWH in the city.

present demand of 48,600 ha m, only 37,374 ha m is supplied from both surface and groundwater resources. The shortage in supply of water amounts to 11,26 ha m, which means nearly 22 lakh urbanites do not have access to drinking water or face water scarcity.

Another study conducted by the Centre for Symbiosis of Technology, Environment and Management (STEM)[25] estimates that 40% of Bengaluru's population is dependent on groundwater sources. There has been a phenomenal growth of borewells from 5000 to around 4.08 lakh over the last three decades, drawing 750 MLD (Sawkar 2012b: 15), which is 3.7 times more than the recharge from the city's annual rainfall of 900 mm (Balasubramanian 2013). In addition, to meet the domestic, commercial and industrial needs, about 7300 ha m/yr. of groundwater is extracted from about 2 lakh unregistered private borewells (Hedge and Chandra 2012: 1103) (Picture 6.4).

The Table 6.11 shows that the south zone records the highest groundwater consumption with 31% of borewells followed by the west zone with 26% of borewells the second highest consumer. While the west zone is the highest consumer of surface water, the central zone is the lowest.

6.3.4.1 Groundwater Extraction and Capacity to Recharge

Groundwater extraction by BWSSB amounts to 28% (CGWB 2008: 13; GoK 2009: 85). Dependency on groundwater (3568 tube wells supply about 57 MLD of water) is high, particularly in the newly added regions of CMCs and TMCs and Bengaluru Rural. Both Bengaluru Rural and Urban, particularly newly added regions like 7 CMCs, 1 TMC and 110 villages, predominantly rely on groundwater resources. Nearly 3568 tube wells supply about 57 mld of water to expanded areas (GoK 2009: 86) (Table 6.12).

The problem of water shortage in the city is overcome by tapping groundwater sources, amounting to 282 mld from more than 7000 public borewells in Bengaluru City. In addition, an estimated 120 mld is drawn from an approximately 80,000 private borewells in the city. Similarly, studies have clearly pointed out that about 40% of potable drinking water to the city of Bengaluru is met by extraction of groundwater (GoK 2009: 85). Households in slums and informal settlements constitute 20% of the city's population Schenk (2001), 17% of which rely on groundwater for both drinking and domestic consumption.

Similarly, an estimated 70,000 industries in Bengaluru Metropolitan Region (BMR) and a total of 99 industrial units are heavily reliant on groundwater resources (GoK 2009: 90). Further dependence on groundwater is very significant in the peripheral areas of BBMP as they are yet to receive formal drinking water supply. The total groundwater extraction (see Table 6.13) in BMR area amounts to 1216 MLD as against the recharging capacity of 221 MLD, while the extraction in BBMP

[25] Visit http://www.rainwaterharvesting.org/Crisis/Urbanwater-scenario.htm for details on water scarcity in Bengaluru.

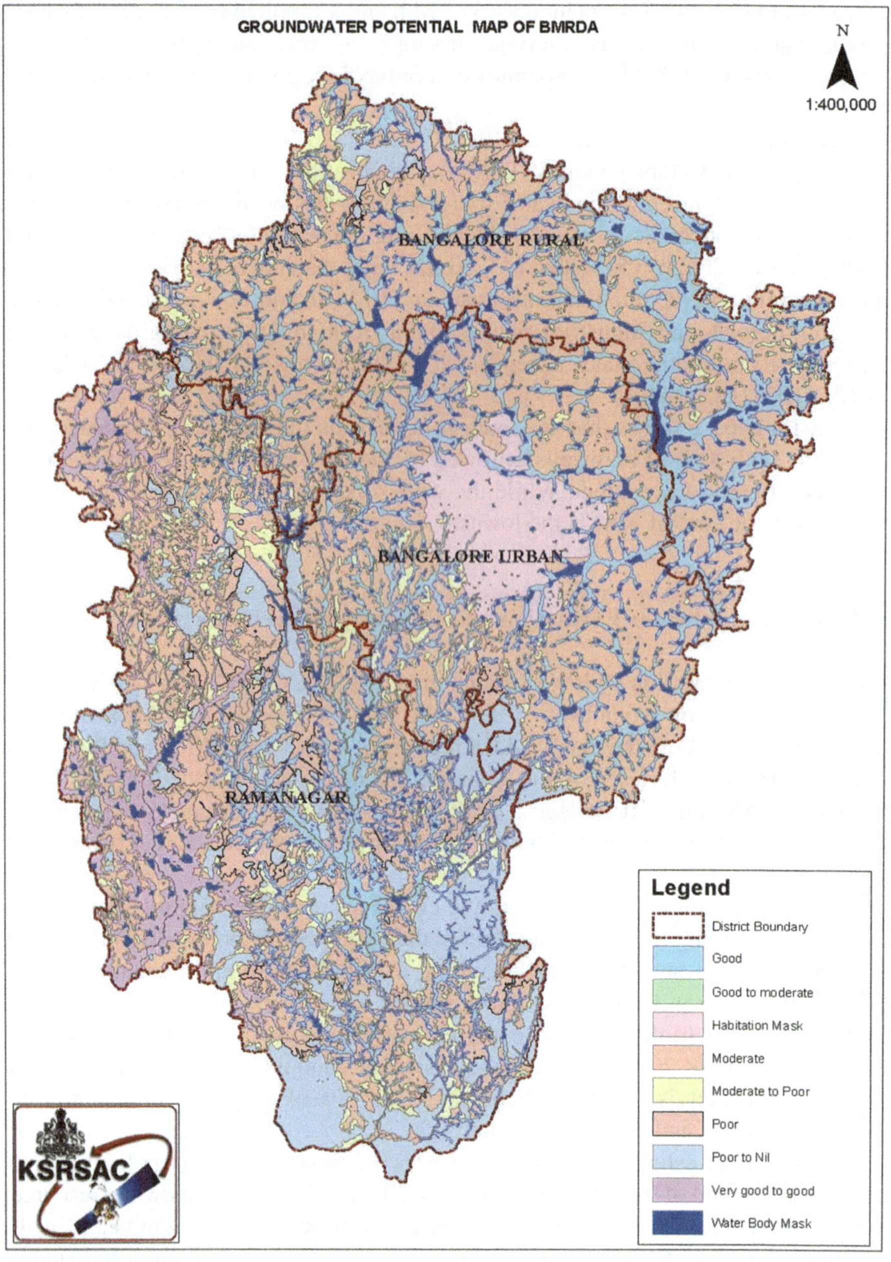

Source: GoK (2009)

Picture 6.4 Groundwater potential of BMR

Table 6.11 Zone-wise piped water supply and groundwater consumption

Zone	Surface water (MLD)	Groundwater (MLD)
Central	67.10	38.91
North	210.46	87.08
West	184.89	**149.45**
East	169.19	50.46
South	133.106	**176.00**
South-east	104.79	**67.80**
Total	**869.54**	**569.70**

Source: Sawkar (2012a, b: 17)

Table 6.12 Groundwater drawing

Sources	Estimated number	Water drawn (MLD)
BWSSB borewells	7000	282
Private borewells	80,000	120 (estimated)

Source: GoK (2009: 85)

Table 6.13 Groundwater extraction in BMR

		Water drawn		Capacity of recharging	
Area/district	Area in acres	MLA	MLD	MLA	MLD
Bengaluru urban	528,559	207,000	567	–	–
Bengaluru rural	574,520	–	–	–	–
Ramanagara	879,201	–	–	–	–
BMR area	**1,982,280**	**444,000**	**1216**	**81,100**	**221**
BBMP area	**197,684**	**103,000**	**282**	**71,000**	**195**

Source: GoK (2009: 85)
Empty space signifies data not available on extraction and recharging capacity

area comes to 282 MLD as against the recharging capacity of 195 MLD (GoK 2009: 85). The prevailing scenario simply suggests that there is an alarming gap between groundwater extraction and the capacity to recharge, going by the non-availability of official data or records. The highest number of dug wells is found in Bengaluru North Taluk.

A study conducted by Vishwanath S. (Rainwaterclub) states that, as of January 31, 2010, there were 7206 borewells in three CMCs (CMC 1, 2 and 3). Similarly, there are about 149, 302 borewells in BMR[26] (see Table 6.14).

Table 6.14 shows that borewells are more in number in the south zone followed by the west zone. The south-east zone has a moderate number of borewells followed by the central zone with the least number of borewells in the city. The over-extraction of groundwater beyond rechargeable capacity has led to semi-critical, critical, over-exploited watersheds.

[26] Constituting east, west, north, south and south-east regions.

Table 6.14 No of Borewells in Bengaluru (as of January 2010)

Zone	No of Borewells (as of January 2010)
Central	7206
North	16,126
West	**27,675**
East	9346
South	**32,593**
South-east	12,555
Total	**10,5501**

Source: www.rainwaterclub.org

Table 6.15 Groundwater draft by various sectors in Bengaluru City

Sectors	Number of wells	Methodology	Quantity of water extracted in MLD
Domestic use	2,61,573	From survey and assuming 1000 L per day per tube well	261
Nondomestic use	65,393	–	–
Irrigation	28,250	KPTCL	–
Agencies	13,235	Concerned depts	156
Industries	4400	DAG	46
Parks	413	Survey	15
Commercial establishments	432	Survey	11
Tankers	100	Survey	162
Others (institutes, offices, hotels, hospitals, etc.)	18,643	Assuming 5000 L per tube well	95
Total groundwater extracted	3,26,966	–	746

Source: Raju et al. (2008: 24)

Similarly, groundwater extraction from various sources is presented below (see Table 6.15).

The availability of groundwater depends largely upon the depth which, in turn, is determined by topography, rainfall and rock type. The ground drawn and depth in some areas of Bengaluru are presented in Table 6.16.

A study report on the status of groundwater resources in Bengaluru Urban district, prepared by the Central Ground Water Board (2008), reveals that three districts, namely, Bengaluru North, Bengaluru South and Anekal, have been notified as overexploited zones of groundwater resources. It has been clearly pointed out that the city and its surrounding areas have reached the saturation level[27] of 100% beyond

[27] According to the 'Ground Water Estimation' report (2011), Taluks sucked are Doddaballapur, Hoskote, Nelamangala and Devanahalli in Bengaluru Rural District; Anekal and Bengaluru East, North and South taluks in Bengaluru Urban district; Chikkaballapur, Chintamani, Gauribidanur, Gudibande and Sidlaghatta taluks in Chikkaballapur district; and Bangarpet, Kolar, Malur, Mulbagal and Srinivasapura taluks in Kolar district. Ban drilling of borewells (Mukherjee 2014).

Table 6.16 Depth of borewell, DTW and drawn down in Bengaluru

Areas	RL (m)	Depth drilled (m)	DTW (RL) (m)	Drawn down RL (m)
CAR Sirisi circle	882	264	877	849
Bengaluru university	876	120	860	854
CPWD, Koramangala	900	215	879	859
Kannamangala	875	225	852	838
Vidyanagara	893	209	878	–
Hesaraghatta	879	241	866	857
Kumbalgodu	810	149	776	–
Yelahanka	893	200	881	867
Chandapura	920	203	899	879
Sarjapura	920	172	900	894
Somanahalli	–	210	–	–
Attibele	920	241	895	882

Source: Sawkar (2012a, b: 14)

which water cannot be tapped. According to 'Groundwater Resource Estimation', 2011, a report prepared by the Central Ground Water Board reveals that the saturation level for India was 61% in 2009, 68% for Karnataka and 136% for Bengaluru. A 2011 report puts the level at 64% and 141% for Bengaluru (Mukherjee 2014).

A study report prepared by the Dept of Mines and Geology on groundwater resource of Karnataka, 2004, points to alarming levels of groundwater extraction and exploitation as against natural recharge. Particularly, the rate of extraction exceeds the rate of recharge in BMR and BBMP areas. The Table 6.13 shows that the level of groundwater extraction in Bengaluru Urban district amounts to 567 MLD as against BMR's entire drawing estimated to be 1216 MLD. Similarly, BMR's groundwater utilisation rate is 1216 MLD as against the recharge rate of 221 MLD. The recharge in BBMP area is approximately 195 MLD as against the withdrawal of 282 MLD (GoK 2009: 85). The status of overexploitation of groundwater in BMR for 1999 is presented in Annexure.

6.3.4.2 Some of the Key Environmental Issues with Respect to Groundwater Extraction

(i) The entire Bengaluru terrain is a hard rock formation with groundwater depth reaching more than 1200 feet. Exploratory wells give information on the depth of water levels. Sometimes, the depth reaches to more than 1500 feet.

(ii) A majority of the people, particularly in the newly merged areas, own borewells. The situation turns worst during summer while mild during rainy season. Since drinking water is not available in the periphery region, the local community resorts to groundwater tapping which is usually unauthorised. Groundwater levels are very low from 1200 to 1500 feet.

(iii) Lack of homogeneity of groundwater aquifers is the major problem. Success depends on the availability of water.
(iv) In addition, illegal groundwater extraction has led to a flourishing groundwater market, particularly in the newly merged city periphery region. It is only water tankers selling groundwater that cater to the day-to-day needs of the exploding population. It is estimated that about 3000 tankers, with a capacity of 3000 litres each, sell about 85 lakh litres per day (CGWB 2008: 13). As a result of such an unabated overexploitation of groundwater (estimated to have crossed 196–200%), there seem to be no groundwater resources left for future development.
(v) In fact, the rapid urbanisation processing along with an increase in the conversion of agricultural lands to nonagriculture purposes has led to a reduction in the number of lakes and tanks, the main sources of groundwater recharge. A study by the Central Ground Water Control Board (2008) has identified three significant reasons behind the overexploitation of groundwater resources: (i) sewerage and industrial pollution; (ii) man-made pollution, leading to high nitrate concentrations; and (iii) unabated overexploitation of groundwater resources.
(vi) Man-made pollution, such as sewerage diversion to lakes, particularly Vrishabhavathi Valley, is the result of a haphazard urbanisation. In addition, industrial pollution has crossed threshold levels, particularly in the industrial belts of Bengaluru City such as Peenya, Rajajinagar and Hoskote, which indicate high concentration levels of chloride and magnesium.
(vii) There is a lack of strong augmentation, recharge and recycling measures practised by the formal and informal sectors such as public and private agencies.
(viii) Lack of a monitoring system on the loss of water and measures to reduce it.
(ix) There is no mechanism to monitor groundwater extraction through a drilling and recharge structure. Devanahalli and its surroundings have been declared as grey areas due to a decline in the groundwater table to alarming levels.
(x) An inadequate storage and distribution network in most of the Taluk Panchayats (TPs) and City Municipal Councils (CMCs).
(xi) Lack of sufficient awareness regarding depleting water tables, while there is a need for water harvesting measures and groundwater recharge (BMRDA 2009: 159).
(xii) Nitrate levels of groundwater exceed the standard limits of WHO, constituting public health risk (CSD 2008: 30).
(xiii) Generally, people lack awareness with regard to effective groundwater management and regulations. Monitoring and maintenance do not happen at the community level, particularly in respect of government buildings.
(xiv) Lack of an effective groundwater management is affecting groundwater resources (Picture 6.5).

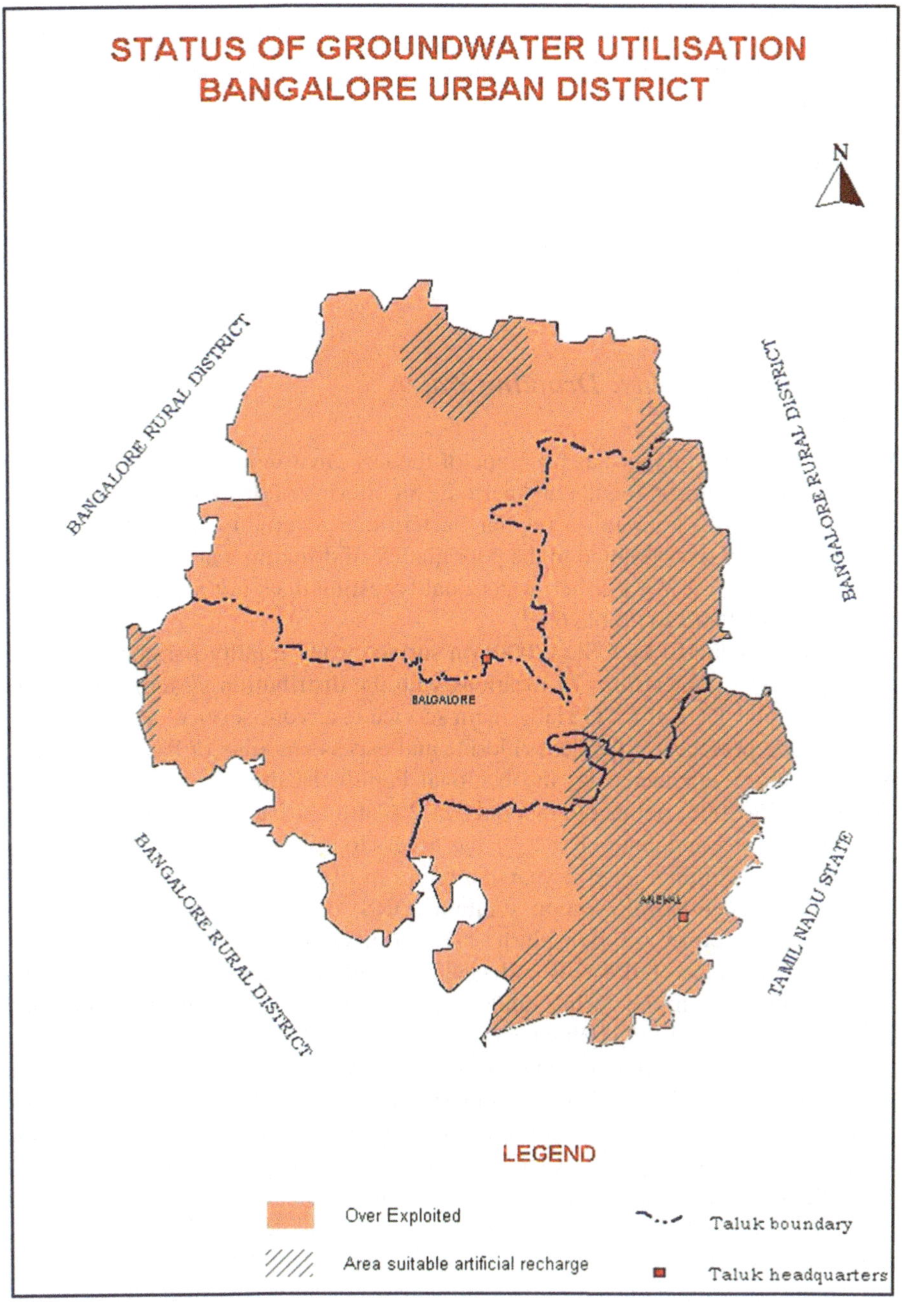

Source: CGWCB (2008: 25)

Picture 6.5 Status of groundwater utilisation in Bengaluru Urban district

Table 6.17 Citizen survey on water quality

Water quality issues	Percentage of piped water	Percentage of public tap
Timings of water supply	79	26
Adequacy of water	70	64
Accuracy of billing	90	–
Clarity of water	98	99
Odour-free water	81	83

Source: BBMP (2009: 68)

6.3.5 Water Quality: Drinking Water

Maintaining water quality has been one of the key environmental factors affecting the urban population. A regular and periodic monitoring of drinking water quality is done by BWSSB. Although a regular bacteriological quality testing is done by BWSSB, studies have pointed to the poor quality of drinking water all over the city (Sawkar 2012a, b: 877). Further, water quality testing does not meet the standards of WHO and CPHEEO (Table 6.17).

A study conducted by CSD (2008) on surface water quality parameters on the basis of water samples from 79 locations with the distribution systems of the city (which include TK Halli, TG Halli, main service reservoir, service reservoir (SR)) indicates the presence of residual chlorine and excess amounts of bacterial counts. Another survey conducted by the National Family Health Survey in Karnataka (NFHS, 2005–2006) points out that only 71.2% of the urban population has access to drinking water and that only 82.7% has access to toilets. The survey also reports that 67.3% of children aged 3 years suffer from diarrhoea in urban areas, while 38.5% have received rehydration solution (ORS) (GoK 2009: 152). As per the Environment Report Card, Bengaluru (2012), a survey conducted by CSD on water quality (16 samples of BWSSB collected from different service stations), a high coliform content is found during summer months of April–June, mostly in east and west zones (CSD 2012: 26) (Fig. 6.6).

Similarly, the water quality monitoring on 44 lakes reveals that most of them are not conducive for drinking or other domestic use (GoK 2009: 52). Due to the lack of an effective sewerage and solid waste management in the city, most of the untreated waste and sewerage reach water bodies like tanks and lakes severely polluting the local ecosystem. This proves a gradual degradation of the biosystem and the spread of contamination to groundwater. Several large lakes like Hebbal, Bellandur, Vrishabhavathi, Agraharam, Hulimavu, Madiwala and Doddabeguru remain severely contaminated with the presence of bacteria and other physiochemicals at more than permissible limits, threatening human health (GoK 2009: 53–55) (Picture 6.6).

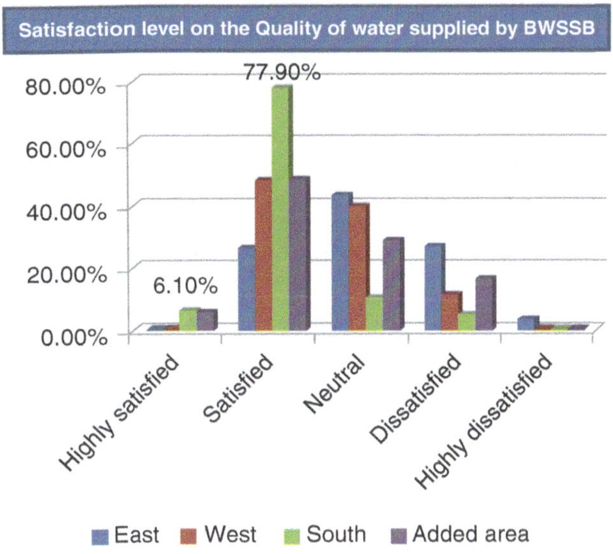

Source: CSD (2012: 26)

Fig. 6.6 Levels of satisfaction regarding the quality of water supplied by BWSSB

Picture 6.6 An overflowing manhole contaminating wells in Shivajinagar

Source: GoK (2009)

Fig. 6.7 Coliform organism count in piped water supply

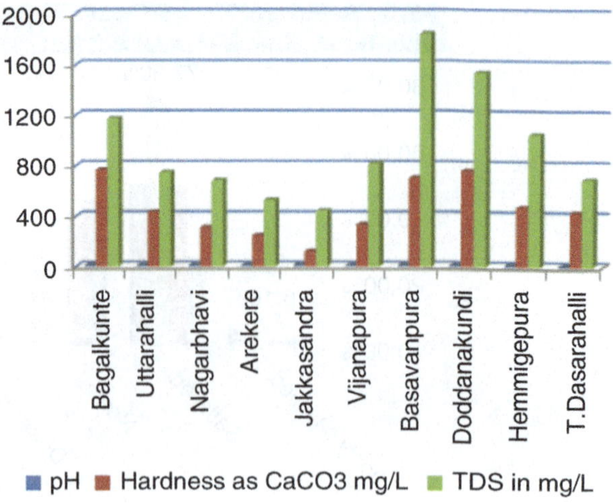

Source: CSD (2012: 27)

6.3.5.1 Groundwater Quality

Studies have pointed to an increased groundwater contamination and pollution due to rapid urbanisation, population growth, poor sewerage system and insufficient disposal methods of waste (GoK 2009: 92; CSD 2008: 29; Raju et al. 2008: 7; Sawkar 2012a, b). Groundwater quality is a major problem plaguing urban areas such as Bengaluru. Borewells having a more chloride content are a natural phenomenon, but the percolation of nitrate contents into deep groundwater is due to (i) urban waste, (ii) diversion of sewerage and (iii) application of fertilisers by farmers.

A study conducted by the Dept of Mines and Geology (2006) on the quality of water shows that 58% of the groundwater is not potable due to the presence of nitrates, phosphates, iron and high hardness values. For instance, Vrishabhavathi river catchment area[28] records highly polluted groundwater values (CSD 2008: 6; GoK 2009: 92). The condition is even more worse in CMCs and TMCs which rely completely on groundwater with high levels of nitrate even as blue baby syndrome is very common in these areas (CSD 2008: 6). Most of the major rivers and lakes like Vrishabhavathi, Hemavathy, Arkavathi, Bellandur, Hebbal, Agraharam, Hulimavu, Madiwala, Doddabeguru, etc. are contaminated beyond human use. One of the most serious problems associated with groundwater is the lack of an effective management system with no records maintained by both the government and private agencies regarding usage and contamination (Fig. 6.7).

[28] Peenya Industrial Area, Nandini Layout, Kamakshipalya, Industrial Area, Rajajinagar Industrial Area, Mysore Road, Nagarabhavi 1st Stage, Mahadevapura, Thimmpasandra, Vittasandra, Doddavangala, Bellandur village, Shampura, Bharathinagar and Geddlahalli. High nitrate concentrations are also found in Sulthanpalya, Cholanayakanahalli, Hebbal, Srirampura, Doddabanaswadi, Jayanagar 4th Block and Banashankari.

In Bengaluru North Taluk, only small portions in south-western part and north-eastern part, groundwater is affected by high nitrate contents, while a larger part of Anekal taluk, isolated patches in the eastern part of Bengaluru north and south-west and eastern parts of Bengaluru South Taluk are found to have chloride in the range of 250–1000 mg/l (CGWB 2008: 12).

Similarly, a study conducted by CSD (2008: 40) on groundwater quality in 2003 consisting of a total of 918 samples[29] collected from different borewells in 100 wards of Bengaluru City shows that nearly 51% of the water samples are not fit for drinking due to the presence of either chemicals or bacteriological parameters. Particularly, the condition is worse in the periphery of Bengaluru City (CMCs and TMCs) which is highly dependent on groundwater for all purposes. The samples show the presence of high nitrate levels besides the blue baby syndrome, most commonly suffered by newly born babies. As per a study jointly conducted by Public Health Institute and the Department of Mines and Geology, GoK (2011), out of 2209 groundwater samples, 52% of borewell water and 59% of tap water in Bengaluru are not potable due to the presence of 8.4% and 19% of *Escherichia coli* bacteria, respectively (Balasubramanian 2013). In addition, nitrate content in excess of 29% and 10% of iron (while permissible limit being 8.5%) are found to have crossed the permissible limits (CSD 2012: 28). A study by CSD (2012) on groundwater monitoring in some of the selected locations of the city (see Fig. 6.5) shows the presence of high TDS which indicates the presence of inorganic and organic solids in excess of the prescribed standard limits (500 mg/l), thus rendering it unfit for drinking purposes. Similarly, a CSD study (2012), on the basis of nine samples covering all eight zones of BMR (see Fig. 6.8 below), finds six zones contaminated with coliform content, making water unfit for drinking. Both east and west zones receive contaminated water from BWSSB.

6.3.6 Packaged Drinking Water

With a population of 5,686,000, Bengaluru happens to be India's fifth largest city. As per the estimates of Bengaluru Water Supply and Sewerage Board (BWSSB), the total demand for water is 840 million litres per day (MLD) (assuming a population of 6 million and a supply rate of 140 litres per capita per day [lpcd]).[30] The corresponding demand-supply gaps are 135 and 495 MLD.

According to BWSSB sources, the city is desperately short of water. It is reported that 870 MLD of water supplied every day reaches the core areas and at least 72 of the 225 wards in former CMCs and TMCs. But the government is least concern

[29] These samples are tested against 15 parameters namely calcium, magnesium, sodium, potassium, total iron, carbonate, bicarbonate, chloride, nitrate, sulphate, total dissolved solids, specific conductance, hardness, fluoride and PH.

[30] The demand works out to be 1200 MLD, at the standard rate of 200 lpcd set by the Bureau of Indian Standards [BIS] for water supply in urban areas.

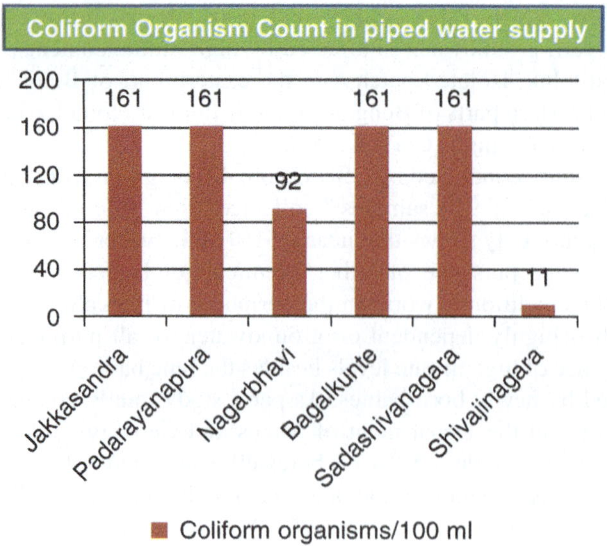

Source: CSD (2012: 27)

Fig. 6.8 Coliform organism count in piped water supply

about the extent of impact that can be caused due to over-felling of trees and constructing of concrete roads which reduce the water holding capacity of the soil which, in turn, reduces groundwater and thereby drinking water.

Several packaged drinking water samples are available for urban citizens. Both authorised and unauthorised packaged drinking water are serving the city of Bengaluru. There is a flourishing business going on of packaged drinking water around peri-urban areas of the city. In most of the agricultural lands surrounding the city, groundwater is pumped out for package industries. More than hundred packaged water units thrive around the city supplying water, while half of the packaged water units do follow the quality control norms[31] set by the government of Karnataka (Kidiyoor 2013. About 51 in Bengaluru and 568 in Karnataka are illegal units (Ashwini 2012; Kidiyoor 2013). Most of these illegal units supply water without purifying causing serious health issues. The consumption of non-purified water might lead to health problems like liver damage, Hepatitis A and E (*ibid)*. A study of eight packaged drinking water samples by the Public Health Institute (of the Dept of Health and Family Welfare) in June found seven sample (i.e. 87.5%) substandards, not meeting the prescribed specifications of the Bureau of Indian Standards (BIS) (Ashwini 2012). Similarly, an independent study, conducted by the Centre for Science and Environment (CSE), found deadly pesticides in most of the bottled water samples sold in India, not in conformity with the European norms (CSE 2013.

[31]Recently in 2013 it was found that 51 units had not obtained the required permission and license as per the ISI and BIS norms.

Table 6.18 Distribution of water bodies in Bengaluru Metropolitan Region (BMR)

Type of lakes	Bengaluru Development Authority		Bengaluru Metropolitan Region	
	Number of lakes	Water spread area (ha)	Number of lakes	Water spread area (ha)
Minor (less than 2)	285	194.0	1260	864
Small (between 2 and 8)	185	833.1	847	2809.3
Medium (between 8 and 25)	94	1220.0	469	4258.0
Large (between 25 and 50)	29	1003.4	117	2742.1
Very large (more than 50)	15	1322.3	96	7586.0
Total	**608**	**4572.8**	**2789**	**18,260.2**

Source: GoK (2009: 48)

The Bureau of Indian Standards (BIS), which issues license to the packaged drinking water industry, has recently issued 148 licenses to packaged drinking water companies in Karnataka of which 100 are based in Bengaluru. These companies have their own borewells and treatment plants, but the BIS monitoring system that involves inspecting the plants twice a year shows that 20–30% of them flout directions. According to a leading producer of packaged mineral water, the demand is increasing in east and south of Bengaluru[32]. Interestingly, the BIS officials point out that there could be more than 100 illegal units operating in and around Bengaluru, supplying 'unsafe' drinking water (Prasad 2008).

6.3.7 *Governance of Urban Lakes*

Surface water also constitutes several lakes and tanks in and around the city of Bengaluru. The city of Bengaluru is known for its lush green space and wetlands, promoting a salubrious climate through the year. This fact is supported by Mathur and Cunha (2006) book on Bangalore. Lakshman Rau Committee[33] reports that there were 278 lakes around Bengaluru City, covering more than an area of 40 ha during 1961–1972 (Thipaiah 2009: 5; GoK 2009: 81). But their number subsequently decreased to 81 as per the Lakshman Rau Committee Report (1986), a decline of 35.09% (Thipaiah 2009: 5). In the name of development, nearly 79% of the city's water surface has been lost (Siraj 2014). The distribution of water bodies in BMR is presented in Table 6.18.

[32] [http://waterbangalore.blogspot.com/2009_03_01_archive.html].

[33] The city's lakes were for the first time gazetted from 1988 with the commissioning of N Lakshman Rau, Committee.

Table 6.19 Management of lakes in Bengaluru

Organisations	No of lakes
BBMP	59
BDA	123
Lake development authority	4
Karnataka Forest Department	5
Minor irrigation department	18

Source: Lake Development Authority, Bengaluru

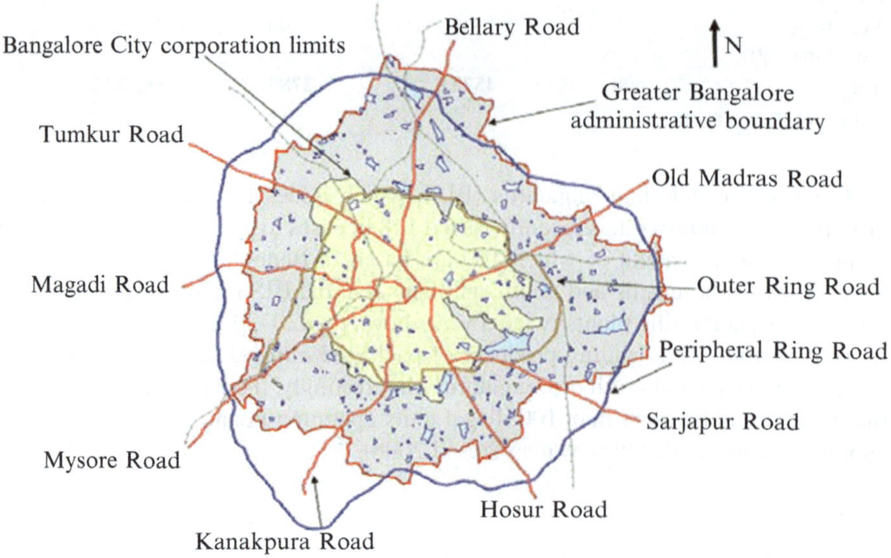

Source: Ramachandra and Kumar 2010

Picture 6.7 Greater Bengaluru with 265 water bodies

As per the official records of Lake Development Authority (LDA), there are about 2789 lakes (spread over 18,260.2 ha) in and around the city. There are 117 large lakes and 96 very large lakes covering around 25 ha and 50 ha, respectively. While BBMP manages 59 lakes, BDA 123 lakes and LDA 4 lakes, Karnataka Forest Department manages 5 lakes, and the Department of Minor Irrigation manages 18 lakes[34] (see Table 6.19, Picture 6.7).

[34]Details of Lakes were collected during an interview with Mr. Sivanna, Chief Executive Officer, LDA, on April 24, 2014.

6.3.7.1 Loss of Water Bodies: Lakes and Tanks

Rich water bodies of lakes earned the city title ‘Garden City’ as part of an ecological heritage of the city. Nearly 2/3 of the water bodies, particularly lakes, are lost due to a rapid urbanisation process propelled by migration and industrialisation; several lakes and tanks have disappeared or vanished (Sherwood Institute n.d). Most of the lakes have been either encroached upon or polluted beyond recognition, adversely impacting public health, biodiversity and local environment. As per the NK Patil report (2011), out of 386 lakes, the status of 121 lakes is unknown. The report further acknowledges the fact that 100 lakes have disappeared, majority of them have been converted into commercial purposes (like bus stations, roads, layouts, garbage and waste dumping sites or truck stands, etc.). Many lakes are contaminated due to discharging of untreated sewerage and industrial effluents, deforestation, denudation of wetlands, excessive use of chemical fertilisers and pesticides, mining and extraction of groundwater aquifers and waste dumping, spreading diseases and damaging the local ecology of lakes, aquatic life and bird habitats (GoK 2009: 47; Venkataraman 2000). The Justice N.K. Patil Committee report (2011) highlights that lake beds in the conurbation area are used for commercial and residential sites and layouts. Further, lakes are also contaminated due to discharging of domestic effluents and sledge (Picture 6.8, 6.9, 6.10, 6.11, 6.12, 6.13, 6.14, 6.15, and 6.16).

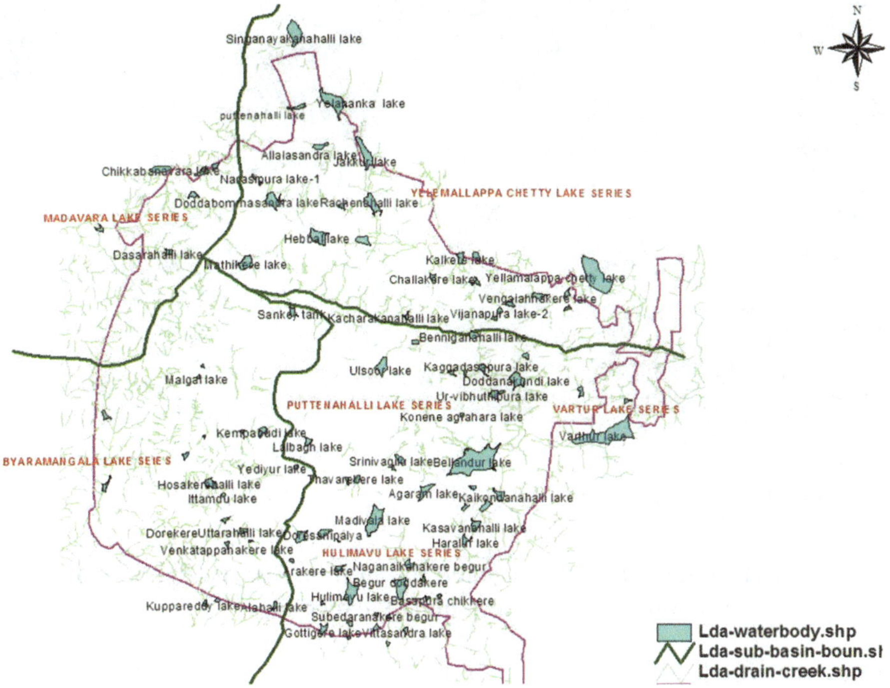

Source: BWSSB 2014

Picture 6.8 Bengaluru cascading lake series

Source: BWSSB 2014

Picture 6.9 Highly polluted Biratena Agrahara Lake

Source: BWSSB 2014

Picture 6.10 Weed-infested Jaraganahalli Lake

6.3.7.2 Urban Lake Rejuvenation and Restoration Efforts: Public-Private Partnership (PPP) or Community Partnership

Consistent efforts are made to rejuvenate and restore the visible lakes by NGOs (national/international/local). Community-based organisations (CBOs), particularly around the lakes, are initiating serious efforts towards restoration. Several PILs (public interest litigations) have been filed in the High Court of Karnataka seeking

Source: BWSSB 2014

Picture 6.11 Highly polluted Challakere Lake

Source: BWSSB 2014

Picture 6.12 Contaminated Madivala Lake

Source: BWSSB 2014

Picture 6.13 Encroached Singasandra Lake

Source: BWSSB 2014

Picture 6.14 Encroached Konappana Agrahara Lake

Source: BWSSB 2014

Picture 6.15 Encroached Kacharakanahalli Lake

Source: BWSB 2014

Picture 6.16 Conservation and Rejuvenation of Subedaranakere Lake

Table 6.20 Lakes taken up for redevelopment by different organisations

Organisation	No of lakes	Share (%)	Area (ha)	Share (%)	Status of development
BBMP	55	29.10	1895	26.61	21
BDA	123	65.08	4426	62.15	12
KFD	5	3.17	476	6.68	1
LDA	4	2.65	325	4.56	3

Source: BWSSB et al. (2014)

action towards restoration and protection of the existing lakes. Particularly, the role of ESG (Environment Support Group) is laudable in this direction. The reports of both N Lakshman Rau Committee (1985) and Justice N.K. Patil Committee (2011)[35] took notice of the extent of deterioration of Bengaluru's wetlands while recommending for the preservation and development of lakes in the city. Subsequently, as per a study conducted by Bruhat Bengaluru Mahanagara Palike (BBMP) (2007, 17 lakes were selected for rejuvenation and restoration (GoK 2009: 50) in the BBMP area. A further concerted effort towards lake rejuvenation made was the constitution of Lake Development Authority (LDA) in 2002 for conserving the lakes of Bengaluru. Under a combined initiative of LDA, BDA and BBMP, 25 lakes have taken up for rejuvenation. Further, LDA has taken up four lakes, using funds from the National Lake Conservation Programme. The Department of Forest has taken up 17 lakes for development and 11 lakes for maintenance besides developing a Master Plan for five lakes in the city. By 2007, BBMP had developed four lakes (BDA 2007: 43–44).

Similarly, BDA has taken over 123 lakes from BBMP for rejuvenation (see Table 6.20). The overall development of lakes has been undertaken in three phases, involving 'rejuvenation and restoration of lakes[36]'. The task under rejuvenation of lakes involves checking the inflow of surface run-offs by providing silt traps and wetland ponds so as to arrest the debris, organic and inorganic substances, so that fresh water is impounded in the lakes. This facilitates an increase in groundwater tables and a good environment by way of restricting both air and water pollution. Under Phase I, 13 lakes; Phase II, 32 lakes; and Phase III, 78 lakes have been undertaken for rejuvenation. While the rejuvenation of Phase I, 13 lakes is completed, the other two phases are going to be undertaken based on the availability of funds and necessity or urgency of works. The status of rejuvenation and water recharge before and after rejuvenation is presented below (Picture 6.17).

[35] The Lake Development Authority (LDA), an autonomous body, was constituted in 2002 as the nodal agency for the convergence of BBMP, BDA, BWSSB, BMRDA, Zilla Panchayat and Department of Forest (GoK 2009: 64). The main aim of the LDA is to reinvigorate and rejuvenate the existing lakes within BMR, thereby boosting aquifers and protecting biodiversity of water bodies.

[36] Details regarding lake rejuvenation were collected in an interview with Mr. Shivashankar, EOI, Dept of Town Planning, BDA, Bengaluru, on April 25 2014.

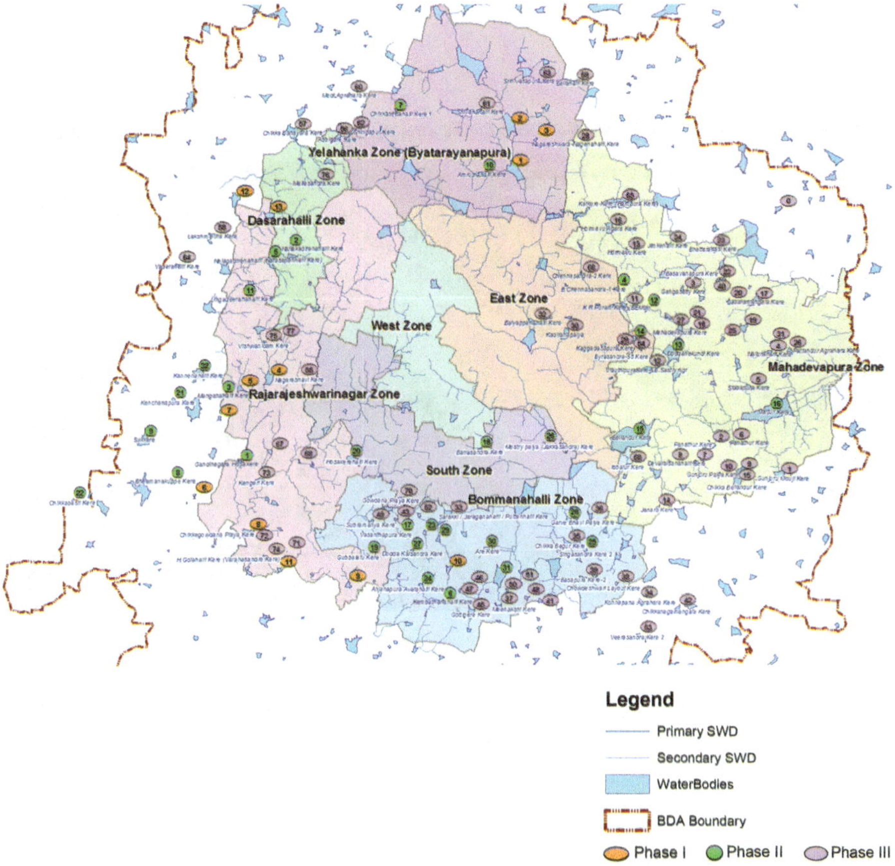

Source: BDA, Bangalore

Picture 6.17 123 Lakes rejuvenated by BDA

Table 6.21 Lake redevelopment and sources of funds

Organisation	No of lakes	Loan sanctioned (crore)
KUIDFC (under megacity scheme)	17	46.55 (32.65)
Government of Karnataka	20	25 each for 2 years
BBMP own resources	24	50

Source: BWSSB et al. (2014)

As regards the lake development projects, out of 16 NLCP-assisted LDA projects, 11 NLCP projects have been completed. A lake prioritisation list is prepared according to the NLCP guidelines. Further, a technical committee scrutinises DPRs before the submission, and GIS readings are collected for all the water bodies in Bengaluru district. So far, GIS reading for 500 lakes has been collected. In addition, the collection of basic data and digitalisation of lakes as per revenue records are planned (Table 6.21, Pictures 6.18, 6.19, 6.20, 6.21, 6.22, and 6.23).

Source: BDA, Bengaluru

Picture 6.18 A completely rejuvenated Malathahalli Lake

Source: BDA, Bengaluru

Picture 6.19 A completely rejuvenated Ullal Lake

Source: BDA, Bengaluru

Picture 6.20 A completely rejuvenated Konasandra Lake

Source: BDA, Bengaluru

Picture 6.21 Rejuvenated Jakkur Lake

Source: BWSSB et al. 2014

Picture 6.22 B. Hosahalli Lake before rehabilitation

Source: BWSSB et al. 2014

Picture 6.23 B. Hosahalli Lake after rehabilitation

Table 6.22 Total expenditure incurred on lake management and rejuvenation by BBMP up to March 2014

Sl No	Year	Fencing	Development	Total amount (in Rs. Lakhs)
1	Up to 2008–2009	0.00	0.00	0.00
2	2009–2010	422.64	424.70	847.34
3	2010–2011	1586.45	1328.18	2914.63
4	2011–2012	1480.71	1813.88	3294.59
5	2012–2013	755.18	2534.13	3289.31
6	2013–2014	26.06	2839.11	2865.17
Total	**Total**	**4271.04**	**8940.00**	**13211.04**

Source: BBMP, Bangalore

Similarly, up to March 2014, 44 lakes[37] have been rejuvenated[38] by BBMP. While three BBMP lakes, namely, Kaikondahalli Lake, Ulsoor Lake and Chinnappanahalli Lake, are maintained through a 'tripartite[39]' partnership, i.e. PPP agreement, seven lakes have been adopted by registered NGOs/RWAs/charitable trusts/corporations/corporate houses/business houses, etc. The contract is to last for 5–15 years. Under PPP, two schemes were introduced, namely, (i) Expression of Interest and (ii) Adopt-a-Lake. Although in 2009 36 lakes were given to private companies under 'Expression of Interest', only four lakes have been developed, i.e. Nagavara, Vengaiahnakere, Hebbal and Agara. Similarly, under 'Adopt-a-Lake' scheme (in 2009), 15 lakes were given to private companies, but only five have been developed – Sheelavanthakere, Kelaginakere, Mahadevapura lake, Kundanahalli lake and Challakere. Under this scheme, the companies that have adopted are required to make yearly payment to LDA as per the lease agreement (Thipaiah 2009: 23).

Total expenditure incurred on lake management and rejuvenation by BBMP up to March 2014 amounts to Rs. 13211.04 lakh. Nearly 49 lakes are being developed with fencing sewerage diversion and STPs (sewege treatment plants) fixed for four lakes, while six lakes are developed with 'neighbourhood parks' with the assistance of RWAs (Residential Welfare Associations). The total lake waste water treated works out to nearly 2.5 MLD. Total expenditure incurred for lake management and rejuvenation by BBMP up to March 2014 is Rs. 1,32,11.04 lakhs (see Table 6.22).

Recently, in 2014, LDA has taken up a number of sustainable initiatives for conserving and protecting the ecology of the city. One such effort is a baseline survey conducted throughout the city listing lakes of present and past. This survey has compiled all the details on the number of lakes, present status, encroachment, eviction, survey numbers, etc. In addition, a model DPR (detailed project report) has been commissioned under LDA by way of inviting national level tenders. This DPR

[37] Data on urban lakes in Bengaluru was collected from Lake Dept., BBMP, Bengaluru. An interview was held with Mr. B.V. Satish, Chief Engineer, BBMP, Bengaluru, on May 20, 2014.

[38] See Annexure for Tables 6.62 and 6.63 on the Basic status of Lake Rejuvenation by BDA.

[39] A Tripartite Agreement has been reached between BBMP-RWAs-IT/corporate sector and local councilors for maintenance of urban lakes.

Table 6.23 Action taken on lakes/proposed

Status	No of lakes	DPR cost (in Rs.)	Expenditure (in Rs.)
Lakes developed	21	143.47	58.27
Lake development in progress	8	28.66	4.85
Tender in progress	6	24.61	0
Tender floated	15	128.97	1.99
DPR under preparation/submission	5	8.43	0
Total	55	334.15	65.72

Source: BWSSB et al. (2014)

will set a model for the management and protection of the city's lakes under LDA. The DPR contains all the details on lake management issues such as (i) recreational activities, (ii) community participation, (iii) maintenance issues, (iv) revenue generation, (v) lake advisory committee, etc. At present, LDA has taken up for the management of Agara, Madivala and Puttenahalli lakes at Yelahanka. Details of progress and proposed lake redevelopment are presented in Table 6.23.

6.3.7.3 Community Participation in Lake Management

To enhance and strengthen the organisational powers of LDA, Karnataka Lake Development Authority Bill 2014 has been submitted to the Government of Karnataka (GoK) for making LDA the sole authority in the management and protection of the city's lakes. A three-member 'technical committee' has been formed consisting of academicians from IIM, Bengaluru, for the preparation of a DPR on the city's lakes. Besides, a voluntary pressure is exerted on LDA for the restoration and rejuvenation of lakes in the city. Under Karnataka Lake Conservation and Development Authority Act 2014, Lake Conservation and Development (LCD) has been constituted in 2015. The Karnataka Lake Conservation and Development Authority Rules has come into existence from 2016.

A minimum of 10–12 complaints have been received from the public and the NGOs concerned, demanding restoration and rejuvenation of polluted lakes in the city. A voluntary pressure from the local community is gaining momentum in the city of Bengaluru for restoration and rejuvenation of lakes. For instance, the community had approached the Lokayukta for the restoration of JP-Nagar-Puttenahalli Lake. Immediately, thereafter, the Lokayukta directed BWSSB, BDA and LDA to initiate the required action. In the second instance, the ex-military staff approached LDA for restoring Arkehalli Lake. Similarly, apartment dwellers were willing to participate in the restoration of lakes adjacent to their apartments as part of enhancing their aesthetic values. For example, Seetharampalya Lake restoration was an initiative taken by the apartment community and was willing to donate Rs. 1 crore. The proposal was cleared by BDA. Although private builders were interested in restoring lakes with their own funds by installing STPs, they wanted to divert recycled waste later into lakes. For example, Mantri Developers wanted to restore

Agara lake, but permission was not given. In yet another instance, apartment dwellers wanted a connecting road constructed between the apartment complex and Madivala Lake but were not permitted.

6.3.7.4 Constraints Involved in the Management of Urban Lakes

Many constraints are faced by the local bodies in the protection and preservation of urban lakes. They include the following:

- The most significant of them being the lack of a separate act or laws for the protection and restoration of lakes in the city. As a result, numerous local bodies such as BBMP, BWSSB, BDA and LDA have their own set-ups for the protection of lakes under their jurisdictions. Further, different administrative jurisdictions such as BBMP, BDA and LDA (like Bengaluru Urban Agglomeration, Bengaluru Metropolitan Area and Bengaluru Metropolitan Regional Authority 1985 further complicate the process of institutional intervention for the protection of lakes.
- Varying figures and estimates of the different departments on the distribution of BMA have further problematised the issue of conservation and protection of vast resources of urban lakes. Further, the lack of consolidated official statistics on lakes and their status further complicates the management and protection of the city's lakes.
- At present, only four lakes are under LDA for management and protection.
- Before LDA was formed, all the city lakes were under the management of the Forest Department. But later, both BBMP and BDA took charge of some lakes for managements.
- The criticism is that LDA is a 'toothless body'. Although LDA has been constituted with the sole authority for the protection and management of city lakes, one of the serious and major constraints facing LDA is that it has absolutely no powers in terms of sufficient funds, adequate staff, infrastructure and other facilities to effectively implement any policy or to take necessary actions for the protection and restoration of the city's lakes. At present, LDA is seriously understaffed with only two engineers and deficient funds. Therefore, LDA has completely failed in its efforts to protect and manage the city's lakes.
- Particularly, the BBMP political representatives influence the restoration efforts and is even claimed that they are the main encroachers of the city lakes.
- Illegal encroachments by private developers and politicians are one of the serious concerns as far as the protection and restoration of lakes in the city are concerned. Besides, many slums have encroached upon lakes in the city. Nearly 30 acres of Kowdenahalli Lake area has been encroached. And so far, BBMP has strived to reclaim nearly 85 acres of the encroached lake.[40]
- Sewerage-induced pollution is a major problem facing the city's lakes. There is no effective monitoring system to prevent the diverting illegal sewerage into the lakes.

[40] Data collected from Dept of Lakes, BBMP, Bengaluru, May 2014.

- Sand quarrying is serious issue affecting the lakes.
- Lack of community participation is mainly affecting the restoration efforts.
- Haphazard development of the city has grossly affected the city's lakes.
- Blocking storm water drains and diverting the city's sewerage from industries and other bodies such as bus depot have enormously polluted the city's lakes.
- Once, the city was known for its small 'Kuntes',[41] but now all of them have completely vanished. No single Kunte is visible in the city vicinity, with almost all of them encroached upon.

6.3.8 Sanitation

Access to sanitation[42] is one of the Millennium Development Goals (MGDs) of the United Nations. A large number of people in Asian cities do not have reliable sanitation facilities, particularly access to toilets; instead they rely on either public toilets or open spaces. Such poor sanitation practices lead to serious health hazards (UN-HABITAT 2010: 148). Public toilets comprise 'Nirmala Bengaluru'[43] 'Sulabh-Sowchalaya' and toilets maintained by BBMP. Most of the public toilets are 'pay-and-use' toilets in the city. As per the data available with BBMP,[44] there are just about 500 public toilets (for an estimated population of 96.2 lakhs), out of which only 200 are functional which means that for every 19,000 people, only 1 toilet is available. Moreover, there is a need for 25,000 additional public toilets for the growing population (Jain 2013). According to a Janaagraha report, covering a total of 369 arterial and collector roads, 90 out of the 198 wards do not have public toilets (Kumar 2013). These toilets are often poorly maintained without regular cleaning. Often most of the public toilets do not have a proper water facility with women and children in particular from low-income groups and slums bearing the brunt of unhygienic conditions.

Bengaluru Agenda Task Force (BATF) constituted by the Government of Karnataka (GoK) in 2000 managed to garner public support, and with the involvement of corporate houses, nearly 100 public toilets were constructed, called as 'Nirmala toilets'. Households from slums were charged with Rs. 20 per month instead of the regular 'pay-and-use' toilets. Private companies were also involved, through contracts, in operating and maintaining toilets. In Phase I, 27 'Nirmala toilets' were constructed at a cost of Rs. 10 lakh per toilet block. By 2005, BCC (Bengaluru City Corporation) constructed nearly 50–60 toilets (Sharma n.d.).

---

[41] Interview with Shivanna, Chief Executive Officer, LDA, Bengaluru, on April 25, 2014. Kunte means 'small water body'.

[42] Sanitation constitutes access to toilets, sewerage and drainage facilities.

[43] Bengaluru Agenda Task Force (BATF) implemented 'Nirmala Bengaluru', by way of constructing pay-and-use public toilets.

[44] BBMP is obliged to ensure access to public toilets as per the Karnataka Municipal Corporation Act, 1976 (The Hindu 2013a, b).

A report card survey published by the Public Affairs Centre (PAC) point out that, while service delivery grossly varies across zones and wards, overall, the satisfaction level is higher in east zone than the other zones. Also, the performance in south zone is good with respect to slum households (Shekar and Shah 2006). As per the Revised Master Plan 2015, over 50% of the households in slums are not connected to either latrines or drainage (BDA 2007: 54).

The sanitation condition in the periphery and conurbation areas of Bengaluru is worse, with only 47% of the households having access to toilets, while 19% share toilets and 35% defecate openly. In the green belt area, only 26% have access to toilets and a staggering 70% defecate in the open spaces (GoK 2009: 94). The households in both conurbation and green belt areas are not connected to the sewerage network. A poor provision of sewerage and sanitation facilities spreads communicable diseases like cholera, typhoid, hepatitis, polio, diarrhoea, cryptosporidiosis and ascariasis (GoK 2009: 93).

6.3.8.1 Access to Public Toilets

Public toilets comprise 'Nirmala Bengaluru'[45] 'Sulabh-Shauchalaya' and toilets maintained by BBMP. Most of the public toilets are 'pay-and-use' toilets in the city. As per the data available with BBMP,[46] there are just about 500 public toilets (for an estimated population of 96.2 lakhs), out of which only 200 are functional which means that for every 19,000 people, only 1 toilet is available. Moreover, there is a need for 25,000 additional public toilets for the growing population (Jain 2013). According to a Janaagraha report, covering a total of 369 arterial and collector roads, 90 out of the 198 wards do not have public toilets (Kumar 2013). These toilets are often poorly maintained without regular cleaning. Often most of the public toilets do not have a proper water facility with women and children in particular from low-income groups and slums bearing the brunt of unhygienic conditions.

Bengaluru Agenda Task Force (BATF) constituted by the Government of Karnataka (GoK) in 2000 managed to garner public support and with the involvement of corporate houses, nearly 100 public toilets were constructed, called as 'Nirmala toilets'. Households from slums were charged with Rs. 20 per month instead of the regular 'pay-and-use' toilets. Private companies were also involved, through contracts, in operating and maintaining toilets. In Phase I, 27 'Nirmala toilets' were constructed at a cost of Rs. 10 lakh per toilet block. By 2005, BCC (Bengaluru City Corporation) constructed nearly 50–60 toilets (Sharma n.d.). A report card survey published by Public Affairs Centre (PAC) point out that, while service delivery grossly varies across zones and wards, overall, the satisfaction level is higher in east zone than the other zones. Also, the performance in south zone is

[45] Bengaluru Agenda Task Force (BATF) implemented 'Nirmala Bengaluru', by way of constructing pay-and-use public toilets.

[46] BBMP is obliged to ensure access to public toilets as per the Karnataka Municipal Corporation Act, 1976 (The Hindu 2013a, b).

Table 6.24 No. of public toilets in Bengaluru constructed and maintained by BBMP

Sl No	Zones	Total number of public toilets	BBMP-maintained toilets	Private-maintained toilets
1	South	161	29	132
2	West	214	82	132
3	East	151	115	36
4	Dasarahalli	2	2	0
5	Rajarajeshwari Nagar	24	10	14
6	Bommanahalli	2	0	2
7	Mahadevapura	19	3	16
8	Yelahanka	14	6	8
Total		587	247	340

good with respect to slum households (Shekar and Shah 2006). As per the Revised Master Plan 2015, over 50% of the households in slums are not connected to either latrines or drainage (BDA 2007: 54).

The sanitation condition in the periphery and conurbation areas of Bengaluru is worse, with only 47% of the households having access to toilets, while 19% share toilets and 35% defecate openly. In the green belt area, only 26% have access to toilets and a staggering 70% defecate in the open spaces (GoK 2009: 94). The households in both conurbation and green belt areas are not connected to the sewerage network. A poor provision of sewerage and sanitation facilities spread communicable diseases like cholera, typhoid, hepatitis, polio, diarrhoea, cryptosporidiosis and ascariasis (GoK 2009: 93).

Public Toilets by BBMP

Public toilets are the most important part of any social infrastructure. BBMP is also responsible for setting up public toilets at various places in Bengaluru. According to section 58(4) of the Karnataka Municipal Corporation Act, it is an obligatory duty of BBMP to provide public toilets. The data collected from BBMP indicated that Bengaluru has 587 public toilets (Table 6.24).

The newly added wards, consisting of 110 villages around the city, do not have a single public toilet. In older wards (located in the core city) where public toilets exist, some were constructed by Infosys Foundation, while some are under MLAs and MPs funds, and few more are constructed by BBMP. The maintenance of these toilets has been given to contractors, while the BBMP monitors it on a daily basis.

Another issue of concern is insufficient/lack/disconnected public toilets and availability of water supply. BWSSB has disconnected water supply as BBMP has not paid water bills due to insufficient funds. In addition, most of the public toilets are suffering from problems like unhygienic conditions, no tap or water, no latch on the door, stained toilet bowl and the remains of a sink and beer bottles and cigarette butts thrown around.

Plate 6.1 Nirmal Toilets in Bengaluru

Public Toilets: Nirmala Bengaluru by Infosys Foundation

Nirmala Bengaluru Movement focused on providing access to toilet facilities for the urban poor, particularly women and children living in slums. Besides, Nirmala Bengaluru Movement was also a public sanitation movement aimed at creating a larger awareness about public sanitation, cleanliness and hygiene among citizens. It also focused on instilling civic sense and identifying of Nirmala Bengaluru with good sanitation and hygiene practices. There are about 77 Nirmala Shauchalayas in Bengaluru across three zones – 33 in the east zone, 24 in the west zone and 20 in the south zone. The Infosys Foundation funded the construction of these toilets (Plate 6.1).

The aim was to provide users with a unique experience of hygiene termed as 'the Nirmala experience' through providing the best ambience and service delivery for ensuring revenue generation and sustainability.

The toilets were based on a 'pay-and-use' concept, at affordable rates, so as to extend it to the whole of Bengaluru City. However, at a later stage, user charges were waived off for encouraging public to use toilets, and also the management of the Nirmala Bengaluru toilets was handed over to Hype Integrated Communications, a private company on (BOOT) while the company could exhibit advertisements. The movement intended to build a strong network for future sanitation programmes by setting a benchmark in the provision of sanitation facilities at prominent locations, based on scientific area analysis for catering to different requirements of people based on user profile and demand.

The processes were well planned with an emphasis on (i) plans to have a comprehensive documentation of the program and systems for easy replicability and scalability; (ii) inter-stakeholder coordination expediting construction and delivery of infrastructure services; (iii) community participation across slums at different levels, i.e. construction, monitoring and maintenance of toilets; (iv) awareness creation regarding the existing spatially distributed toilet facilities as part of catering to the needs and demands; (v) Shuchi Mitras (citizen volunteers) and NGO involvement in maintenance and monitoring; (vi) establishment of the brand image of Nirmala Bengaluru in the city for a precise recognition of the logo/symbol and facility in the vicinity; and (vii) increasing the expectations of citizens of the city with a specific reference to sanitation and hygiene maintained in public toilets in association with the brand image of Nirmala Bengaluru (Plates 6.2 and 6.3).

6.3.8.2 e-Toilets[47]

As a novel measure to improve the public sanitation services, BBMP has proposed to install e-toilets across the city. One hundred such toilets are to be installed across the city (The Hindu 2013b). The recent initiative by the Bruhat Bengaluru Mahanagara Palike in installing e-toilets is viewed as a welcome initiative by majority of the users. In view of the increasing demand and need for public toilets in Bengaluru, e-toilets[48] were installed in several locations of Bengaluru City. e-Toilets are state-of-the-art toilets, with several outstanding and fully automated features – (a) flushes automatically, (b) automatically cleans up the entire toilet after five uses, (c) intelligent lighting and fan systems (switches on and off as a person enters and leaves the toilet premises), (d) waste processing and (e) portable design to fit busy

---

[47] e-Toilets are made out of stainless steel with a coin-operated entry system. They consist of an auto-flush and floor wash system. These toilets have a battery with an inbuilt water tank.

[48] The Kerala State Women's Development Corporation (KSWDC) has bagged the prestigious World CSR Congress Responsible Business Award 2014 for promoting the user friendly 'she toilets' for women.

Plate 6.2 Location before and after the construction of Nirmala Toilets

Plate 6.3 Nirmala toilets in Bengaluru

Indian towns and cities. e-Toilets do not need manpower for cleaning the toilets, a major constraint encountered in the previously installed public toilet initiatives.

Safety measures are also included to prevent damage and encroachment of public toilets. e-Toilets are equipped with GPRS device to alert the officials concerned in case of theft or trespassing. In case an attempt is made to break the automatic door lock or if two users enter a single unit, the alarm will ring, drawing the manager's attention. A supervisor and two technicians are employed for managing each e-toilet. The staff will be trained to ensure smooth functioning of the project.

e-Toilets are designed to avoid water wastage as every flush or floor wash requires only 1.5 litres of water. It is also equipped with a bio-membrane reactor that helps recycle the water and reuse it for flushing and cleaning. Solar panels are available for alternate energy needs. In the areas where drainage tank facility is not available, e-toilets can provide a green eco-friendly solution to manage the waste and its disposal like the bio-membrane tank system and the water recycling unit. It is user-friendly and indicates with a red light if the toilet is in use. e-Toilets are aesthetically appealing and convenient to be installed in busy places.

BBMP is funding the project with a total cost of Rs. 4 crore and has engaged Eram Scientific Solutions Pvt. Ltd., a Kerala-based company which is authorised to install e-toilets, at 75 major locations and crowded areas of the city. The company, Eram Scientific Solution, is responsible for installation and supervision of electric and sewage connections for the project and also maintenance for 2 years. After 2 years, it will be handed over to BBMP for maintenance. Each e-toilet costs Rs. 3.7 lakh, but with the construction cost of sewage, water and power, the total cost of installing each toilet will amount to about Rs. 5 lakhs. However, so far, finding a suitable space/land and a sufficient access to water has remained a constraint.

Eram Scientific Solutions, in collaboration with California Technology, USA, is also working on a research project to produce electricity through urine stored in these e-toilets. Besides, the firm has also collaborated with the Duke University, USA, to install waste treatment plants with human waste collected, processed and used for plantations. This is already in practice across 400 e-toilets and working well in Kerala and a few in Maharashtra (Nagpur) and Tamil Nadu.

In 2013, three e-toilets (two in Krishna Rao Park, one in Dodda Ganapathi Temple premises) were launched in the city on a pilot basis. With subsequent positive feedback received from the public, and with BBMP Mayor's concern, the corporation is keen on expanding the services to other locations. The process is underway, and at present there are 12 e-toilets already working in the city, while eight are under installation. A pilot visit was carried out to e-toilets located at various locations, and it was observed that e-toilets are popular among a large number of users. For example, the e-toilet in Gandhi Bazar serves street vendors, shop keepers, auto drivers and customers. Women vendors in particular find it most useful as it is cost-effective, hygienic and convenient to use (Plate 6.4).

Toilets installed in Krishna Rao Park have been popular among a large number of people who visit the park for walking and jogging in the mornings and evenings, in particular senior citizens (Plate 6.5).

e-Toilet located at a petrol bunk is also used by customers and passers-by. The owner of the petrol bunk opines that the toilet consumes less space and hence is beneficial to be installed in several such spot and attracts people to this petrol bunk, given the toilet access (Plate 6.6).

Plate 6.4 e-Toilet in busy street of Gandhi Bazar, Bengaluru

Plate 6.5 e-Toilet in Krishna Rao Park, Bengaluru

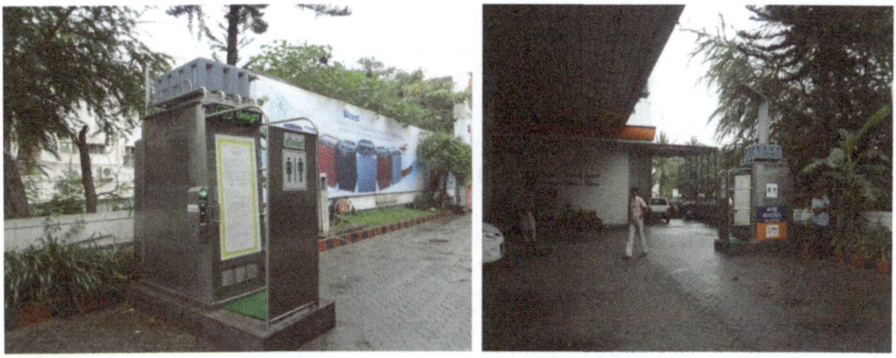

Plate 6.6 First e-toilet in Bengaluru at Annapurna Service Station Southend Circle, Bengaluru

6.3.8.3 Access to Toilets in Slums

About 98% of the urban households have access to clean toilets, but according to a study carried out in 2006, only 47% of the households in slums of India have access to sanitation (UN-HABITAT 2010: 148). Similarly, a study shows that effective sewerage connections have not increased correspondingly to the demographic explosion in Indian cities/urban centres (ibid). Out of a vast population in Bengaluru (78.28 lakhs), nearly 20% constitute of slum population. Slum population is composed of early and later migrant population from different districts within Karnataka and outside Karnataka state. They usually settle down in informal settlements or temporary shelters without an adequate access to basic services which include water supply and sanitation. As per 2001 census, Bengaluru Metropolitan Region has over 733 slums, while according to Karnataka Slum Development Board (KSDB) Karnataka Slum Clearance Board (KSCB), there are 598 slums, and going by the Revised City Development Plan (CDP) (2009: 21), there are a total of 640 slums in BMR. Studies have shown that 17 of the slum dwellers (i.e. every third in recognised slums and 700,000 in unrecognised slums) in the city do not have access to in-house toilets and usually rely on public or shared toilets or open spaces (GoK 2009: 94). Those with toilets are not effectively connected to sewerage with flow either open or connected to the running adjacent drainage channels. Similar conditions exist with regard to drainage facilities. Slum households are not connected to an underground drainage system (UGD) with household sewerage flowing through adjacent open space or open public drainage. A study by Manasi et al. (2016) has highlighted that Bengaluru City faces severe challenges in providing sanitation infrastructure for the urban poor. Lack of toilet usage was largely due to technical discrepancies, behavioural concerns, space issue, water scarcity and poor maintenance of toilets.

Environmental Implications of Non-accessibility of Toilets

1. A study by WHO (2004) reports that in the developing countries, annually, 1.8 million people die due to diarrhoeal diseases with 90% of them being children under the age of five (GoK 2009: 93).
2. There is a lack of data on the state of sanitation in unrecognised slums for the entire city of Bengaluru.
3. The existing public toilets are inadequate besides being very poorly maintained.
4. Lack of cleanliness in the existing toilets adversely impacts health and hygiene of the surroundings.
5. Storm water drains are poorly maintained.

Table 6.25 Sewerage treatment in Bengaluru City

Year	Population	Demand in MLD	Supply in MLD	Shortage in MLD
2001	53.79	870	540	330
2010	75.00	1125	900	225
2015	88.00	1500	1470	30
2021	100.00	1800	1470	330
2036	125.00	2500	1470	1030

Source: Kumar (n.d)

6.3.9 Waste Water Management

Bengaluru is located on a natural undulated terrain and slope. This facilitates the easy flow of sewerage water. The principal valleys of the city include (i) Vrishabhavathi, (ii) Koramangala and (iii) Challaghatta in the south and five minor valleys. The city generates around 3000–4000 t/day of USW which is 1000 mld waste water accounting for 80% of its daily water consumption from both surface and groundwater sources (GoK 2009: 95). Although Bengaluru City (including) covers an area of 800 km^2, only 290 km^2 is connected to sewerage facility. According to the Revised Master Plan 2015 (by BDA), the number of households connected to the sewerage network increased from 157,000 in 1991 to 344,000 in 2001. Yet with the merger of the extensive area of Bengaluru Rural into the erstwhile BMP area, nearly 70% of the BMP population has access to a drainage system, constituting 40% of the LPA (local planning authority) area (BDA 2007: 45).

At present, the core city area of 226 sq.km (i.e. erstwhile BMP) is completely connected to a sewerage network. By 2012, there were about 6,56,000 sewerage connections in the city. At present, the sewerage network covers 4300 km of BBMP area and a proposed network of 2300 km of the former CMC and TMC areas. To prevent blockages, BWSSB has set up 106 service stations with 27 subdivisions (CSD 2012: 17). According to the CAG report (Comptroller and Auditor general of India), 2010, the existing sewerage network covers only 40% of BMR, and that sewerage received amounts to only 47% with the remaining sewerage water ending up in storm water drains and lakes, contaminating groundwater (DNA 2011).

At present, BWSSB treats around 900 mld of waste water (see Table 6.25) and provides tertiary treatment to 73 mld (with four tertiary treatment plants). Both major and minor treatment plants cover an area of 240 km (Kumar n.d). BWSSB has 12 secondary treatment plants, treating about 40% (i.e. 721 mld) of BMR sewerage (GoK 2009: 96). In addition, there are 626 secondary level sewerage treatment plants (STPs) in Bengaluru City, the highest number in India. A report by KSPCB points out that though these secondary level STPs treat 834 million litres (MLD) of waste water per day, much of the treated water flows into the public open drains or underground drains. Although these STPs work at 81% of their capacity, there are no facilities in place to make use of the treated water (Rohith 2013).

The newly merged regions of Greater Bengaluru do not have access to a sewerage network. However, BWSSB under Cauvery Project IV, Phase II has initiated measures to connect CMCs to drinking water facility along with a sewerage network. The three drainage channels, namely, Vrishabhavathi, Hebbal and Koramangala-Challaghatta valleys, with sewerage treatment plants have a total capacity of about 203 MLD (BMRDA 2009: 161). Out of 1400 MLD of water supplied to the city, 70–80% of BWSSB's water ends up in sewage (Balasubramanian 2013; Sawkar 2012b; GoK 2009: 95). BWSSB has 14 secondary treatment plants (STPs) and 4 TTPs (tertiary treatment plants) with a total capacity of 794 MLD, while 10 more STPs are planned by BWSSB with a capacity of 339 MLD at a cost of Rs. 5400 million. However, only 30–40% of the treatment capacity is utilised (ibid; CSD 2012: 17) (Table 6.26).

At present, 10 ULBs (urban local bodies) in BMR are not connected either to an underground drainage system (UGD) or sewerage system. The existing sewerage system is not functioning to a desired capacity. Most of the times, untreated sewerage flows into natural water courses, polluting lakes, nallahs, tanks with silt and sludge (BMRDA 2009: 161; CSD 2012: 17).

6.3.9.1 Environmental Impacts of Waste Water Treatment

1. A report by KSPCB on STPs in Bengaluru (2013) points out that most of the harmful sludge or slurry generated after sewerage treatment is indiscriminately allowed into water bodies that include lakes, tanks and public open drainage or sewerage. In most of the private STPs in apartments (95%), the filter press is not operated at all nor is it commissioned (Rohith 2013).
2. One of the significant observations made by the KSPCB report is that the treatment capacity of private STPs treatment capacity is better with total capacity of 113 mld at 75% designed capacity, while public STPs of BWSSB (14 STPs) function with a total capacity of 721 mld and an operational capacity of only 63.5% (Rohith 2013). Only 302 mld gets treated, which means around 60% of the sewerage water in Bengaluru City remains untreated (Angad 2012).
3. Nitrate from untreated sewerage (400 million litres per day) is the single most polluter of groundwater contamination in Bengaluru. Most of the untreated sewerage (more than 60%) is allowed into storm water drains or gutters and surface water bodies like lakes and tanks which not only causes pollution but also is mostly unhygienic and unhealthy. Particularly, untreated domestic sewerage is the largest source of pollution in the city (Vishwanath 2010).
4. The treated wastewater from Vrishabhavathi STP is discharged into the stream, which is already contaminated. Since the treated wastewater is mixed with the stream water, it gets contaminated again. Around 120 MLD treated sewage is let into Vrishabhavathi stream, thus contaminating the same.
5. The natural stream that carries wastewater from Peenya, Rajajinagar, Kamakshipalya and Mysore roads is not treated; instead it runs through Kengeri before joining to Byramangala tank.

Table 6.26 Key issues in sewerage connection and environmental effects in BMR

Key issues	Environmental effects
Inadequate coverage	Only 40% of the area is covered with sewerage connections
Environmental concerns	Pollution of water bodies – Lakes, tanks and groundwater system Unhygienic surroundings lead to the spread of diseases and various other health-related ailments through mosquitoes
Sewerage entering water bodies like lakes/tanks, etc.	Public health problems – Spreading communicable and non-communicable diseases Loss of ecology and biodiversity of the locality Blocks drainage and storm water drains
Insufficient capacity of sewerage: (both trunks and mains) both primary and secondary	Untreated sewerage enters into river streams, thereby polluting water bodies (like rivers Arakavathy and Vrishabhavathi, etc.) Severely affects the residential surrounding by entering into low-lying areas and subsequently leading to floods Damages sewerage lanes causing floods
Silting up of sewers	Sewerage flow from manholes in residential areas causes floods and health problems including loss of life
Direct connection of sewers from slums and low-lying areas to (primary and secondary drains) storm water drains.	Flooding in slums and low areas Back flows during rainy season when storm water drains are full leading to the spread of various diseases and health problems
Silt, grease and floating debris (plastics, papers, etc.) into open drains and treatment plants	Inefficient treatment plants; untreated sewerage contaminates drinking water besides spreading diseases Free flow of sewerage on streets and manholes causes severe problems to both residents and animals
Encroachments of sewer lines and manholes	Sewage overflows into residential areas (slums, low-lying areas) Sewer cleaning and removing silt are very difficult Health issues due to the spread of mosquitoes and unhygienic conditions

Source: BBMP (2009: 70)

6. Many sewers are in critical condition due to corrosion, requiring immediate replacement. The areas where there is no underground sewerage (UGD) system, let waste water into water bodies like streams and lakes, causing a serious damage to the environment.
7. Effluents discharged into storm water are mostly untreated or poorly treated, adding to environmental pollution. The natural ecology of water bodies, animals and birds consequently gets damaged.

6.3.10 *Drainage*

The city of Bengaluru is located 900 m above the mean sea level. The total Bruhat Bengaluru Mahanagara Palike (BBMP) area covers 710 sq.km (i.e. core city 225 sq.km and peripheral area 485 sq.km). The total watershed or catchment area comes to 1016.50 sq.km. Further, the total length of SWD in BBMP area constitutes 840 sq.km (core area 240 sq.km and peripheral area 600 sq.km). The city's drainage runs parallel to the subparallel formed by natural streams, forming a dendritic pattern (CSD 2012: 18). The drainage extending to a length of 2000 km is at least half a century old, while the total length covers 6000 km of the city (Ray 2013). Although storm water drains of BMP cover 240.40 km under greater Bengaluru, the total area has increased to 869.83 km, covering all the eight zones (CSD 2012: 18). The primary drain length is 415 sq.km, and secondary drain length is 425 sq.km (BWSSB 2014). The bustling population of nearly 9 million of the city is clearly exposed to an inadequate drainage network.

6.3.10.1 Impact of an Inadequate Drainage Network

(i) Most of the garbage dumped by the commercial and industrial establishments reaches into sewer drains or storm water drains, clogging or leading to a blockage, thus exposing the urban residents to its foul smell and odour.
(ii) Often, residents living in the low-lying areas of the city like Church Street, Shivajinagar, Chickpet and Koramangla and other 25 areas are exposed to environmental hazards like frequent floods and bear the consequences which affect their health.

6.3.11 *Solid Waste Management (SWM)*

The city of Bengaluru has grown tremendously both demographically and spatially. The city's population has grown from 56.86 lakh in 2001 to 84 lakh in 2011 which includes the population of newly merged areas of 7 CMCs, 1 TMC and 111 villages. The decadal growth rate of population amounts to as high as 61.36% for 1991–2001 (Ramachandra and Bachamanda n.d.). Consequently, the solid waste generation has increased and is a serious concern for ecology and environment. Solid waste consists of various streams such as municipal solid waste (household, commercial and industrial), biomedical waste (hospital and dispensaries), industrial waste (industries) and electronic waste (electronic goods which include computers and mobiles). The city's spatial expansion beyond the old corporation limits and the inflow of migrant population have increased the generation of waste (Table 6.27).

Table 6.27 Solid waste management in Bengaluru

Bengaluru City	Statistics	Solid waste
Area	800 sq.km	–
Population	78 lakhs	–
Households	25 lakhs	Per capita generation of domestic waste amounts to 350 gram per day 50% of the total generation of waste
Commercial properties	3.5 lakhs	Markets 20% (84 tpd) of total waste Commercial and institutions, 17% (96 tpd) Others, 9%
No of zones	8	–
No of wards	198	Waste generation 3000 tpd Segregation of waste at source, only 10%

Source: Solid Waste Cell (2007), BBMP, Bengaluru

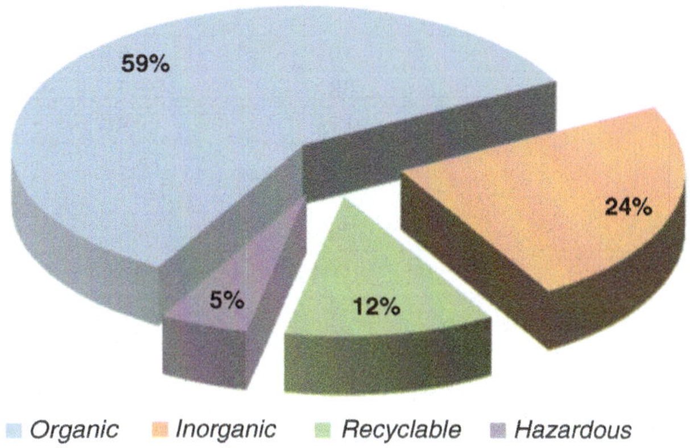

Source: GoK (2009: 311)

Fig. 6.9 Various streams of waste generated in BRM

As presented in Fig. 6.9, inorganic waste generation constitutes 24% followed by 5% of hazardous waste. Thus, nearly 30% of waste generated is not properly collected or recycled.

6.3.11.1 Generation of Municipal Waste

The total amount of MSW has increased from 650 tpd (in 1988) to 1450 tpd (in 2000) and to 2200 tonnes tpd (by 2007), while the per capita generation of municipal solid waste has gone up from 2500 tonnes to 5000 tonnes per day over a span of 10 years. This has increased the per capita waste generation from 0.16 (1988) to 0.58 kg/day (2009) (Varshney 2012; BBMP 2012; Chanakya et al. 2010; Saldanha et al. 2012). There has been a rapid increase in the generation rate from 1450 tpd

Table 6.28 Generation of municipal waste in Bengaluru City

Bengaluru City	Statistics
Area	800 sq.km
Population	78 lakhs
BBMP area and Anekal taluk	93.15%
Bengaluru rural districts	3.14%
Ramanagara	3.67%
Landfill	95.2%

Source: GoK (2009: 247)

Table 6.29 Waste generation and collection efficiency in BMR (Urban and Rural) 2009

Bengaluru Urban	Generated in Mt./d	Collected in Mt./d	Collection efficiency
Bruhat Bengaluru Mahanagara Palike (BBMP)	2374	2300	96.9
Bengaluru rural			
Anekal	10	6	60.0
Doddaballapura	38	21.7	57.1
Devanahalli	10.7	8	74.8
Nelamangala	16.2	15.5	95.7
Ramanagara			
Chanapatna	24	20	83.3
Kanakapura	19	12	63.2
Magadi	12.4	9	72.6
Ramanagara	39	30	76.9
Total	2543.3	2422.2	95.2

Source: GoK (2009: 311)

(in 2000) to 3600 tpd (in 2008–2009) (Chanakya et al. n.d.) and to 4000 tpd (CSD 2012: 19). Nearly 600 tonnes of waste is picked up from KR Market.[49] The generation of municipal waste in BMR is presented below (see Table 6.28).

Details of waste generation and collection efficiency in BMR (both urban and rural) are presented below (see Table 6.29).

Municipal waste consists of waste generated from households, markets, hotels, restaurants, streets, etc. While studies claim that there is no accurate estimation of waste generated by different sources, a study carried out the erstwhile BMP area in 2001 and quantified the waste generated by various sources, as presented in the Table 6.30.

Similarly, since the formation of BBMP with the annexation of additional areas – 7 CMCs and 1 TMC – into the city corporation limits, there is an additional burden on BBMP with respect to handling of MSW. The tables (see Tables 6.31 and 6.32) below present the current scenario with respect to waste generation and collection efficiency.

[49] Interview with Joint Commissioner, Solid Waste Unit, BBMP, Bengaluru, May 20, 2014.

Table 6.30 Composition of municipal solid waste (MSW) generation in core city area (BMP)

Source	Quantity (TPD)
Residences	1562
Markets	84
Hotels and restaurants	96
Total	1742

Source: BBMP (2009: 88)

Table 6.31 MSW generation and collection efficiency in the former CMCs and TMCs (in 2001)

Bengaluru Urban	Generated in Mt./day	Collected in Mt./day (%)
Yelahanka	61	80
Byatarayanapura	75	80
KR Puram	75	80
Bommanahalli	141	80
Dasarahalli	131	80
RR Nagar	38	79
Mahadevapura	81	80
Kengeri	30	80
Total	**632**	**Average: 80**

Source: BBMP (2009: 89–90)

Table 6.32 MSW practices in BBMP area

Areas	MSW practices	Key issues
Collection	Near 100% collection efficiency in the core area	Segregation is practiced only in a few areas
	100% door-to-door collection in residential localities of the core area	
	Private participation in several wards	
Transportation	Private participation in 182 wards of the core area	No transfer stations available
	With cover vehicles being used in most areas	
	Compactors and mechanical sweeping of roads proposed	
Treatment	Existing treatment capacity being 700–800 TPD (biggest plant KCDC -350 TPD)	Treatment capacity inadequate (shortfall of more than 1000 TPD) to treat the entire waste generation
	Compost plants of 1000 TPD and waste-to-energy plants of 1000 TPD being developed with private participation on BOT basis	
Disposal	Engineered sanitary landfills being developed with private participation on BOT basis	Expected to be operational only by 2008
		Waste currently dumped on roadsides and low-lying areas

Source: BBMP (2009: 88–89)

6.3.11.2 Municipal Waste Collection

Administrative wards are further divided into 294 health wards for the proper management of sanitation facilities. BMP has introduced PPP (public-private partnership) services for door-to-door collection of household-generated garbage. While 98 wards are assigned to private agencies on a contract basis, the remaining 112 wards are managed by BMP. Nearly 98% of the BMP wards are covered under door-to-door garbage collection system, while 70–90% in major cities of India (Ramachandra and Bachamanda n.d.: 38). The collection of municipal waste through community bins is most common in low-income and slum settlements. Street sweeping/cleaning method also constitutes one of the waste collection methods, and many local bodies allocate 30–50% of solid waste budget to street cleaning/sweeping (ibid). The BBMP has 25 collection centres in the city.

6.3.11.3 Municipal Waste Transport

At present, transfer stations are not available, but large-scale trucks are used to carry loads of garbage to dumping sites. The city's (former BMP area) 70% of the waste generated is transferred to dumping sites and private sector manure companies (Ramachandra and Bachamanda n.d.; Abide 2010: 83). Most of the trucks are push-carts (No. 11000), open trucks (No. 600), tractor trailers, tipper trucks (No. 650) and dumper placers; the city of Bengaluru has about 13 dumper placers (*ibid* 40).

6.3.11.4 Municipal Waste Disposal Methods

Most of the waste generated and transported does not reach dumping sites. Many times, truckloads of garbage spill onto the streets, and informal ragpickers[50] collect waste feeding intermediaries who, in turn, recycle these waste materials and use as raw materials (Ramanchandra and Bachamanda n.d.: 40; Chanakya et al. n.d.: 2). Large portion of the total urban solid waste is dumped on 60 shifting open sites, posing serious environmental problems to the city including the (Chanakya et al. n.d.: 2), newly added regions (*ibid*). Besides, the city has 28 assembly segments consisting of wet-processing units. The total generation of waste comes to around 4000 tonnes per day, while only 1000 tonnes of waste is treated. Nearly 1500 ragpickers contribute significantly towards segregation of waste to the extent of about 100 tonnes.[51]

---

[50] Ragpickers play a significant part in the economy of recycling process in India (Ramachandra and Bachamanda n.d.: 40).

[51] Data collected from Solid Waste Unit, BBMP, Bengaluru.

Table 6.33 Key issues and environmental effects of SWM in BMR

Key issues	Environmental effects
Inadequate awareness	Alhough household segregation has been made compulsory in BMR, there is a poor compliance due to lack of awareness Such large quantities of non-biodegradable waste poses severe threats to health
Available transport vehicles are poorly managed and equipped	Spilling of domestic waste on streets and surroundings leads to unhygienic conditions, posing threats to both humans and animals
Absence of adequate landfills for treatment	Unabated dumping of waste on roads and streets, especially low-lying areas, poses severe environmental threats
Poor regulation mechanisms and policies dealing with waste reuse and recycling	Illegal dumping of waste in the sewers and water bodies like lakes and tanks threatens local biodiversity and ecology

Source: BBMP (2009: 91)

6.3.11.5 Some of the Key Issues in Municipal Waste Management

(i) Although the core city area (former BMP area consisting of 100 wards) boasts of 100% door-to-door garbage/waste collection efficiency, both conurbation and green belt areas do not have waste collection facilities.

(ii) Waste generation amounts to about 3056 tonnes per day in BMR, of which only 69% is collected (Abide 2010: 83). The existing treatment capacity is only 400 MT/day, while the total waste generated comes to about 2300 MT/day (as of 2011) (Ramachandra and Bachamanda n.d.), whereas the city has treatment and disposal facilities only for 2000 and 1600 MT, respectively (BDA 2007: 45; GoK 2009: 312). Thus, 60% of the waste generated in the city ends up on unknown sites.

(iii) Although eight landfill sites have been identified (GoK 2009: 312), none of them are available for treatment and management. Therefore, there is an insufficient land available for treatment and management of waste generated in BMR. Mandur organic firm collects 150–200 tonnes of municipal waste, while RAMKEY firm is shut in Malavalli dumping site (Table 6.33).

(iv) Most of the lakes, tanks or public/open drains remain clogged and polluted due to seepage and dumping of untreated waste.

(v) Although waste segregation has been made mandatory for households (since 2011), it is not effectively practised due to lack of awareness. Hence, the waste reaching landfill sites is not segregated, leading to various problems. Trucks carrying waste do not have separate sections for segregated waste resulting in the mixing of waste. Most of the trucks carrying waste are overloaded, scattering waste on the way spreading foul odour and leakage on the streets.

(vi) Mixing of plastic waste with municipal waste produces harmful effects on the environment.
(vii) Nearly 17% of the total slum households in the city do not have access to door-to-door waste collection facilities. Often, the poor households dump their waste on streets and adjacent vacant lands. As a result, most of the slum households are exposed to serious health problems.
(viii) Conventional methods of burning waste – both paper and plastic – at street corners cause air pollution besides increasing the danger posed to the environment with emissions of smoke and toxic matters.
(ix) Most of the ragpickers suffer from respiratory problems. A study conducted in 2003 shows that 75% of ragpickers suffer from upper and lower respiratory symptoms (Bhattacharya 2005; Ramachandra and Bachamanda n.d.: 40).
(x) Dumping of waste on streets and road sides has led to many public health issues. For instance, 313 dengue cases have been reported in Bengaluru between January 1, 2014 and June 2, 2014 with 82 cases in May alone (Kamath 2014).

6.3.12 Hazardous and Toxic Waste Management

Household-generated hazardous waste consists of four subcategories: (i) toxic waste like paints, cleaning agents, solvents, insecticides and their containers; (ii) biowaste like used syringes, expired medicines, thermometers, used cosmetics, etc.; (iii) e-waste consisting of waste from electronic goods like computers, mobiles, etc.; and (iv) plastic waste. The city of Bengaluru, the main hub of 'Asia's IT/BT, rapid industrialisation and large-scale infrastructure projects, is choking due to the production of 64,379 million tonnes of hazardous waste by 1702 industries every year. It is revealed that Bengaluru generates about 300 barrels of waste oil which does not reach the refineries or recycling plants (Ray 2011). Nearly 40–70% of the waste constitutes the city's biomedical waste (BBMP 2009: 90). As per the Hazardous Waste Management Rules, 2003, an amended version of the Hazardous Waste (Management and Handling) Rules, 2000, the management of hazardous waste at present is the combined responsibility of BBMP and KSPCB.

6.3.12.1 Hazardous Waste Generation

The total hazardous waste generated in BMR, by a total of 1042 industries, amounts to newly 23,756 metric tonnes per annum (MT/a). Further incinerable, recyclable and landfill wastes constitute 2811 MT/a (11.8%), 11,135 MT/a (46.9%) and 9810 MT/a (41.3%), respectively. The total estimated number of industries in the city comes to 70,000 of which approximately 30,000 generate hazardous waste (GoK 2009: 239) (Table 6.34).

Approximately 147 industries have been identified as hazardous waste-generating industries in Peenya industrial area alone. Bengaluru South constitutes the highest

Table 6.34 Hazardous waste generation in BMR

Bengaluru City	Statistics
Area	800 sq.km
Industries	70,000
Bengaluru City	13.2%
Bengaluru east	12.4%
Bengaluru west	10.8%
Bengaluru north	11.1%
Bengaluru south	47.5%
Bengaluru south II	3125 MT/a
Peenya industrial area	5.1%

Source: GoK (2009: 239)

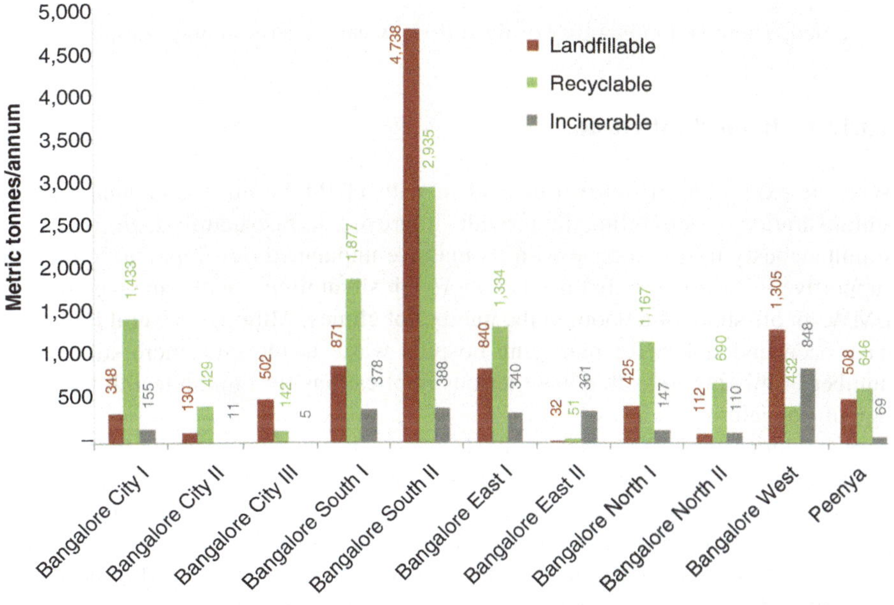

Source: GoK (2009: 239)

Fig. 6.10 Hazardous waste generation (in metric tonnes) per annum

hazardous waste-generating area. Similarly, 30 illegal dumping sites were identified during 2003–2004 (Fig. 6.10).

6.3.12.2 Environmental Implications of Hazardous Waste

1. Most of the hazardous waste gets mixed up with municipal waste. Disposal is a major problem, hence causing serious health and environmental problems in the city. Household level segregation does not happen.

2. Due to a rapid industrialisation of the periphery and green belt area, there is an increase in the generation of hazardous waste around the city which lacks effective monitoring and capacity to recycle or reuse it. Most of the newly emerging industries, particularly small-scale ones, do not follow environmentally norms and standards.
3. Karnataka State Pollution Control Board (KSPCB) is in charge of monitoring and effective implementation of rules and regulations with respect to hazardous waste generation and illegal dumping by industries. But with KSPCB, there is no comprehensive data or records maintained regarding hazardous waste generated stored. Besides, most of the illegal dumping of hazardous waste reaches lakes, tanks and vacant sites.
4. Hazardous waste carries toxic pollutants contaminating water bodies and wetlands which include lakes, tanks and groundwater resources due to illegal dumping without any strict vigilance exercised by the local authorities concerned. There is no inventory of hazardous waste generated and dumped.

6.3.12.3 Biomedical Waste

With the expansion of tourism in tandem with IT/BT boom, the Bengaluru City administration is struggling to provide improved infrastructure facilities while simultaneously trying to cope with its massive unplanned development. One such supportive infrastructure that needs a thorough streamlining is the management of BMW, an off-shoot of a boom in the number of clinics. Although several initiatives have been undertaken for managing hospital waste in the city, there still exist a number of missing links that have serious implications for human health and ecology of Bengaluru.

Bengaluru is one of the top medical tourism destinations in India with professional experts, technological sophistication and health-care services that easily match the best in the world. Its reputation as a global technology hub and cosmopolitan city has made it a prominent health-care destination for foreigners from developed as well as developing countries. Medical tourism is slated to become a 2.3 billion dollar industry next only to IT and BPO (The Times of India, Medical tourism, 28 Feb, 2007).

6.3.12.4 Generation of Biomedical Waste

The state of Karnataka generates the highest quantity of biomedical wast generated at least e,[52] amounting to 62,241 kg/day (as per the MoEF figures for 2011). In BMR, biowaste generation amounts to 193.5 g/day/bed, which is 25 MT per day

[52] BMW (infectious waste) is defined as a solid waste which is generated during diagnosis, testing, treatment, research and production of biological products for humans and animals. It includes needles, syringes, laboratory samples, cultures, live vaccines, bodily fluids and so on.

Table 6.35 Generation of biowaste in BMR up to 2009

Category of hazardous waste in BMR	Solid waste in Mt./a	Liquid waste Mt./a
Huge	432.8	8800
Extra large	173.4	2360
Large	167.3	1900
Medium	98.6	570
Small	95.4	660
Very small	399.2	1250
Total	1366.7	15,540

Source: GoK (2009: 244); Note: BMR: Bengaluru Metropolitan Region

Table 6.36 Generation of biomedical waste in Bengaluru

Institution type	No of institutions	Number of beds	Waste generated (kg/day)
Major hospitals (500 and above)	12	7533	3766
Major hospitals (200–499 beds)	15	4868	2434
Less than 200 beds	608	5849	2924
Non-bed health-care establishments (such as clinics, labs, blood banks, dispensaries, medical centres)	683	0	100
Total	**1318**	**18,250**	**9224**

Source: BBMP (2009: 90)

(Kehle et al. 2008: 92; GoK 2009: 244). Table 6.35 provides details of solid waste and liquid waste generated in BMR. Till 2009, the city generated at least 1350 tonnes of biomedical waste per year (GoK 2009: 156).

A study by TERI indicates that the generation of BMW in Bengaluru accounts for 0.5% of the total waste generated. Medical waste from households constitutes 0.5% of the total waste generated (Manasi, Umamani and Latha 2014) (Table 6.36).

While trying to understand the status of BMW, studies have quoted varied figures on the quantity of waste generated. A study by Pruthvish et al. (1997) indicates that Bengaluru generates 132,500 kg of health-care waste per day, while the health-care facilities generate 5100 kgs daily. Government hospitals generate far less BMW than corporate hospitals. In government hospitals, such as Victoria, it is around 0.5–0.8 kg per bed per day, whereas in corporate hospitals, it is 1–1.5 kg. Currently, Bengaluru generates 40 tonnes of biomedical waste daily. The waste meant for incinerators amounts to a mere 2%, while the infectious waste meant for autoclaving amounts to about 15%. General garbage makes up for over 75% of the waste (Manasi, Umamani and Latha 2014).

6.3.12.5 Environmental Impact of Unregulated Biowaste Generation

1. Most of the biowaste is dumped or untreated or unregulated, leading to serious environmental problems. The problem of mixing both solid and liquid biowastes compounds the environmental problem. Another alarming factor is that only two-thirds of the waste, i.e. 14 tonnes out of the total medical waste generated and 40 tonnes in Bengaluru every day, is treated scientifically by the recycling units. The rest gets mixed up with municipal waste. Besides, these units use only 60–70% of their capacity (BBMP 2009; GoK 2009).
2. The generation capacity of biowaste has increased manyfold with the emergence of unregulated industries, clinics and private hospitals in the BMR.
3. Instances of irregularity in the collection observed at times add to the spreading of diseases. The waste is not collected regularly by eco-industries, which lead to the spread of infections. Dental clinics are a major source of infection because of the waste generated related to the oral treatment. Therefore, it is necessary to collect the waste daily from clinics too.
4. Most of the unregulated industries and clinics do not follow norms and standards of biowaste as they are not registered with KSPCB. In the absence of a strict vigilance, there is every responsibility of biowaste being dumped into water bodies and vacant sites.
5. Poor segregation methods due to the lack of awareness, followed by private clinics, significantly impact the environment due to dumping of hazardous materials.
6. Many industries (large/medium) and hospitals do not have proper treatment plants (Manasi, Umamani and Latha 2014). They usually resort to illegal dumping of untreated biowaste into water bodies or public (open/underground) drainage/sewerage systems, inevitably leading to their clogging besides spreading to the surrounding areas and the environment. Instances of illegal dumping in Bengaluru have drawn the attention of the media several times. For instance, hospital waste found dumped near Avalahalli, off Kanakapura road, Mavallipura, causing serious health and environmental hazards, has been reported by the media.
7. Particularly, with regard to the hospital waste, a poor management has led to the spread of infectious diseases like Hepatitis B and C from HIV/AIDS (Manasi, Umamani and Latha 2014).
8. Occupational hazards are clearly associated with chemicals, drugs and unauthorised repacking and sale of disposable items and expired drugs.

6.3.13 Plastic Waste

The use of plastic is considered inevitable in our daily routine. Plastic is used excessively for domestic, commercial and industrial purposes. Stringent measures have been introduced to prevent the use of plastic. For instance, the Recycled Plastics Manufacture and Usage Rules, 1999, later amended in 2003, restrict the usage of plastic carry bags and containers with a thickness of 20 microns and less.

6.3.13.1 Generation of Plastic Waste

Out of 5000 tonnes of waste generated every day, 3000 tonnes constitute dry waste including 15–20 tonnes of plastic (Balakrishna 2013). Another source (IBNLIVE.com n.d) points out that the city generates 9000 tonnes of plastic waste every day. There are about 118 plastic manufacturing units in BMR. In addition, the city generates about 150 tonnes of lightweight plastic every day (of less than 20 microns) (Anil 2013).

6.3.13.2 Collection and Transport of Plastic Waste

Despite regulations, the usage of plastic is causing enormous ecological and environmental problems. Varied types of plastic waste[53] are formed due to its excessive usage. (i) Foremost, plastic waste is dumped in garbage on streets, open water bodies like lakes, tanks and sewerage networks causing blockage and clogging, spread of foul smell and thereby harm to health; (ii) households or waste pickers pool up the plastic waste and burn them causing serious toxic and furan emissions into the environment; (iii) many times plastic which is accessible on dump yards, in garbage bins and on streets amounts to fodder for cows, street dogs and other animals which might cause serious consequences; and (iv) plastic poses a serious threat by decreasing permeability and air circulation making composting plants highly sluggish (GoK 2009: 252). There is a lack of separate disposal method practices for plastic waste.

6.3.13.3 Serious Concerns for the Environment Due to Plastic Waste

1. Plastic waste is non-biodegradable and hence poses a grave threat to the environment.
2. Excessive usage of plastic results in dumping on streets and water bodies causing colossal damage to the environment. It is estimated that 0.25 million tonnes of carbon dioxide is released into the atmosphere (Nayak 2013).
3. Sporadic burning of plastic waste on streets and in other usage leads to emission of toxic fumes into environment.
4. There are rampant illegal manufacturers of plastic without following norms reach markets for commercial purpose.

[53] The Bengaluru Mahanagara Palike will use plastic waste to asphalt over 40% of the roads under a World Bank scheme (Yasmeen 2005).

6.3.14 E-Waste

e-Waste is one of the latest and fastest growing waste streams in the world. According to Greenpeace (international environmental group), in recent years, 20–50 million tonnes of e-waste is generated each year. e-Waste management is one of the key areas of environment in urban areas. Both formal and informal sectors are involved in the generation of e-waste. Of the top ten e-waste generator cities, Mumbai ranks first followed by Delhi, Bengaluru, Chennai, Kolkota, Ahmedabad, Hyderabad, Pune, Surat and Nagpur (MoEF 2009). According to the MAIT-GTZ e-waste assessment study, based on the data by hardware manufacturers, the annual generation of e-waste in India works out to 400,000 tonnes with 19,000 tonnes being recycled (Business Standard 2012). For the regulation of e-waste, E-waste (Management and Handling) Rules 2011 was promulgated, making the producers liable for recycling and reducing e-waste in the country. According to the ASSOCHAM report (2013), electronics and durables sector in India was valued at 340 billion in 2012, and in the same year, India produced over 8 lakh MT of e-waste (Chhakchhuak and Jain 2014).

Increased e-waste, one of the recent outcomes of the IT boom, is seen as a major threat to the already deteriorating environment in Bengaluru. With little awareness among the majority about the magnitude of the problem, e-waste has been accumulating almost unhindered as one of the most serious management challenges in the recent times. Experts have cautioned against the potentially harmful impacts of e-waste on the environment and human health, hence the need for its safe disposal. Home to more than 1200 foreign and domestic technology firms, Bengaluru figures prominently in the danger list of cities faced with e-waste hazards (Habib Beary 2005). As IT firms continue to inundate India's technology hub, Bengaluru City is beginning to choke, thanks to e-waste generated.

6.3.14.1 Generation of E-Waste

The city of Bengaluru is the main IT/BT hub of Asia, consisting of 928 IT companies which rose to 1850 companies in 2005 (GoK 2009: 254). The city generated as high as 8000 TPA in 2009. Bengaluru produces 200,000 tonnes of e-waste a year (businesstoday.intoday.in) Manasi (2013).[54]

There are several estimates with regard to the quantity of e-waste generated in Bengaluru. Discussions with PCB officials indicate that 13,000 tonnes of e-waste was generated in 2009, which excluded household appliances. As estimated by E-Parisara, Bengaluru generates 12,000 tonnes per year and India 330,000 tonnes per year with another 50,000 tonnes being illegally imported. The secondary market for old PCs accounts for 40 tonnes per hour, while it is 4000 tonnes per hour for the

[54] Visit the article titled 'Cleaning E-wasteland: Cerebra will extract precious and other metals from mountains of e-waste (http://businesstoday.intoday.in/story/companies-that-are-making-wealth-from-waste/1/195163.html).

Table 6.37 Generation of e-waste in Indian metropolitan cities

Metropolitan cities in India	E-waste generation (in tonnes)
Mumbai	61,500
New Delhi	43,000
Bengaluru	37,000
Chennai	30,700
Kolkata	23,000
Hyderabad	16,000

Source: GoK (2009: 254)

whole world. Manufacturers and assemblers generate about 1800 tonnes of electronic scrap every year. According to another estimate, about 1000 tonnes of plastics, the same equivalent of iron, 300 tonnes of lead, 0.23 tonnes of mercury, 43 tonnes of nickel and 350 tonnes of copper are generated as e-waste in Bengaluru, and this figure might increase tenfold by 2020 with the city generating one-third of the State's e-waste (Manasi 2013). The data from State of Environment Report 2009 shows Bengaluru as the third highest generator of e-waste (see Table 6.37).

The ASSOCHAM report (2013) points out that e-waste[55] comprises steel (50%), plastics (21%), non-ferrous metals (13%) and other constituents (16%) (Deccan Herald 2013). Both Public and private industries generate more than 70% of e-waste, while households contribute 15%. Household materials like television sets, refrigerators and washing machines are some of the major e-waste contributors, amounting to 20%, while mobiles contribute 2% (The Hindu 2013a). On the whole, the city generates one-third of the state's e-waste.

6.3.14.2 Environmental Impacts of E-Waste

1. Although the city produces 37,000 tonnes of e-waste, only 12.5% is recycled. There are only two recycling plants; however, there are many unregistered backyard recyclers (nearly 70% is handled by the unorganised sector and only 30% by the organised sector firms/units registered with the Pollution Control Board) using crude methods, thereby exposing workers to health hazards (GoK 2009: 254; Deccan Herald 2013).
2. In the absence of properly regulated recycling facilities, e-waste generation is expected to grow by 2110 lakh MT.
3. Sound practices of e-waste recycling are a major area of concern in the absence of an effective enforcement of rules and regulations. Illegal dumping and recycling methods are key environmental problems.
4. Chemicals released by e-waste such as beryllium which is found in motherboards and cadmium in chip resistors and semiconductors are toxic and can lead

[55] The ASSOCHAM report (2013) states that Indian organised e-waste market for 2012 was from the order of Rs. 1 billion to Rs. 2 billion. There are around 72 formal e-waste recycling units in India with a recycling capacity of 1.75 lakh tones per annum (Deccan Herald 2013).

to cancer. Similarly, chromium in floppy disks, batteries and computer monitors and mercury in alkaline batteries and fluorescent lamps can pose server health threats (BBMP 2009: 90–91).

6.3.14.3 Disposal and Treatment of e-Waste

The Pollution Control Board, in association with GTZ, a German Engineering Firm, has set up an e-waste disposal landfill site, with an investment of Rs. 55 crore, at Dobbaspet about 50 km off from Bengaluru near Tumkur road. The landfill has been operating since February 2009, despite encountering opposition from environmental activists, local residents and politicians, fearing hazardous impacts on health and environment. The landfill, known as 'treatment, storage and disposal facility' (TSDF), is spread over 93 acres with a potential storage capacity of 40,000 tonnes of industrial and biomedical waste every year and an actual storage capacity of 8,00,000 tonnes – enough to last for two decades. Industries registered with Karnataka (more than 2000) have to pay between Rs.1500 and Rs. 2500 for each tonne of waste for using TSDF. More than 8000 tonnes of hazardous waste is sent to the site, and companies like BHEL, Volvo and Toyota have used this facility (Manasi 2013). KSPCB has authorised a few agencies for e-waste processing and recycling. However, a large portion of the e-waste is handled by the informal sector (CSD 2012: 20).

6.3.14.4 Environmental Impacts of Hazardous Waste (HW)

As per the HAWA survey (2004), there are 25 illegal HW dumping sites in BMR, three of which are more than 20 acres across used for dumping mixtures of paint sludge, ETP sludge, spent solvents, used oil and incineration ash, affecting the local environment and communities, particularly women and children.

6.3.15 Industrial Waste

An estimated total of 70,000 industries operate in the Bengaluru Metropolitan Region (BMR) and 162,000 micro-, small and medium enterprises are operating in BMR. There are about 2100 IT-based and 200 BT- and 245 IT-enabled service industries located in BMR. Of 130 highly polluting industries in Karnataka (as of 2007), 21 are located in BMR. The environmental impact of 2345 IT/BT industries is yet to be established. As per KSPCB classification, nearly 40% of the industries classified as red and orange constitute together 2900 industries that are considered as major polluters in BMR.

6.3.15.1 Environmental Impact of Industrial Waste

1. Water and sewerage facilities for industries under the former BMP area are taken care of by BWSSB. While only 20% of the industries are supplied by majority of industries, those in the periphery and green belt area of BMR rely on groundwater (that amounts to 509 mld) and dump the industrial waste into running drains and sewers. An estimated 50% of industries depend on borewells, and nearly 30–40% of industries outside the BMR dump their waste into running sewerage, soak pits, storm water drains or open lands. Sewerage seeping from soak pits contaminates the environment. Organic matter discharged severely impacts biodiversity. An estimated 5–15 mld of industrial waste generation contaminates the soil environment. Exposure to industrial effluents degrades land, thereby affecting fertility.
2. At present, there is lack of infrastructure for the collection of industrial waste. Many unauthorised industrial dumping sites contaminating groundwater have sprung up, causing environmental damage. For example, eight unauthorised dumping sites are present in Peenya industrial area alone.
3. Many private industrial areas are located along open drains with serious environmental implications.
4. Industries constitute major polluters emitting toxic chemicals, lubricants, hazardous compounds, oil and grease, dyes, suspended solids and non-biodegradable matter into the environment. Industrial sewerage generation amounts to 56 million litres per day (MLD). The presence of heavy chemicals and solids affect plant metabolism and reduce the growth in urban areas.
5. High levels of chemical compounds threaten the extinction of certain animal and bird species.
6. Untreated waste water from industries causes water-borne diseases like diarrhoea, affecting the central nervous system.
7. Industrial greenhouse gas emissions from fuels and grid-supplied electricity in BMR amount to 2 million tonnes per year.

6.3.16 Slaughter House Waste

Slaughter[56] house-generated waste has created major environmental problems in the city. There are three main organised slaughtering houses in Bengaluru, which are meant for cattle, sheep, goats and pigs, while there are no organised shops found for poultry. More waste is produced while slaughtering cattle, buffaloes, sheep and goats because of stomach and intestinal waste, more a roughage. India has the world's largest population of livestock. According to the Ministry of Food Processing, a total of 3616 slaughter houses slaughter over 2 million cattle and

[56] The slaughter houses come under the purview of the animal husbandry division of the Ministry of Agriculture mainly for the purpose of funding the expansion and modernisation activities (www.urbanindia.nic.in).

buffaloes, 50 million sheep and goats, 1.5 million pigs and 150 million poultry annually, for domestic consumption as well as export purposes. The waste generated here is liquid and solid in nature. Slaughtering of animals generates waste consisting of nonedible organs, stomach contents, dung, bones and sludge from waste water treatment. The Central Pollution Control Board brought out 'Draft Guidelines for Sanitation in Slaughter Houses' in August 1998 (www.cpreec.org).

The state of Karnataka produced 81,057 metric tonnes of meat in 2001 with the city of Bengaluru contributing 18% of the total slaughter waste (14,277 metric tonnes). The regulation of slaughter waste comes under the purview of Municipal Waste (Handling and Management) Rules 2000 (GoK 2003: 136; Kehle et al. 2008: 94). Slaughter waste is similar to biowaste in terms of spreading infections and generating biohazards.

6.3.16.1 Environmental Impacts of Slaughter Waste

1. Slaughter waste mainly consists of hide, hair, undigested and digested food particles from food, bones and meat. The waste is highly putrescible and could have pathogens which cause zoonoses (Chandrappa 2012: 49). The leftovers of carcass of slaughter houses are feeds for street dogs which tend to get infected with *E. coli* and salmonella, while the surroundings of slaughter houses generate foul biogas (Times of India, October 2, 2010[57]).
2. The internal parts of animals like cow, goats etc. are sold at cheaper prices which can lead to health complications, and animals feeding on them can cause serious infections.
3. The street dog menace in Bengaluru City is attributed to feeding on the waste generated by slaughter houses. These dogs are a potential threat to the community and passers-by, especially at night (dnaindia.com).[58]
4. Slaughter houses in the city do not adhere to environmental regulations or norms. The city lacks effluent treatment plants for slaughter houses. The effluents usually consist of blood, animal fat, intestinal parts and pieces of inedible parts.
5. Usually slaughter houses due to their unregulated nature suffer from very poor hygiene standards and dispose of waste and highly polluted effluents, thus posing major health and environmental hazards.
6. Unauthorised and illicit slaughtering of animals has also increased manifold with the related problems threatening the environment. The illegal activities of the meat industry do not meet the standards for discharge of effluents as laid down and notified under the Environment (Protection) Act, 1986.

---

[57] Visit Indiatimes.com for an article titled 'BBP needs slaughterhouse', October 2, 2010. (http://articles.timesofindia.indiatimes.com/2010-10-02/bangalore/28269036_1_slaughterhouse-salmonella-intensive-care).

[58] Refer the article titled 'Slaughterhouse waste posing health hazard', August 28, 2014. Access the link www.dnaindia.com

7. The waste generated by slaughter houses and packaging houses is similar in terms of chemical concentrations. The principal deleterious effect of this waste on streams and water bodies is deoxygenation.
8. Most of the times, the animal waste from slaughter houses gets mixed up with domestic waste, posing serious threats to the environment. Internal waste and body fluids of the animals mixed with pond water and groundwater can adversely affect human health in terms of causing skin disorders.

6.3.17 *Urban Land Utilisation/Use*

The city of Bengaluru has grown spatially from 28.85 sq. km in 1901 to 151 sq. km in 1981[59] and to 530.85 sq.km with an area of 561 sq.km in the year 2001 (Ravindra 1996: 57; BBMP 2006: 15). The Urban Agglomeration of 1981 (which included 17 towns and 79 outgrowths) covered an area of 365.65 sq. km[60] (Ravindra 1996: 57). The total geographical area of BMR constitutes 8022 km^2 (GoK 2009: 222). For an orderly growth of the city, zone delineation and the permissible land uses within the zone and the respective regulations for land use are determined by the Local Planning Area (LPA), declared under the Karnataka Town and Country Planning Act, 1961 (BDA 2010).

At present, the Bengaluru Urban Agglomeration (BUA) has grown from 226 to 800 sq. kms incorporating 7 CMCs, 1 TMC and 111 villages. The rapid urbanisation process and the resultant urban sprawl have detrimental effects on the environment and ecology surrounding the city. A study of Ramachandra and Mujumdar (2009) clearly indicates a 466% increase in the built-up area from 1973 to 2007. Further, the study points out a sharp decline of 61% in the area covering water bodies due to encroachment (54%), 66% of the lakes being sewerage fed, 14% encroached upon by slums and 72% of catchment area loss. Meanwhile, the vegetation cover decreased by 32% from 1973 to 1992, by 38% from 1992 to 2002 and by 63% from 2002 to 2007.

According to the BMR Structure Plan (2011), there are five area planning zones (APZs)[61] and six interspatial zones (IZs) for spatialising the region's land and water resource management. Table 6.38 shows the categories of land utilisation, the area covered by each for 2000 and 2008, the percentage of each category vis-a-vis the

[59] The entire Bengaluru metropolitan region has been classified into five zones, namely (i) core area, (ii) pericentral area, (iii) recent extensions, (iv) new layouts and (v) green belt and agriculture areas (BBMP 2006: 16).

[60] According to the plan, metropolitan area drawn for 1984 covered an area of 1279 sq.km that included the conurbation area of 440 sq.km (Ravindra 1996: 57).

[61] The APZs and IZs serve as simple tools for land planning and maximising the utilisation of the existing and available infrastructure and also for minimising the need for conversion of agriculture land to urbanisation (BMRDA 2009: 116).

Table 6.38 Land utilisation changes between 2000 and 2008

Land utilisation categories	2000		2008		% of Change
	Area in sq. km	As % of total BMR	Area in sq. km	As % of total BMR	
Agriculture land	5323	66.49	5288	66.06	−0.65
Built-up land	723	9.03	920	11.49	27.18
Forest	830	10.36	831	10.38	0.15
Grassland/grazing land	22	0.27	5	0.07	−74.80
Waste lands	690	8.61	599	7.48	−13.14
Water bodies and wetlands	340	4.25	287	3.58	−15.62
Others	79	0.99	76	0.94	−4.47
Total	8006	100.00	8006	100.00	

Source: BMRDA (2009: 117)

total area of BMR and the percentage change of land utilisation between 2000 and 2008 for each category.

An analysis of the land utilisation changes based on the Table 6.38 clearly shows a decrease in the agricultural lands by 0.65% for the period between 2000 and 2008 (amounting to 34 sq.km) and a lesser decrease in grasslands/grazing lands by 74.80% (16 sq.km). Similarly, water bodies have reduced by 15.62%, i.e. 53 sq.km, and the wetlands by about 13.14% (91 sq.km). But the built-up area has increased by 27.18% (i.e. 197 sq.km).

Bruhat Bengaluru Mahanagara Palike (BBMP) was formed in 2007 by amalgamating the erstwhile Bengaluru Mahanagara Palike (BMP), the surrounding 8 smaller urban local bodies and 111 villages. At present, BBMP spreads over an area of 800 sq.km (see Table 6.39) consisting of 198 wards.

The city of Bengaluru is divided into eight zones consisting of three zones from BMP and five zones from newly formed areas of BBMP: (i) the core area consisting of the traditional business areas, the administrative centre and the central business district; (ii) the pericentral area with older, planned residential areas, surrounding the core area; (iii) the recent extensions of the city (past 5–7 years) flanking both sides of the Outer Ring Road; (iv) the new layouts that have developed in the periphery of the city, with some vacant lots and agricultural lands; and (v) the green belt and agricultural areas on the city's outskirts including small villages. Table 6.40 shows the land use pattern in BDA (Bengaluru Development Authority).

The city has been growing in all available directions along with major corridors. The projected land use pattern for 2015 has been assessed in the Master Plan 2015 of BDA. It is clear from Table 6.41 that the proposed urbanised area amounts to only 300 hectares (23%), while the actual urbanised area covers 512 hectares (39%).

The CDP (City Development Plan) and BDA have come up with land use projections for 2007–2015 (see Table 6.42).

Table 6.39 Bengaluru at a glance

Parameters	BMP	Bommanahalli	Byatarayanapura	Dasarahalli	KR Puram	Mahadevapura	RR Nagar	Yelahanka	Kengeri	Total	110 Villages
Area sq.km	226.2	43.6	47.0	38.0	21.3	46.2	66.0	38.8	34.0	561.0	239
Population (2001)	4,303,033	243,870	210,007	309,956	198,991	163,486	111,553	99,993	44,995	5,685,884	304,855
Gender ratio	915	867	908	844	911	866	879	863	936	906	NA
Literacy levels (%)	86	81	82	96	87	83	78	84	84	86	NA
Number of households	1,225,307	65,885	52,813	91,071	50,186	44,927	30,073	28,953	14,319	1,603,534	NA
Developed properties	521,939	41,371	25,800	26,759	44,824	18,186	16,715	24,110	6009	725,713	NA
Vacant land (ha)	57,993	37,193	10,167	20,689	15,176	24,575	27,300	7555	5115	205,763	NA

Source: BBMP (2009: 12)

Table 6.40 Land use pattern in BDA (up to 2009)

Land use category	Area in hectares	% Use
Residential	16,042	14.95
Commercial	1708	1.59
Industrial	5746	5.36
Parks and open space	1635	1.52
Public and semipublic	4641	4.33
Transportation	9014	8.40
Public utility	192	0.18
Water sheet	4066	3.79
Agriculture	64,243	59.88
Total	107,287	100

Source: BBMP (2009: 15)

Table 6.41 Projected land needs in the Bengaluru Metropolitan Area

Land use category	Area in hectares	% Use
Existing urbanised area	512	39
Proposed area to be urbanised as	300	23
Housing	135	
Hi-tech development	25	
Other industries	15	
Logistics	13	
Large-scale facilities	24	
Office spaces	2.5	
Other facilities	85.5	
Inside peripheral road	270	–
Outside peripheral road	30	–
Green belt	270	20.7
Inside peripheral road	40	–
Outside peripheral road	402	–
Agricultural land	174	13.5
BMICPA	50	3.8
Total	1306	100

Source: BBMP (2009: 51), (sourced from BDA Master Plan 2015)

Table 6.42 Land use projections for 2007–2015, BDA

Master plan	ODP	CDP	CDP (R1)	MP (R2)
Plan period	*1961–1976*	*1984–2000*	*2000–2011*	*2005–2015*
Projected population (lakhs)	19	70	70	88
Local planning area (LPA) (Sq.km)	500	1279	1279	1306
Conurbation (sq.km)	264	440	564	812
Green belt (sq.km)	236	866	742	494
Existing land use (Sq.Km)	142	202	284	420
Existing density (PPH)	116	144	143	138
Proposed density (PPH)	72	159	125	108

Source: CSD (2008: 50)

Table 6.43 Land use changes over time (1991–2011)

Land use category	Area in Km^2		Percentage	
	1991	2011	1991	2011
Residential	98.88	243.7	34.8	43.2
Commercial	6.8	16.4	2.4	2.9
Industrial	20.4	38	7.2	6.8
Parks and open space	21.3	77.9	7.5	13.8
Public and semipublic space	26.2	49.1	9.2	20.7
Transportation	21.4	117.0	31.5	20.7
Unclassified	89.5	22.1	7.5	3.9
Total	**284.0**	**564.7**	**100**	**100**

Source: GoK (2009: 102)

As per the State Environment Report of Bengaluru (2009), urban land use has changed drastically due to rapid urbanisation and growth of urban sprawl. Table 6.43 presents the land use changes over the years from 1991 to 2011.

The urban land has been put to intense economic activities, increasing from 284 km^2 in 1990 to 740 km^2 in 2007. Consequently, while the built-up area has increased substantially, the wetlands and water bodies have reduced. There has been a substantial increase in the residential areas from 99 km^2 (35%) in 1991 to 244 km^2 (43%) in 2011. Both industrial and commercial area use have increased from 7 km^2 to 17 km^2 (an increase from 2% to 3%). Similarly, there has been a drastic increase in the land use for transportation from 21 km^2 in 1991 to 117 km^2 in 2011 (GoK 2009: 102).

6.3.17.1 Encroachment of Green Belt

Green belt area has been encroached upon for urban activities by converting agriculture lands to nonagricultural purposes. Although accurate data is not available regarding the exact area of encroachment, *The Hindu* and the statistics available with the government indicate that over the years, particularly between 2004–2005, about 6295 acres of land was converted from agriculture to industrial and other housing purposes (mainly for developing layouts) in five taluks of Bengaluru Urban district. A study by Thippaiah (2009: 2) clearly points out that the net sown area in Bengaluru Urban district which was 1,02,331 ha in 1986–1987 declined to 85,575 ha in 2000–2001.

Table 6.44 presents the extent of green belt encroachment.

A study conducted by the Indian Institute of Science (IISc) using ISRO's remote sensing data on land use and environment of Bengaluru has revealed the extent of encroachment and loss of urban land for various purposes over the years, i.e. from 2000 to 2006 (BMRDA 2009: 265) (Table 6.45).

The Revised Structural Plan by BMRDA 2030 (BMRDA 2009: 265) clearly states that the city's green cover has been declining at an average rate of 30 sq.km

Table 6.44 Green belt encroachment

Green belt region	Agriculture to nonagriculture (in acres)
Anekal taluk	2174.33
Bengaluru north (additional)	*1382.28*
Bengaluru south	1080.21
Bengaluru east	909.5
Bengaluru north	*748.19*
Conversion of agriculture land to industrial purpose	
Anekal	49.14
Bengaluru north	6.07
Bengaluru east	98.15
For housing purposes	11,126.27
For other purposes	600.17

Source: BMRDA (2009: 265)

Table 6.45 Land use changes (2000–2006)

Land use category	2000	2006	% of Change
Built-up area	186.42	301.27	+ 61.6
Vegetation	859.24	684.85	−20.30
Water bodies	42.03	32.05	−23.7

Source: BMRDA (2009: 265)

per annum. On the other side, the built-up area has increased from 180.42 sq.km to 301.27 sq.km (61.6%). Besides, while the city has lost 23.7% of water bodies at an annual rate of 1.7 sq.km, similarly, it has lost 20.30% of vegetation.

According to the BDA official statistics, Outline Development Plan (ODP), City Development Plan (CDP) and Revised Master Plan (2015), the green belt area has reduced by 262.5 sq.km, as presented in Table 6.46.

From Table 6.47, it is very clear that urbanisable area of Bengaluru has drastically increased from 220 sq.km (approved by Outline Development Plan in 1972) to 800 sq.km by 2007 as per Revised Master Plan 2015 prepared by BDA. However, the green belt region has reduced from 830 sq.km in 1984 (approved by City Development Plan) (CDP) to mere 419.50 sq.km, as per the Revised Master Plan 2015. Thus, the green belt has lost nearly 262.5 sq.km between 1984 and 2007.

A similar conclusion can be drawn from Table 6.41 that the vegetation cover has substantially reduced from 68.27% (in 1973) to just 25.2% (by 2007), while the built-up area has increased from a mere 7.97% (in 1973) to a whopping 45.19% (by 2007). Visibly, a satellite imagery-based study acknowledges the fact that between 1973 and 2007, the vegetation cover has shrunk by 30% and a shrinking of water bodies by 61% (Subramanian 2014).

Table 6.46 Urbanisable and green belt area as per ODP, CDP, RCDP and RMP 2015

Area	ODP approved on 22.5.1972 area in sq.km	CDP approved on 12.10.1984 area in Sq.km	Difference (2–3)	RCDP approved on 05.01.1995 area in sq.km	Difference (4–5)	RMP 2015 approved on 25.06.2007 area in sq.km	Difference (6–7)
Conurbation area (urbanisable)	220	449	229 (+)	597	148(+)	800	203(+)
Agriculture zone (green belt)	280	830	550(+)	682	148(−)	419.50	262.5(−)
Total local planning area	500	1279	779(+)	1279	–	1219.50	59.5(−)

Source: Dept of Country and Town Planning, BDA, Bengaluru
Data collected from EOI, Dept of Town and Country Planning, BDA, Bengaluru, as of April 24, 2014

Table 6.47 Shrinking vegetation and built-up area

Year	Vegetation		Built-up area	
	Area (ha)	(%)	Area (ha)	(%)
1973	46,639	68.27	5488	7.97
1992	31,579	46.22	18,650	27.30
1999	31,421	45.99	23,532	34.44
2000	31,272	45.77	24,163	35.37
2002	28,959	42.39	26,992	39.51
2006	19,696	28.83	29,535	43.23
2007	17,298	25.32	30,876	45.19

Source: Subramanian (2014)

6.3.17.2 Land Acquisition

Land acquisition in and around Bengaluru City has been the most contentious issue. Thousands of landowners, particularly farmers, and residential owners have lost their lands in the city and catchment areas around the city. The lands have been acquired by the Karnataka Housing Board (KHB), Karnataka Industrial Areas Development Board (KIADB) and Bengaluru Development Authority (BDA) for the development of large-scale infrastructure projects (like urban corridors, fly-overs, special economic zones and residential apartments). Land acquisitions around Rajarajeshwarinagar, Bommanahalli and Yelahanka, etc. for Bengaluru Mysore Infrastructure Corridor Project (BMICAPA) have been the most controversial involving role of political parties and politicians. Resistance movements erupted, time and again, against illegal land de-notification for acquisition (Kumar

2012) with farmers and landowners refusing to part with their lands without an adequate compensation package and rehabilitation. In fact, the role of politicians[62] and land mafia is held responsible for land-related and housing scams in the city. Similarly, for the construction of Arakavathy Layout, 2750 acres of land covering 16 villages was acquired by BDA. But there was a huge uproar over the compensation and acquisition as the authorities concerned had not sought consent of the local landowners. The matter was taken to the Supreme Court of India, which upheld the BDA acquisition, and the government of Karnataka came out with an amicable compensation package for the landowners (Rao 2013; Kushala 2013). Due to such an indiscriminate conversion and de-notification of land, many illegal layouts have sprung up around the city outskirts. According to unofficial statistics, there are more than 108 illegal or unauthorised layouts under BDA jurisdiction (Ravikanth 2013).

6.3.17.3 Environmental Impacts of Land Acquisition

1. Agricultural lands and catchment areas consisting of various water bodies are being indiscriminately converted to residential/commercial/infrastructure projects.
2. With the emergence of information technology industry, land is increasingly converted into special economic zones and corridors like IT and BT, affecting the local ecology and environment (Yousaf 2012).
3. Such a fast land acquisition process has not only affected the local farmers but also uprooted them from producing agricultural goods to consumption. Besides, land price has exponentially increased in the city (Prasad and Khan 2013).

6.3.18 Forest Region

Compared to other metro cities of India, like New Delhi, Hyderabad, Chennai, Mumbai, etc., the city of Bengaluru is rich in forest cover and greenery. Bengaluru Rural district has a forest cover of around 45,000 acres spread across Nelamangala, Hoskote, Doddaballapur and Devanahalli taluks that are mostly raised under social forestry. However, the forest region has been under a tremendous pressure in the recent years due to encroachment and illegal occupation for industrial, commercial and residential purposes, and as a result, the city is losing its green cover region. The notified forest area in the city limits constitutes just about 2%.

Table 6.48 presents forest cover under Bengaluru Rural.

According to the Karnataka State Remote Sensing Applications Centre, the total area under vegetation was around 12.4% (in 2000–2001) which has drastically reduced now. As per the Forest Survey of India report (2009), the forest cover in the Bengaluru City is only around 6.80% (The Hindu 2011).

[62] For instance, Katta Subramanya Naidu and his son Katta, MLA Ashok Kumar, B.S. Yedurappa and his sons and S.N. Krihnayya Shetty and others were involved in land scams (Mondal 2013).

Table 6.48 Forest cover in Bengaluru Rural

Forest Cover	Acres
Nelamangala	10,000
Hoskote	12,000
Devanahalli	8000

Source: Srivatsa (2008)

Table 6.49 BBMP's tree planting budget allocation

Year	Budget allocation	Actual budget released
2010–2011	39.23	9.83 (25%)
2011–2012	38.23	18.04 (47%)
2012–2013		6.42
2013–2014		3.91 (13%)

Source: Chaturvedi (2014)

Tree planting in the city of Bengaluru has enormously come down due to callousness on the part of local bodies. Besides, the budget outlay of BBMP for tree planting has substantially decreased from 47% during 2011–2012 to mere 13% in 2013–2014. At present, due to a low budgetary allocation, BBMP has invited local communities like NGOs or interested individuals to plant saplings. Documents available with BBMP clearly show that for the past 4 years, BBMP has substantially neglected its greening activities (Table 6.49).

Table 6.49 clearly indicates that the budget allocation and the subsequent actual release have been gradually declining from year to year. For the year 2010–2011, the BBMP's budget allocation amounts to Rs. 39.23 crore towards sapling and planting, while the actual amount released comes to only Rs. 9.83 crore, i.e. only 25% of what the budget allocated. However, for the following year, 2011–2012, BBMP allocated 38.23 crore and actually released Rs. 18.04 crore, amounting to only 47% of the budgetary allocation, almost twice the amount released for the previous year. But, thereafter, the amount released has steadily declined; in 2012–2013 a mere Rs. 6.42 crore was released towards greening activities while only Rs. 3.91 crore in 2013–2014, just 13% of the budgetary outlay. The BBMP officials categorically point to the fact that since 4 years, planting of saplings has been stopped.

6.3.19 Emerging Issues in Land Utilisation Pattern

(i) The emergence of a large metropolitan area with a population of more than 9 million has given rise to myriad environmental resource-use problems. The problems include water supply, sanitation, waste management, air/noise pollution, electricity supply, depletion of groundwater and vanishing water bodies and wetlands.

(ii) Rapid urbanisation has led to an unabated encroachment of urban land, thereby increasing the built-up area. Particularly in the core city area, water bodies and valleys have been encroached upon.
(iii) The green belt zone covering BMP/core city area has been encroached upon for varied purposes like building corridors, special economic zones, IT/BT industrial zones, satellite townships, etc., thus corroding the valuable natural resources.
(iv) Development of (i) Bengaluru-Bidadi corridors, (ii) Bengaluru-Nelamangala corridor, (iii) Bengaluru-Tumkur corridor, (iv) Bengaluru-Mysore corridor, etc. has increased the intensity of urbanisation process. Such an indiscriminate urban sprawl has adverse implications for the surrounding natural resources and agricultural lands.
(v) Spatial expansion of the city has severely threatened the available water resources due to a combination of factors such as encroachment, depletion of catchment area, urban sprawl eutrophication, etc. In addition, an indiscriminative groundwater extraction has led to the severe depletion of aquifers. This trend is closely related to an increase in the built-up area based on a growth model followed by the local government which has severely affected water bodies like lakes and tanks (Ramachandra and Mujumdar 2009: 48).
(vi) Sewerage and industrial pollution has led to high levels of groundwater contamination with both chemical and biological bodies.
(vii) The continued indiscriminate mining and query activities surrounding Bengaluru City have serious long-term implications for the environment and ecology of the city.

6.3.20 Urban Aesthetics of Ecology

The city of Bengaluru has to live up to the standards of 'International City'. The aesthetic value of the city lies in the understanding of the characteristics of our ecology and environment. The aesthetics canvass of the natural environment can be divided into three components, namely, (i) architecture and the city, (ii) special places and homes and (iii) landscape, gardens and countryside (Kaminska 2008: 176). Therefore, from an environmental aesthetics perspective, the built-in environment is measured by applying qualitative measures to determine the value of visual and non-visual aesthetic features as also the tools for interpreting the active nature of perceptions *(ibid).*

The urban landscape of Bengaluru is undergoing a rapid transformation that includes public space like traffic, litter and noise, urban design of buildings, sidewalk/pedestrian space and visual space and landscape of the city like gardens and parks and the rural side.

The city of Bengaluru, the very epitome of 'paradise' and a 'global city', is facing a serious challenge of 'capture/commodification' of urban public spaces by a host of illegal hoardings, flex boards, cutouts and film posters. Visual pollution has been one of the serious threats to the urban aesthetics of the city. The clutter occurs at various levels: (i) Hoardings of advertisements. (ii) Display of flex boards on occasions like marriages, birthdays of politicians, obituaries, etc. (iii) The menace has been further compounded by obscene film posters pasted on walls, metal sheets of metro works and huge hoardings on the roadside lanes, electric poles and tree trunks. (iv) Display of personal advertisements across billboards.

According to a recent estimate, there are 1997 hoarding spaces (including 381 self-hoardings), and more than 2000 hoardings (Vittal 2010; Times of India 2013) rented out for commercial advertisements alone. The development of huge malls (29 big ones), a symbol of hyperbole culture, has triggered a further competition for space along with IT firms in the city. Each year, the global companies in the city allocate for advertisement budgets. Thus the orderliness of the city aesthetics has been seriously compromised by illegal and unscrupulous hoarding mafia practices assuming alarming proportions.

6.3.20.1 Urban Aesthetics Vs Environmental Impact

Urban citizens are the victims of various kinds of pollution such as (i) air, (ii) noise, (iii) water and now (iv) visual environmental pollution. Public space is a democratic space, and every citizen has a right to participate for recreation and acquiring knowledge. The menace of hoarding cluttering space on busy hubs in Bengaluru City is a violation of our constitutional right to public space and a pleasant visual treat of the city landscape. In 2005, some of the roads in the city were made hoarding-free, but this measure seems to be no solution to a thriving illegal hoarding mafia/business. Similarly, there are no laws to clamp down on the number of hoardings to be put up. Visual pollution can have a tremendous effect, particularly on the young and general population in the city. Both school-going children and college students are exposed to indecent and obscene posters, particularly showing women in poor light. This may be one of the strong reasons for an increase in the crime rate of the city, particularly, instances of rape, rise of exhibitionism like pervert flashes (Chandrashekar 2013). A walk through busy central city areas, particularly, Majestic, among other places, shows huge/life-size hoarding of movies, political parties, etc. severely hampering the pedestrian movement. Many lowly hung hoardings can cause injuries, thus forcing the pedestrians to guard their heads. Such blatant violations of the city rules and by-law regulations with the mushrooming of illegal hoardings in undesignated places seriously hamper in terms of recognising or promoting urban aesthetics, architecture and conservation of historical or heritage structures.

6.3.21 Urban Parks and Open Spaces

Urban parks constitute one of the chief characteristics of a rich biodiversity and are critical to the preservation of the city's valuable ecosystem (Nagendra and Gopal 2010: 2). The city of Bengaluru enjoys a salubrious climate which earned for it the name of 'Garden City[63]'. A pleasant climate supported by its topography and lush green ecology with gardens, parks and water bodies promotes an aesthetic value of the city. Urban parks are spaces for recreational activities in addition to acting as habitats for flora and fauna (*ibid* 2).

Urban parks and open spaces in the city of Bengaluru are regulated as per the Revise Master Plan 2015[64] prepared by BDA. CDP (1995) has earmarked 77.9 sq.km (14%) of land for parks and open spaces, while BDA has earmarked not less than 15% for any new layout (BDA 2007: 44). As per the Revised Master Plan 2015, 22 theme parks and 16 tree parks under 'Gtreener Bengaluru[65]' were to be developed by both BBMP and BDA, and accordingly the Forest Department raised 130 sq.km of plantation by 2007, while BBMP has developed 48% of the 560 parks in the city (BDA 2007: 44).

With a booming population which is likely to reach over 1 crore in the near future, the city is undergoing many changes with respect to its landscape. The city has a wealth of both large and small parks. Some of the large parks like Lalbagh, Cubban Park, Coles Park and Bannerghatta National Park are rich hotspots of biodiversity. Several large parks have been in existence since the eighteenth century (Harini and Gopal 2010: 2). Some of the hotspots like bio-park at Bengaluru University, Gandhi Krishi Vignana Kendra, Pavithra Vana and Sawadurga medicinal plant conservation area are also habitats for several rich species and herbs. Smaller parks are being created by BBMP for recreational purposes in the residential areas of the city post-1990s. Parks[66] in the city can be categorised into (i) during the eighteenth century, (ii) between the 1970s and the 1980s and (iii) small parks created in the residential neighbourhood post-1990s. These established parks differ in their diversity, density and composition of species (Harini and Gopal 2010: 6).

The Department of Horticulture maintains 365 parks (well developed, 55; partially maintained, 130; and undeveloped, 180 under BMP area) (BBMP, 2009: 142). But the official statistics put out by BBMP after a recent survey shows 1067 parks, playfields and open spaces, while 786 are found in the list of published report. Similarly, BBMP's gazetted notification of 2010 records 679 (parks, playfields or open spaces), and the second list consists of only 388 green spaces with ward numbers and names (Navya, 2011). There has been a considerable loss of open space in Bengaluru City, particularly post-1971. Nearly 2.5 icles which has inevitably increased the ve 0% of the existing open spaces were converted into residential areas during 1971–1981 (Mishra 2006) (Table 6.50).

---

[63] The Karnataka Parks, Play-fields and Open Spaces (Preservation and Regulation) Act, 1985, says that the local authority should publicise information about these spaces annually in the gazette and made available to public (Navya 2011).

[64] Refer BDA (2007): Revised Master Plan 2015: Zoning Regulation, Volume 3, Bengaluru: Bengaluru Development Authority.

[65] Refer annexure on the decline of urban parks in the city by 2001.

[66] Bengaluru parks are rich in species with 80 species, recorded from 127 plots of 0.25 ha, each containing 1423 species (Harini and Gopal 2003).

Table 6.50 Regulations and initiatives with respect to open spaces and green areas under BMR

Regulations and initiatives	Green initiatives through parks and open spaces
CDP earmarking 1995	77.9 sq.km (14% of the total land)
BDA	15% of area for parks and open spaces in newly formed layouts
BBMP and BDA –'greener Bengaluru'	22 theme parks, 16 tree parks
The Forest Department	Aiming plantation across 130 sq.km 35 lakh plants already planted
BBMP	Developed 48 per of 560 parks in the city

Source: BBMP (2009: 142)

6.3.21.1 Environmental Impact of Urban Parks[67]

1. In view of the emergence and spread of IT/BT industries and other intense industrial activities, most of the lush green space has been lost in the name of development. Many environmentalists blame BBMP for recklessly cutting down trees for widening roads.
2. There are certain tree species planted which are not suitable to the city environment. Availability of space is the main problem in the core city area for planting trees. Nearly 2 lakh trees are planted per year in the peri-urban and periphery region of the city. At present, trees are planted only in the peri-urban region of the city.
3. There is lack of inventory of these green spaces and biodiversity that the city supports (Navya 2011).
4. A poor maintenance of urban parks and trees (inadequate watering) is leading to their gradual disappearance. Potable water is not used for urban parks. Most of the borewells are not working even at a depth of 1000 feet.
5. The entire city of Bengaluru has been facing a serious groundwater problem. In the entire city, only ten parks have water bodies which can be used for maintenance.
6. Many of them have been encroached upon by in-migrants, land mafia and other infrastructure development agencies.
7. Converting urban parks into tree park is a problem in the BBMP core city area. Despite their being a public property (managed by BBMP), most often, political representatives interfere, encouraging the construction of infrastructure in the urban parks.
8. Tree planting has been stopped since 2013 due to delay in approvals by officials.
9. For the management of urban tree parks in the city, BBMP at present is highly understaffed only with 19 staff members.
10. Finally, budget sanctioning is a major problem.

[67] Some of the details were collected from the Chief Forest Officer, BBMP, Bengaluru, on May 20, 2014.

6.3.22 *Urban Transport*

The city of Bengaluru is known for bustling transport services with long hours of traffic congestion with commuters spending lon 180 cities, yet, critically polluted cities have ris g hours of journey. The city is a designated hub of 'IT/BT' and service sector industries which attract migrants from both within and outside. The city has grown in terms of population from 26 lakhs in 1971 spread over 250 sq.km to more than 9 million spread over 1000 sq.km (Environmental Support Group 2012; Vaidyanathan and King 2011). But the growth of both population and area is not commensurate with the delivery of amiable transport services. The city spans 4300 km of a road network[68] out of which 252 km constitute arterial roads including 81 km in the east sector, 76 km in the western sector and 95 km in the south. Another 100 km of both national and state highways entering the city to the arterial roads carry most of the vehicular traffic (GoK 2009: 172).

The city has been gradually experiencing a vehicular explosion from 400,000 in 1987 to 2.3 million by 2005. The average number of vehicles has registered a six-fold increase from 0.3% (1980) to 1.7% (by 2005). Between 1991 and 2005, the number of registered vehicles rose from 0.68 million to 202 million, constituting an increase of more than 200% (BDA 2007: 38). As a result, Bengaluru accounts for the highest vehicular growth rate among million plus cities in India.[69]

The city has been gradually experiencing a vehicular explosion with 31.37 lakh vehicles which two wheelers constitute 72% (Abide 2010: 35) with an annual growth rate of 14% per annum (MoUD 2008: 16) (see Fig. 6.17). A study by the Ministry of Urban Development (2008: 96) points out that the average trip for the city like Bengaluru has increased by a minimum of 2.4 kilometres (or 33%) and a maximum of 5.5 kilometres (or 86%). The number of registered vehicles has grown from 6.28 lakhs in 1990 to 25.27 lakhs in 2006 and to 37.91 lakhs by March 2011 (Abide 2010: 35; CSD 2012: 21). It is estimated that one in every four persons in Bengaluru owns private vehicles which has inevitably increased the vehicle[70] ownership rates from 58 vehicles per 1000 population in 1981 to 365 vehicles per 1000 population in 2006 (ibid) (Fig. 6.11).

Concomitantly, the vehicular pollution through GHG emissions has increased at alarming levels in the city. As per the State Environment Report of Bengaluru, 2009, PM10 levels (160–180 ug/m^3) are three times that of the Indian National Ambient Air Quality Standards (INAAQS), exposing more than 6 million population to health risks. Annually, vehicular emissions from the transport sector contribute

[68] The different types of roads in Bangalore city are Ring Roads (Core Ring Road (CRR), Outer Ring Road (ORR), Peripheral Ring Road (PRR), Intermediate Ring Road (IRR) and Satellite Township Ring Road (STRR)), expressways (Airport Link Road), highways (national highways, state highways), arterial roads, sub-arterial roads and other link roads.

[69] Bengaluru comes next to New Delhi in terms of having the most number of vehicles per person (32 vehicles per 100 people) (BDA 2007: 38).

[70] Those commuting by cars have almost doubled from 6.7% in 2005 to 12% by 2011 (CSD 2012: 34).

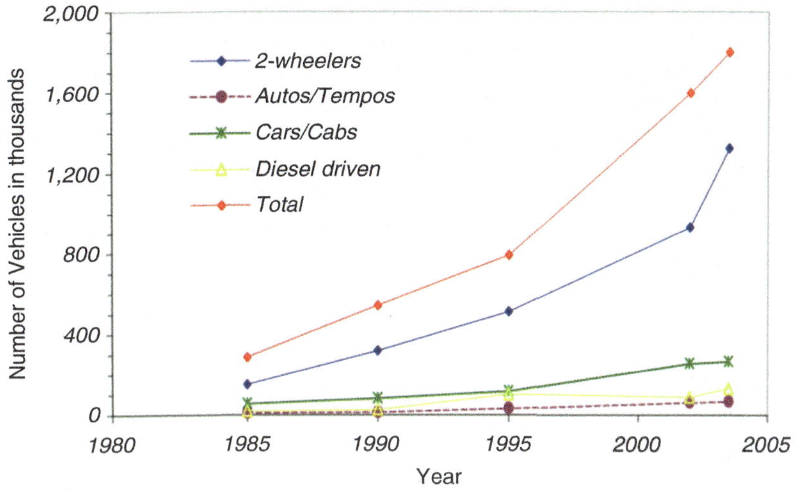

Source: GoK (2009: 171)

Fig. 6.11 Growth of Bengaluru's motor vehicle population

51,221 tonnes of carbon monoxide (CO) and about 2467 tonnes of particulate matter (PM) (GoK 2009: 173). The generation of dust pollution due to traffic which generates particulates of PM 2.5, PM 10 and above is another major concern. At present, one of the single most global concerns has been CO2 emissions from the transport sector that contribute to global warming and climate change.

6.3.23 Bicycle and Pedestrian Mobility

The city of Bengaluru truly lives up to its tag of 'technology' hub by adapting to the trend of motorisation, i.e. through an increase in the use/ownership of two or four wheelers. In addition, flyover construction and metro construction activities are increasingly occupying the city's road space rather than making space available for pedestrians and cycle riders. Increase in the use of motor transport has inevitably disturbed the city's environment and ecological balance. Unlike the Western countries, cities in India do not prescribe pedestrian and urban mobility norms and standards or service network, and as such, the citizen mobility situation has worsened over the years. Urban commuters have no choice but to get increasingly exposed to a host of transport-related negative externalities such as polluted air or noise and road accidents.

However, recently, the Directorate of Urban Land Transport (DULT) (i.e. under Bengaluru Metro Rail Corporation (BMRC)) has identified roads in the CBD area NMT (non-motorised traffic) corridors to be built in the vicinity of MG Road for both pedestrians and bicycle riders. As the first step towards being a cycle-friendly city, on October 27, 2013, the city celebrated a 'bicycle' day, a step closer to 'Cyclovia', with approximately 2 million participants (30% of citizens) using over 120 km of car-free streets. Every last Sunday is designated for celebrating 'cycle day' (Roy 2012; Srinivasan 2013; Walkabilityasia 2013). To reduce the city's carbon footprint, entrepreneurial initiatives are started to encourage commuters to use bicycles for commuting. One such initiative is 'Kerberon' with its sole aim being to connect areas with bicycle paths, thereby reducing carbon footprint (Charan 2011). Cycle days in other parts of the city like HSR Layout and Sanjay Nagar are also popularly followed.

6.3.24 Energy

The transition of Bengaluru from 'Garden City' to being a hub of 'IT/BT' has put a huge stress on the delivery of energy. With the expansion of the city in terms of population and commercial and industrial activities, there has been an unprecedented demand for energy consumption.[71] Bengaluru Electricity Supply Company Limited (BESCOM), a private company, distributes electricity to eight districts, namely, Bengaluru Urban and Rural, Ramanagara, Kolar, Tumkur, Chitradurga, Chikkaballapur and Davangere, covering an area of 41,092 sq.km with a population of over 139 lakh. BESCOM is one of the largest electricity companies with 7 million customers (CSD 2009b: 6). Around 50% of BESCOM's 81 lakh consumers are from Bengaluru City (Ramani 2013).

6.3.24.1 Energy Consumption

Due to the vertical growth of Bengaluru City, the density of power has increased enormously, clearly indicating the growth of power load of both Bengaluru Urban and Rural districts (which was expected to grow at 8.04 per and 9.9%, respectively). But the actual growth registered during the Revised Master Plan 2015 period amounts to 15% to 16%, while the projected demand for the BMR in the 11th Five Year Plan (2007–2012) amounts to 4015 MU (BDA 2007: 50).

At present, the city consumes nearly 30% of the total power generated in the state with negligible savings. While the daily consumption of the city comes to 2000 MW (by 2012) (The Hindu 2012c), *the daily power demand works out to about 2300 MW. The city consumes one-third of the state's* total[72] power generated (Navya 2012). On account of increased demand, BESCOM overdraws up to 100 MW from the southern grid. Domestic connections are dominant in Bengaluru Urban with 85% and Bengaluru Rural and Ramanagara with 75% (GoK 2009: 261). During

[71] At present, 2000 MW is supplied by private power suppliers and another 250 MW to be integrated into the grid (The Hindu 2012c).

[72] Karnataka's average demand comes to 6000 MW per day (Navya 2012).

Table 6.51 Electricity consumption over the years

Category of consumption	Year	Percentage of increase
For entire BESCOM customers	2005–2006	5.15
	2006–2007	21.6
BESCOM customers – Bengaluru urban	2006–2007	6.2
	2008–2009	15

Source: CSD (2009b: 70)

2006–2007, the total electricity consumption in BMR was about 9000 million kWh. Over the years, the power consumption has grown by 14% annually (ibid), resulting in a constant demand-supply gap.

Out of 1.7 lakh industries in Karnataka, 1.56 lakh industries are based in Bengaluru alone of which 10,000 in the city's outskirts and roughly 2000 in the rural areas are fed by BESCOM. Around 48% of power generated is consumed by industries alone (Ramani 2013). The per capita consumption of electricity has increased significantly over the years (see Table 6.51).

6.3.24.2 Transmission and Distribution Loss

At present, the transmission and distribution losses in Bengaluru zone constitute 7%. It is comparatively less because a majority of the consumers are HT and EST (not LT) consumers, and so, not much power is lost through distribution.[73] The transmission and distribution loss of electricity has decreased over the years (see Fig. 6.12), and the loss, if any, could be due to thefts, pilferages and tampering or defective meters.

6.3.25 *Air Pollution*

An increased urbanisation has exposed the urban population to different forms of pollution such as air, noise, water, industrial and construction related. Air pollution[74] can be caused by point sources or non-point sources; point sources include households, industries, hotels or other places where fuels are burnt for cooking or

[73] Details collected during an interview with Mr. Malleshappa, Operation, BESCOM Corporate Office, Bengaluru, on April 25, 2014.

[74] Not even 20 Indian cities follow Euro 4 emission standards for new vehicles, but most follow Euro 3. Euro 4 is 7 years behind European standards, and Euro 3 is behind by 12 years (Hindustan Times 2013).

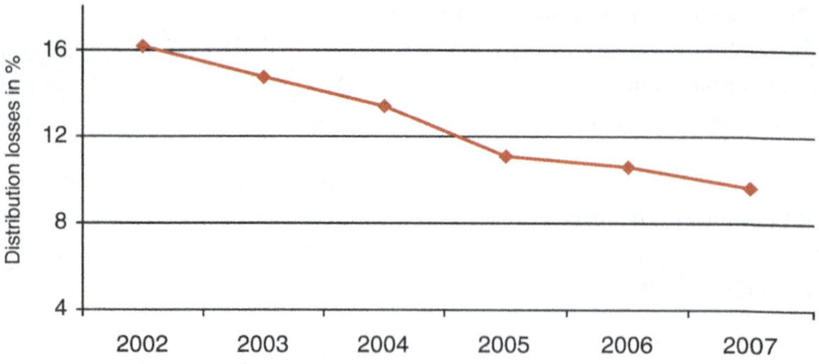

Source: GoK (2009: 266)

Fig. 6.12 Distribution losses in Bengaluru City

heating purposes. Non-point sources include vehicular emissions consisting of exhaust gas, smoke or respiratory particulate matter (CSD 2012: 12).

In 2013, air pollution[75] in Bengaluru caused by vehicles accounted for 44%, road dust for 20%, construction noise for 14% and industrial noise for 14% of the total pollution created (Bhattacharya 2013). This could be because of a tremendous growth in the number of two wheelers from 0.75 million in 1997 to 2.04 million in 2007 (GoK 2009: 134).

Air pollution concentrations are the result of interactions between variations in the physical and dynamic properties of the atmosphere on timescales from hours to days, atmospheric circulation features, winds, topography and energy use. In Bengaluru City, air pollution has reached alarming levels due to an increase in the vehicular population,[76] exceeding 3.7 million, and is consistently increasing at an annual growth rate of 13% (Abide 2010: 35; CSD 2012: 18). A recent study by Energy and Resource Institute (TERI) points out that 42% of Bengaluru's air pollution is caused by vehicular emissions, 34% by dust and 14% by air emissions caused by industrial activities. In 2002, CPCB member secretary attributed 72% of urban air pollution to vehicular emissions (Honaganahalli 2014).

A study conducted by the Centre for Science and Environment[77] on air quality data generated by the Central Pollution Control Board (CPCB) for 2007 under the National Air Quality Monitoring Programme (NAMP) presents deadly revelations about air pollution levels in Indian cities. At least one criterion pollutants exceed the annual average ambient air quality standards in 80% of the cities out of 127 cities/ towns monitored under the NAMP. The levels of SO2 have fallen sharply over the

[75] A study by CSD (2011) states that 'Air pollution is concentration of interactions between variations in the physical and dynamic properties of the atmosphere on time-scales from hours to days, atmospheric circulation features, wind, topography and energy use'.

[76] In 1981 Air Act was enacted to check the problem of air pollution. The Comprehensive Environment Act (1986) also monitors air quality in Bengaluru City.

[77] Access the link for full details on the air quality study (http://www.cseindia.org/node/207).

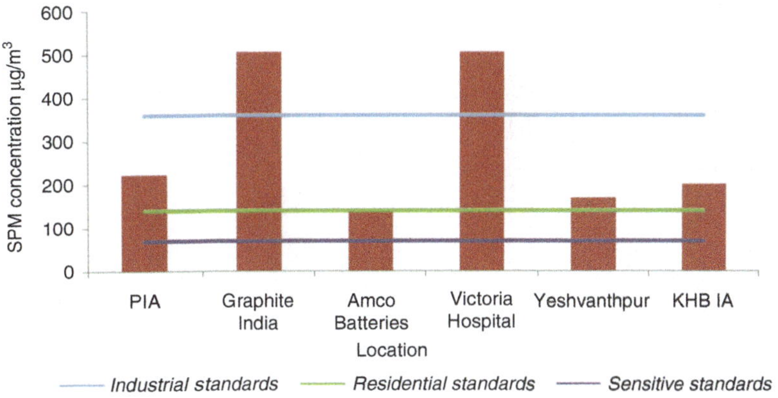

Source: GoK (2009: 122)

Fig. 6.13 Annual SPM concentration in industrial, residential and sensitive areas for/during 2006–2007)

years. Particularly, in south Indian cities like Hyderabad, Bengaluru, Visakhapatnam and Tuticorin, the average PM 10 level has increased since 2000–2007.

Although air quality monitoring has been doubled between 2005 and 2010 from 96 to 180 cities, yet, critically polluted cities have risen from 49 to 89 (Hindustan Times 2013). Under NAMP, Karnataka State Pollution Control Board (KSPCB) has been monitoring ambient air quality at six stations in Bengaluru. Air quality monitoring has been done across classified zones: (i) industrial, (ii) mixed urban and (iii) sensitive zones. As seen, Fig. 6.13 indicates that during 2006–2007 (see Fig. 6.13), the concentration of SPM and industrial limits exceeded both at the industrial area of Graphite and the Victoria Hospital (which is a sensitive zone)[78] area clearly indicating the severity of pollution in terms of causing health problems to urban residents.

6.3.25.1 Respiratory Particulate Matter

The major source of air pollution is respirable dust. As per the data for 2006–2007 (see Fig. 6.15), the concentration of RSPM (respirable particulate matter) exceeds the limits set for sensitive areas, i.e. 50 ug/m^3 at both Victoria Hospital and Graphite India (Industrial Area, the limit being 120 ug/m^3) (Fig. 6.14).

The concentration of RSPM (2007–2008) (see Fig. 6.15) is found highest at the city railway station and SG Halli during peak hours, i.e. beyond the permissible limits (23 h a day).

[78] Which exceeds National Ambient Air Quality Standards (NAAQMS) for sensitive areas of 70 ug/m3 and exceeds even the limit for industrial areas of 360 ug/m3.

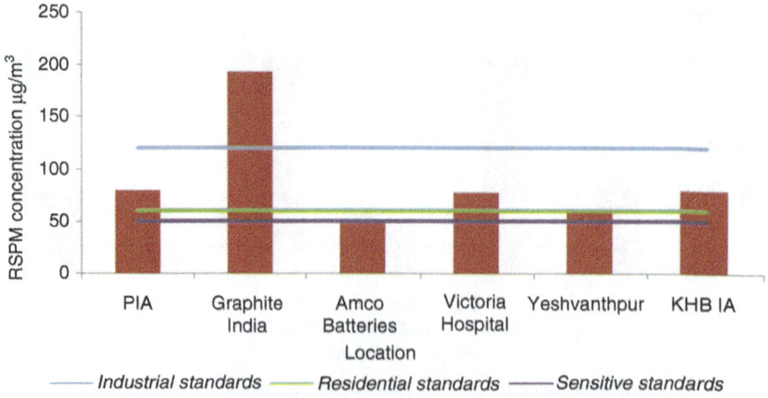

Source: GoK (2009: 123)

Fig. 6.14 Annual RSPM concentration in industrial, residential and sensitive areas for/during 2006–2007)

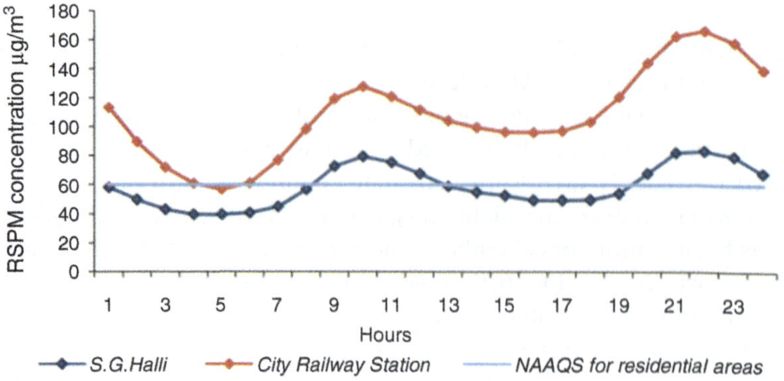

Source: GoK (2009: 123)

Fig. 6.15 Twenty-four-hour RSPM concentration in residential locations for/during 2007–2008

NO2 (nitrogen dioxide) was observed to be highest at SG Halli during 2007–2008 (see Fig. 6.16) which clearly suggests vehicular emissions as the major source of air pollution NO2 (GoK 2009: 124).

A high concentration of carbon monoxide (CO) (see Fig. 6.17) is observed at both the city railway station and SG Halli, particularly during the peak hours (GoK 2009: 124).

6.3.25.2 Industrial Air Pollution

In a study on the levels of SPM with respect to six monitoring locations, three show high levels of SPM, particularly for Graphite India (see Fig. 6.18 below). But in general, the levels of RSPM exceed the prescribed NAAQ in respect of most of the

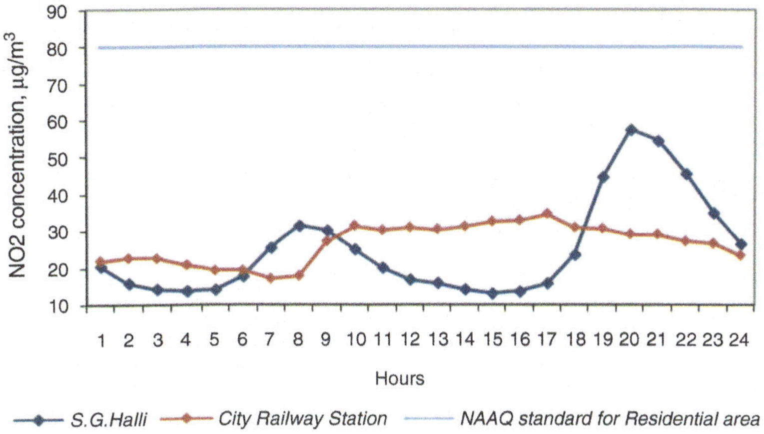

Source: GoK (2009: 124)

Fig. 6.16 Twenty-four-hour NO2 concentration in residential locations during 2007–2008

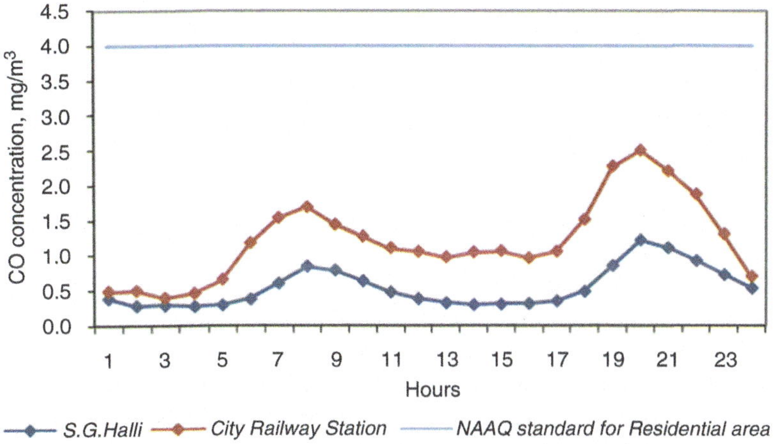

Source: GoK (2009: 124)

Fig. 6.17 Twenty-four-hour CO concentration in residential locations during 2007–2008

industrial areas. The reason attributed is high vehicular traffic, generator sets, industries, incineration, resuspension of traffic dust and use of commercial and domestic fuels (GoK 2009: 128).

Based on the monthly average results on the National Ambient Air Quality Monitoring Programme for 2006–2007, air quality index reveals that of the six industrial locations covered – Graphite India, KHB Industrial Area, Peenya Industrial Area, residential areas of AMCO batteries, Yeshwanthpura, and sensitive areas like Victoria Hospital – Graphite India is most affected by air pollution followed by Peenya Industrial and KHB Industrial Areas. Both residential and

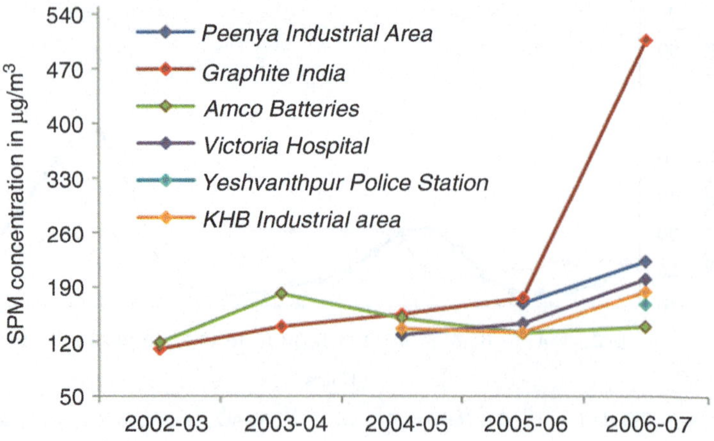

Source: GoK (2000: 128)

Fig. 6.18 Annual average RSPM trends

Table 6.52 Pollution load (in metric tonnes per day) of Bengaluru

CO	NOx	HC	PM
207	29.7	117.4	8.1

Source: State of Environment Report, Bengaluru, 2008

sensitive areas show mixed trends, i.e. from low to moderate pollution levels (GoK 2009: 129).

A study conducted by KSPCB[79] on ambient air quality during 2007 in four major areas of Bengaluru City, namely, City Railway Station (Majestic Area), BTM Circle (near Central Silk Board), Badami House (near Corporation Circle), 11th Cross and Malleshwaram (Yathiraja Mutt), shows that the concentration of PM10 and PM 2.5 and NOx are two to five times above the national limits. Nearly 23–25 metals including toxic metals are found (Table 6.52).

A recent study conducted by Centre for Science and Environment (CSE), in 2013, on the state of air pollution and mobility, highlights the following (The Hindu 2013a; Achanta 2013):

(i) Based on an analysis of air quality data from the Central Pollution Control Board (CPCB), nearly 50% of the southern Indian cities exhibit a high particulate pollution. Fourteen cities including Bengaluru show high levels of particulate matter, i.e. 10 (PM10) with a 41% increase for Bengaluru over the last few years.

[79] Access (http://kspcb.kar.nic.in/breathing_level.pdf) for more details on the study.

(ii) Bengaluru, one of the ten southern cities, indicates rising nitrogen dioxide (NO2) levels. In the recent years, Bengaluru has crossed the standards now and then.
(iii) Of the nine locations in Bengaluru, three show over 50%, exceeding the PM10 24-hourly standard, while seven locations exceed the annual average standard. Of the total locations, Graphite India, Yeshwanthpura Police Station are critically polluted.
(iv) In Bengaluru, vehicles contribute 41% of the particulate matter and 67% of nitrogen oxide to the atmosphere.
(v) This trend of increasing air pollution has, in fact, led to health problems, such as snoring and obstructive sleep apnea syndrome particularly among children aged between 2 and 17.

6.3.26 Greenhouse Gas Emissions (GHG)

The city of Bengaluru is one of the highest polluted among 14 cities in India. High speed and traffic congestion have led to the use of more fossil fuels, thus adding to the problem of greenhouse gas emissions (CSD 2013: 137). The city of Bengaluru accounts for the highest percentage of vehicular population constituting 48% of Karnataka significantly contributing to GHG emissions. In fact, the number of vehicles has crossed 32 lakhs in BMR (Abide 2010: 35).

Emission also is caused by landfilling of municipal solid waste (MSW) which may lead to methane emissions.[80] The figure below (see Fig. 6.19) shows that Bengaluru tops in methane emission as compared to other cities in Karnataka (CSD 2013).

'Environment Report Card', Bengaluru, a study conducted by CSD (2012), covering eight zones of BMR (see Fig. 6.20), reveals that air quality in newly added areas is gradually deteriorating due to an increase in the vehicular traffic. Except for south zone, other zones like east and west and added areas are affected by air pollution and, as a result, are not satisfied with the air quality.

CSD hasmeasured PM in specific areas such as(i) in Marenahalli, where PM exceeds marginally, and (ii) in Bagalkunte, where it is the lowest. The suspended particulate matter is very high in some of the selected areas in Bengaluru; they include the following (Table 6.53).

As per the CSD (2012) study, NO2 levels have significantly increased over the years. The ambient air quality measured by the study across 15 hotspots of the city, such as Anand Rao Circle, Yeshwanthpura Police Station, Mekhri Circle, etc.,

[80] The three main sources of methane emission are (i) solid waste, (ii) domestic waste water and (iii) industrial waste water. CH4 emissions are due to the decomposition of waste in anaerobic condition and N2O cmissions from domestic waste water due to its protein content (EMPRI and TERI 2011: 127).

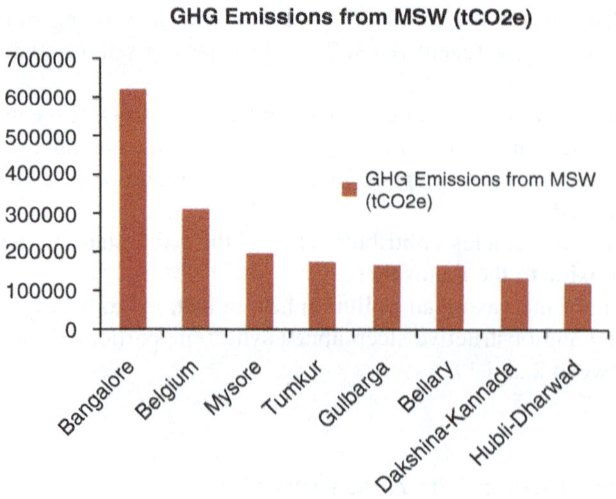

Source: CSD (2013: 72)

Fig. 6.19 Methane emissions from municipal solid waste (MSW)

Fig. 6.20 Satisfaction levels with regard to air quality across eight zones of BMR

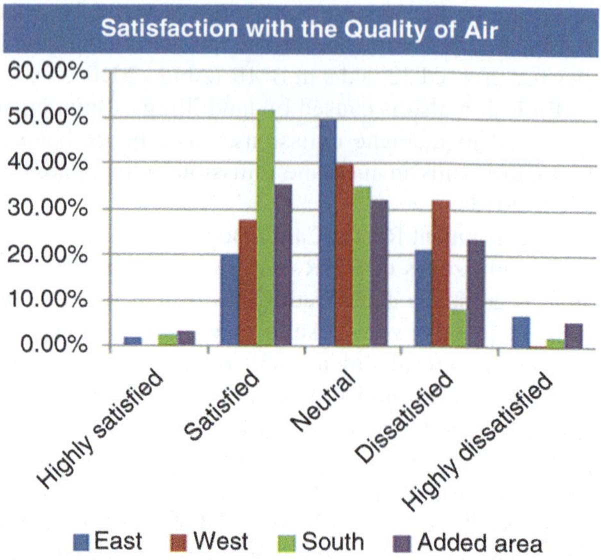

Source: CSD (2012: 29)

amounts to nearly 44%, and in all these places, people are dissatisfied (see Table 6.54) with the air quality.

A recent study carried out by Sridevi et al. (2014) has estimated carbon footprint by calculating GHG emissions in terms of carbon dioxide (CO2) across sectors such as electricity, transportation, industries, agriculture, waste and others of major cities

Table 6.53 Suspended particulate matter (SPM) in the selected areas of Bengaluru City

Selected areas of Bengaluru	Suspended particulate matter
Old madras road	542/549
Russell market	297/280
Lingarajapuram	292/297
Mysore road	282/318

Source: CSD (2012: 31)

Table 6.54 Satisfaction levels with regard to the quality of air across commercial establishments

	Establishment type				
Levels of satisfaction	Function halls	Hospitals	Hotels	Shops	Total
Highly satisfied	0.0	0.0	0.0	0.0	0.0
Satisfied	22.7	44.0	40.7	0.0	26.7
Neutral	9.1	20.0	18.5	51.9	25.7
Dissatisfied	68.2	36.0	25.9	48.1	43.6
Highly dissatisfied	0.0	0.0	14.8	0.0	4.0

Source: CSD (2012: 31)

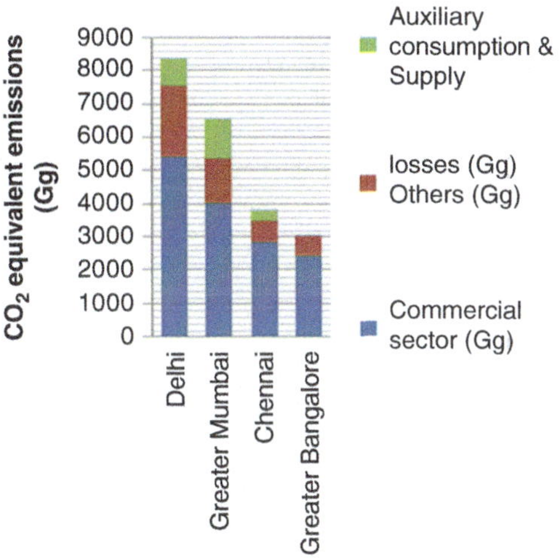

Fig. 6.21 Carbon dioxide equivalent emissions (CO2 eq) from electricity consumption

Source: Sridevi et al. (2014: 40)

like New Delhi, Greater Mumbai, Chennai and Greater Bengaluru. Some of the key findings of the study are presented below (Fig. 6.21).

The study by Sridevi et al. (2014) finds the emissions from the commercial sector (due to power consumption) highest in New Delhi for 2009; emissions from auxiliary consumption and supply loss (857.69 Gg of CO2 eq) are highest among all the cities including Greater Bengaluru (Fig. 6.22).

The study also points out that GHG emissions from vehicles (10867.51 Gg) are highest among all the cities, i.e. New Delhi, Greater Mumbai, Chennai and Greater Bengaluru. But Greater Bengaluru is the second highest after New Delhi (see Fig. 6.23).

Fig. 6.22 Carbon dioxide equivalent emissions (CO2 eq) from the transportation sector

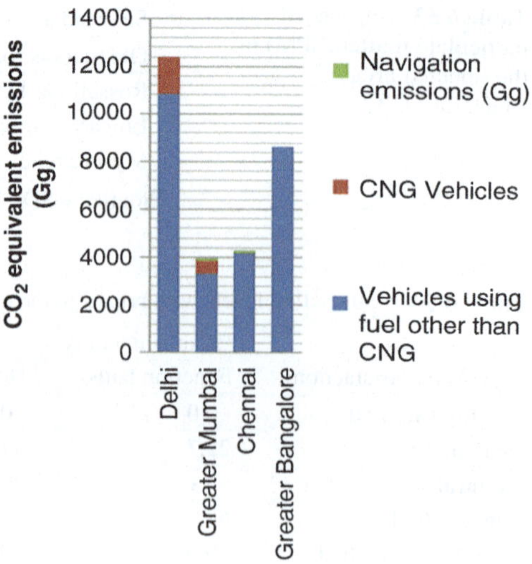

Source: Sridevi et al. (2014: 40)

Fig. 6.23 Carbon dioxide equivalent emissions (CO2 eq) from the industrial sector

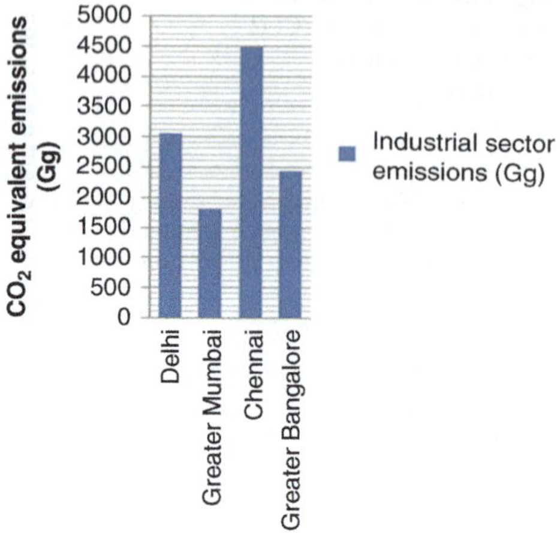

Source: Sridevi et al. (2014: 40)

The study (Sridevi et al. 2014), while calculating GHG emissions from the industrial sector such as iron and steel, cement, fertilisers, chemical manufacturing, etc., located within the respective city boundary, finds Chennai emitting 4472.35 Gg of CO2 eq, the highest emitter followed by New Delhi and Greater Bengaluru (Fig. 6.24).

While calculating the enteric fermentation and manure management which result in the emission of greenhouse gases from animal husbandry, the study (Sridevi et al. 2014) finds both New Delhi and Greater Bengaluru emitting larger amounts of

Fig. 6.24 Carbon dioxide equivalent emissions (CO2 eq) from livestock management

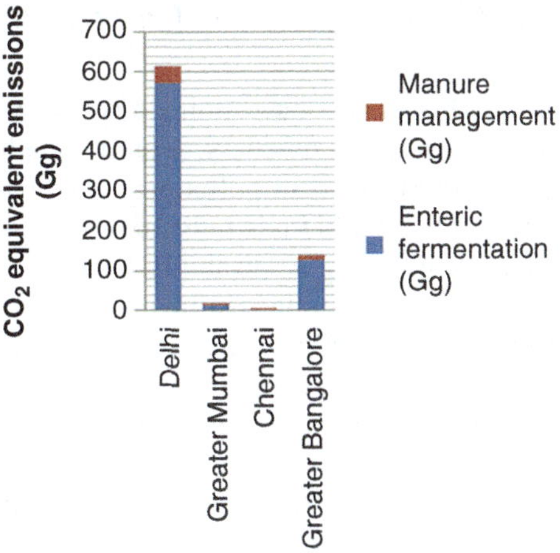

Source: Sridevi et al. (2014: 40)

Fig. 6.25 Carbon dioxide equivalent emissions (CO2eq) from the waste sector

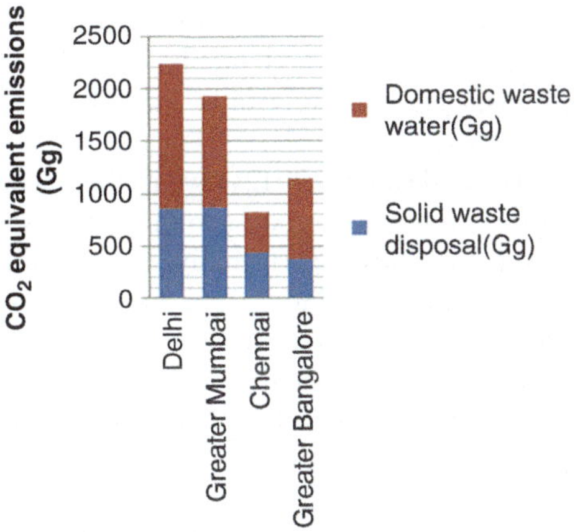

Source: Sridevi et al. (2014: 40)

greenhouse gases as compared to Greater Mumbai and Chennai due to animal husbandry (see Fig. 6.14).

Waste sectors such as municipal solid waste and domestic waste water contribute CH4 and N2O emissions. The study (Sridevi et al. 2014) points out that New Delhi, Greater Mumbai and Greater Bengaluru emit 1378.75 Gg, 1058.09 Gg and 759.29 Gg of CO2 eq, respectively, which are highest emitters of GHGs (Fig. 6.25).

In addition, uncontrolled burning of industrial and domestic waste adds to air pollution. Nearly 64% of 7257 industries registered with KSPCB in BMR operate under the Air Act. However, industries registered with KSPCB constitute only a fraction of 70,000 industries operating in BMR, considered as actual polluters (GoK 2009: 214).

6.3.27 *Impact of Air Quality on Health*

Similarly, studies have pointed to air pollution causing serious health problems among the urban dwellers. A study conducted by Lvovsky (1998) on health damages caused due to air pollution and high levels of particulates with reference to 126 cities of the world finds India, representing 12 largest cities in the sample including Bengaluru, losing 12 disability-adjusted life years (DALYs) per 1000 residents. As regards the economic cost involved (in terms of monetary cost), 9% of the respective income (GDP per capita) is spent on health damage control. This implies that direct productivity loss in the 12 largest cities is as high as 10% of the income generated. In addition, a social cost worth $ 3 million is incurred due to air pollution of which 64% is spent on health cost. Another study by WHO (2003) has estimated that, globally, 800,000 people die prematurely of air pollution-induced lung cancer and cardiovascular and respiratory diseases with 150,000 of these deaths occurring in South Asia alone and an estimated 630 from Bengaluru die prematurely due to air pollution (Honaganahalli 2014).

A CSD study (2012) and 'State of Environment Report' Bengaluru (2009: 130) on indoor[81] air pollution in Bengaluru City points out various sources, such as dust, pollen from trees and plants, volatile organic compounds and indoor cleaning agents, etc., as responsible for causing indoor air pollution and thereby health problems.

The CSD study (2012) reveals that (i) 65% of the respondents residing near main roads or highways report health problems; (ii) more than 75% of the offices situated near main roads are affirmative about being victims of dust pollution; (iii) 2% of the respondents complain of health problems due to the presence of trees and plants near their homes; and (iv) those who have pet animals (13%) complain of health issues. It is found that about 40% of them suffer from either cough or cold, 38% from sneezing and 13.4% from skin allergy, while 35.4% from headaches due to poor indoor air quality.

As per the 'State of Environment Report', Karnataka (2003), the incidence of respiratory diseases in Karnataka (like bronchitis, bronchiolitis, pneumonia and asthma) is increasing at the rate of 1.1% per annum. The vulnerable regions often exposed to industrial pollution happens to be Bengaluru, Bellary, Gulbarga, Chikkaballapur, Udupi, Mangalore and Mysore. Similarly, indoor air pollution is the third most important risk factor particularly affecting the urban poor without access to water supply and sanitation. Besides, 17% of the children underage five face death due to malnutrition (GoK 2009: 130). A study by Dr. Parmesh (2006) reveals that, between 1979 and 2004, the incidents of asthma attacks increased considerably in Bengaluru (GoK 2009: 149). Further, the city of Bengaluru may soon take on a new tag, i.e. 'Lead City,' in view of the presence of lead in the atmosphere being more than the expected levels. It seems that Bengaluru's 'lead scenario' is

[81] Due to indoor air pollution, India accounts for 28% of deaths in the world which is attributed mainly to the use of domestic firewood in rural households (World Bank 2000).

worst in the country which is apparent from the blood samples of youngsters (GoK 2009: 154). As per the 'Environment Report Card', Bengaluru, a study conducted by CSD (2012) covering eight zones of BMR reveals that 35–37% of them suffer from air-borne diseases, more so in the newly merged areas. Another study, carried out by HAWA-GTZ on the potential occupational health hazards of informal e-waste recyclers, finds lead values in the air beyond permissible limits, indicating the health risks associated with the informal e-waste recycling (GoK 2009: 155).

Much of the air pollution in the city is attributed to dust from the city's open spaces, road sides and unpaved footpaths, in addition to the dust generated from construction activities. BESCOM is responsible for 7% of air pollution by not being able to supply adequate power, compelling business and industrial establishments to use diesel–/kerosene-run captive power generator sets which emit fine particulates.

6.3.28 Noise Pollution

With an increase in the vehicular population crossing 3.5 million in Bengaluru City, noise pollution[82] has increased to alarming levels. A study conducted in Bengaluru[83] points out that noise pollution at Victoria Junction is of the highest order, recording between 82 and 86 decibels.[84] The study has further identified noise[85] 'hotspots' which exceed 100 decibels such as K.R Circle, S.M. Junction, Peenya, Silk Board, etc. (GoK 2009: 174).

But noise levels in major commercial areas reach beyond permissible limits: MG road, 105 db; Majestic Bus Sand, 110 db; Majestic Railway Station, 100 db; and Marathahalli,105 db (DNA 2012). Basically, 'the citizens of IT capital, Bengaluru, lack knowledge regarding noise pollution', as observed by Karnataka High Court Chief Justice D.H. Waghela (The New Indian Express 2013). The permissible ambient noise levels in residential areas, as specified by the Central Pollution Control Board (CPCB), are 55 dB (A) during daytime from 6 am to 10 pm and 45 dB at night from 10 pm to 6 am (Fig. 6.26).

A study conducted by CPCB at important locations of the city to assess the magnitude of noise pollution reveals that both at Victoria Hospital and City Market areas, noise levels are beyond the permissible levels. Similarly, a baseline study conducted by Bengaluru Metro Rail Corporation Limited (BMRCL) reveals that

[82] Ambient air quality standards are prescribed with respect to noise levels of industries, commercial, silent and residential zones under The Environment Protection Act 1986, Schedule II (CSD 2009a, b: 5). Similarly, noise standards for automobiles, domestic appliances and construction equipment have been notified in Part 'E', Schedule VI of Environment (Protection) Rules, 1986, as amended on May 19, 1993 (CSD 2009a, b: 9).

[83] State of Environment Report, Bengaluru, 2008.

[84] The permissible limit is 50 decibels.

[85] As specified by the Central Pollution Control Board (CPCB), the permissible ambient noise levels in residential areas are 55 dB (A) during daytime from 6 am to 10 pm and 45 dB during night from 10 pm to 6 am (GoK 2009: 151).

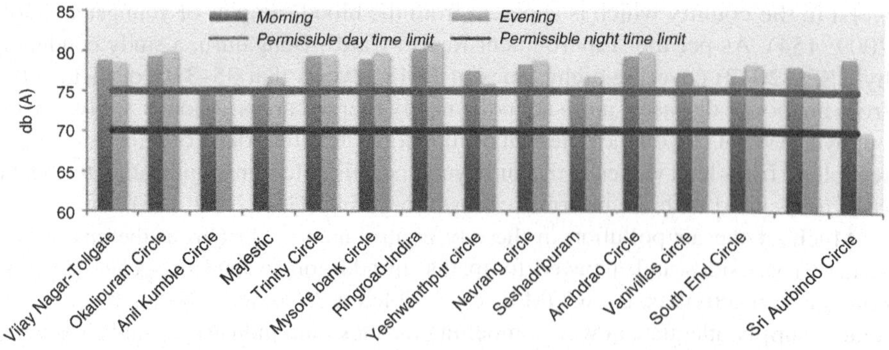

Source: GoK (2009: 131)

Fig. 6.26 Traffic noise monitored by CPCB

Table 6.55 Traffic noise monitored by BMRCL

Locations	L-day	L-night
Tumkur road-ring road junction	80.5	76
Yeshwanthpura Railway Station	76.8	74.2
MKK road-west of chord road junction	76.4	71.7
Majestic circle	78.9	74.0
KR road-Albert victor road junction	75.7	68.7
Harischandra Ghat on MKK road	76.5	70.8
Siddaiah Chowk/circle	76.0	69.3
South end road-RV road junction	76.8	69.7
Byppanahalli depot near Police Station	76.5	69.1
Road-SV road junction	76.6	69.3
CMH Road-100 ft. road junction	76.9	70.5
Trinity circle	76.0	70.5
Queens road-MG road junction	75.3	71.7
Opposite to high court	75.7	71.6
Central Leprosorium – Magadi road	74.8	69.3
Magadi road-chord road junction	76.4	70.3
Magadi road-Mysore road junction	77.9	73.7

Source: GoK (2009: 132)

noise levels at 17 locations (see Table 6.55) in the city have crossed the permissible limits (GoK 2009).

From Table 6.55, it is clear that noise levels measured in the selected areas cross the prescribed decibel levels in the city.

A study conducted by CSD (2009a) points to noise levels in the south zone (see Table 6.56) of Bengaluru City consisting of residential, commercial, silent and industrial areas. The study was conducted in 12 out of 33 wards. The study reveals that the noise levels prevailing in the silent and industrial areas are above the

Table 6.56 Noise pollution levels in the selected areas of Bengaluru City (2006)

Areas	Noise levels (in decibels)
Jayanagar 4th block	82
South end circle	82
JC road	80
Mekhri circle	100
KH road	95
Wilson garden	82
BTM layout	79
Forum mall	78
Brigade road	98

Source: BBMP (2009: 105)

prescribed standards and are almost twice the prescribed standards. In commercial and residential areas while the noise levels are normal during the day, during evening the intensity will increase, contributing to noise pollution. According to the 'Environment Report Card', CSD Bengaluru (2012: 34–35), while 84% have reported noise pollution at intolerable levels, 50% of them have reported noise pollution due to vehicular traffic. The noise monitoring levels at all 8 zones covering 23 residential areas and 15 hotspots in Bengaluru reveal that 75% of them agree that noise levels have increased beyond the permissible limits.

6.3.29 Urban Climate Change

Most of the literature associates urban climate change with uncontrolled population growth and rapid urbanisation that continuously pose dangers to the environment and living conditions of those mostly concentrated in informal settlements and slums. The convergence of urban development and environment at a rapid scale threatens to produce unprecedented negative impacts on the quality of life in the context of urban climate change. The problem is further accentuated by the consequences of key urban challenges which are mostly human induced like (i) greenhouse gas emissions[86] both commercial/domestic and industrial, (ii) innovations in technology and growth of industries which restrict the capacity of mitigation (coping) and adaptation mechanisms, (iii) land use changes, (iv) increase in energy use and (v) tampering with 'eco-zones', particularly the topography consisting of water bodies, biodiversity, coastal regions, wetlands and dry lands (UNHSP (2011: 3).

[86] As against developed countries contributing 47% of global CO2 emissions, the developing countries generate only 25% of the per capita emissions of the developed countries. In fact, three developed countries, namely Australia, the USA and Canada, account for the highest per capita CO2 emissions (UNHSP 2011: 9).

Table 6.57 Rainfall recoded in Bengaluru City (1990–1999)

Year	Rainy days	Rainfall (mm)	Litres/100 sq.m
1990	42	509.40	40,752
1991	65	1338.50	1,07,080
1992	56	844.60	67,568
1993	65	1059.70	84,776
1994	45	587.10	46,968
1995	61	1072.20	85,776
1996	64	1173.30	93,864
1997	52	717.40	57,392
1998	68	1431.80	1,14,544
1999	52	1009.40	80,720
Average	57	974.34 mm	77,987

Source: Ramachandra and Mujumdar (2009: 52)

6.3.29.1 Urban Climate Change in Bengaluru

The city of Bengaluru is one of the fastest-growing Indian metropolitan cities after New Delhi (decadal growth rate being 51.93%). The population of Bengaluru City grew tremendously from 41.60 lakh to 56.86 lakh over the decades 1991–2001, a decadal increase of 37.7%, which is more than New Delhi (BBMP 2006: 14). Bruhat Bengaluru[87] was created by incorporating 7 CMCs (City Municipal Councils), 1 TMC (Town Municipal Council) and 110 villages. The net increase in the population of Bengaluru City was 22% between 1991 and 2001 (BBMP 2006) with the city recording 45% of in-migration (BBMP 2006: 14). The topography of the city constitutes an uneven landscape with hills and valleys. The undulating terrain is highly suitable for the development of water bodies like lakes and tanks around and within the city, the largest of them being Madivala Tank, Hebbal Lake, Ulsoor Lake and Sankey Tank (BBMP 2006: 11) (Table 6.57).

6.3.29.2 Factors Underlying Climate Change in the City

Within and across, the city is witnessing (i) urban sprawl, (ii) increase in population density, (iii) changes in land use pattern and (ii) spatial accumulation in terms of intense economic, social or cultural activities (associated with production, consumption and distribution of industries, particularly services sector, etc.) (Ramachandra et al. 2012: 330) tremendously impacting the local environment. The phenomenon of urban sprawl, which is defined as 'uncontrolled, scattered

[87] In 2007, the Government of Karnataka (GoK) issued a notification for the merger of 7 CMCs, 1 TMC and 110 villages with the erstwhile Bengaluru Mahanagara Palike (BMP) area. The process was completed between April and May 2007, and BBMP came into existence (BBMP 2006: 24).

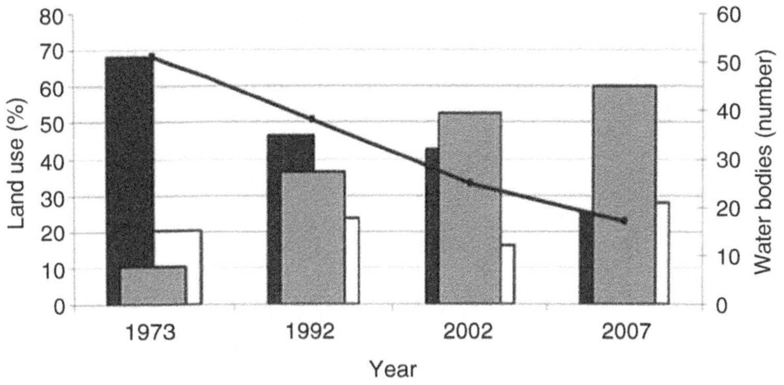

Source: Ramachandra and Mujumdar (2009: 50)

Fig. 6.27 Changes in the built-up area and their impact on vegetation and water bodies

suburbanisation', results in major irreversible changes in land use which impact the carrying capacity of the city, thus depleting the local resources. Nearly 7% of the land (as per the BDA land use pattern) is used for both commercial and industrial purposes in the city. Open spaces have considerably reduced post-1971. Nearly 50% of the open spaces,[88] between 1971 and 1981, were converted into residential areas (Mishra 2006). The city has shown a considerable increase in the number of slums with a concomitant increase in the urban poor population from 4.30 lakh as per 2001 census to 7 lakh, which is 10% of the total population of Bruhat Bengaluru Mahanagara Palike (BBMP) area (up to 2006) (BBMP 2006: 20). Much of the built-up area has been converted into industrial/commercial/residential areas. Already, the man-made wetlands in the BCC and BDA areas have fallen from 262 in 1960 to merely 81 lakes (Mishra 2006). A study by Ramachandra and Mujumdar (2009) on urban floods in Bengaluru clearly indicates the decline of water bodies to 34.48% (during 1973–1992), and therefore, a total percentage decline of 56.90% of water bodies has been reported for the periods 1973–2002. Further, a decline of 70.69% is reported for the periods 1973–2007 in the erstwhile Bengaluru City limits. In addition, the study further indicates a decline of 32.47% (during 1973–1992), 53.76% (during 1973–2002) and 60.83% during 1973–2007. The disappearance and encroachment of water bodies and wetlands in Bengaluru reveal a decline in the groundwater table from 10–12 to 100–200 m over the past 20 years (*ibid*: 48) (Fig. 6.27).

A study by Ramachandra and Mujumdar (2009) points out the loss of aquatic ecosystems (70% decline over the period 1973–2007 in the Bengaluru City limits) and vegetation following an increase in the built-up area (466% over the period 1973–2007) which in turn is the result of an intense urbanisation process.

[88] Open spaces constitute play grounds, parks and vacant lands.

In the context of a rapid process of urbanisation, some of the key problems affecting the city's climate are (i) loss of ambient temperatures due to intense population activities and density and (ii) increase in humidity due to the loss of water bodies and wetlands followed by the conversion of green space and agriculture land to industrial and commercial purposes.

6.3.29.3 'Urban Heat Islands' and Greater Bengaluru City

'Urban heat island' is a conurbation associated rise in the temperatures of the core/inner city/metropolitan area rather than the surrounding rural areas. This phenomenon is caused by changes brought about by rapid urban development that uses materials which significantly retain heat. The spatial expansion is one of the most important reasons behind the creation of urban heat islands (UN-HABITAT 2010: 200). The term 'urban heat island' refers to increased surface temperatures in some pockets of a city, caused by an ever-changing micro-level climate. Some of the significant reasons for the emergence of urban heat islands are an increase in the surface and atmospheric temperatures due to an excessive energy consumption, increase in the presence of artificial materials with high heat capacities and conductivities and a reduction in the vegetation or green cover and water bodies which have the capacity to substantially reduce temperatures through evapotranspiration as also a considerable loss of aquatic ecosystems (Ramachandra and Kumar 2010: 2). Pollution and heat due to vehicular traffic, industrialisation and human activities are the other contributing factors.

The city is witnessing an unprecedented urbanisation process in terms of a demographic explosion and immense spatial transformation. As a result, the process of urbanisation and physical environment cannot be delinked. The studies claim direct linkages between tampering with land and water bodies (see Table 6.58) causing urban heat islands in the city (Ramachandra[89] and Kumar 2010: 2). In fact, due to an unplanned urbanisation, the city has lost coping mechanisms in terms of drastically altering the natural characteristics of catchments or drainage areas through excessive pollution with volumes of untreated solid waste. Similarly, the encroachment of biodiversity-rich wetlands and flood plains has caused a severe damage to the natural ecosystem. Frequency of floods in the city's low-lying areas has increased causing enormous human and material damages (*ibid;* Ravindra 1996) – all leading to the emergence of 'urban heat islands' in the city.

The city of Bengaluru is known for a salubrious climate mostly throughout the year. The temperatures oscillate between 16 and 33° C with an average of 24 °C. A study by Mishra (2009) clearly indicates that though mean monthly maximum temperatures have more or less remained constant, the mean monthly minimum temperatures have increased in the city over the years (see Fig. 6.28).

[89] Dr. T.V. Ramachardra, lead researcher, Centre for Ecological Sciences of the Indian Institute of Science, Bengaluru, observes that an increase of 466% in the built-up area over the past 3 years (from 2009) is not allowing water, both domestic and drainage, to percolate through the soil (Jayashree 2009).

Table 6.58 Loss of water bodies in Bengaluru Metro Region between 1973 and 2007

Year	Bengaluru City		Greater Bengaluru	
	Number of water bodies	Area in (ha)	Number of water bodies	Area in (ha)
SOI	58	406	207	2342
1973	51	321	159	2003
1992	38	207	147	1582
2002	25	135	107	1083
2007	17	87	93	918

Source: Ramachandra and Mujumdar (2009: 9)

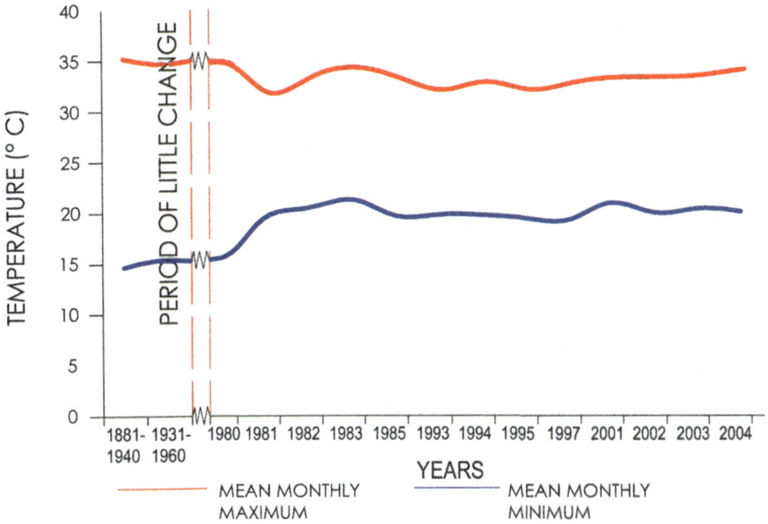

Source: Indian Meteorological Department, New Delhi (Mishra 2006)

Fig. 6.28 Temperature variations in Bengaluru City over the years

The mean annual rainfall comes to 875 mm for 50 days of a year (between the months of July–August and October–November). The city also receives cyclone-induced rainfall between the months of November and December (GoK 2009: 1). The average rainfall works out to 923.7 mm, but the soil receives only 190 mm of water during the rainy season (as against the capacity of 300 mm) leading to water deficiency (Ravindra 1996: 54). Urban experts and environmentalists (Ravindra 1996; Ray 2010; Deepthi 2012) point out and have attributed many reasons for the emergence of Bengaluru as an urban heat island like the failure of an effective town planning and the city's rising temperatures. The fault lies in the unsustainable methods used in the construction of buildings[90] and infrastructure in the city which trap

[90] As per the International Energy Agency, buildings account for 40% of the total energy consumption and 24% of greenhouse gas emissions in the world (UN-HABITAT 2010: 21).

sunlight, causing intense heat. Increased energy consumption has inevitably led to heat radiating outwards. The average temperature of the city has gone up by 2 °C in 1986 to 3 °C by 2010 (Ravindra 1996: 57; Ray 2010; Deepthi 2012). Further, as regards town planning, out of 1200 sq.km, 800 sq. km was planned for green space, but in the subsequent blueprints of town planning, it has got reduced to 500 sq.km (Ray 2010). An alarming reduction in wetlands and local vegetation has had a direct bearing on the local temperatures. Loss of green space in terms of urban parks and vacant lands has led to an increase in the temperature and levels of radiation. The situation is further accentuated by pollution and emissions of GHGs by a growing traffic in the city and faulty architecture that have resulted in soaring temperatures of the city (GoK 2009: 25; Jayashree 2009). The city is witnessing a tremendous increase in vehicular traffic, giving rise to high levels of carbon dioxide. An excessive consumption of fossil fuels (like petrol and diesel) has increased carbon emissions to the extent of 390 ppm (parts per million) between 2010 and 2012 (The Hindu 2012c).

There is a considerable reduction in water bodies like tanks from 200 in 1985 to only 17 visible tanks by 2009. The main reason attributed to the disappearance of the city's rich water bodies is rapid urbanisation. Many lakes have been encroached upon for industrial/commercial and residential purposes such as for the construction of illegal buildings (54%). The field survey observations (between July and August 2007) reveal that nearly 66% of the lakes are sewerage fed and that 14% of them are surrounded by slums and 72% show a loss of catchment area. These catchment areas are used either for dumping municipal solid waste or building debris (Jayashree 2009). Such a loss of catchment areas along with evaporation has led to increased humidity levels in the city.

6.4 Identification of Environmentally Sensitive Zones (ESZs) in BMA

Environmentally sensitive zones have been identified in Bengaluru Metropolitan Area (BMA) based on an analysis of the overlapping sectors, i.e. water supply, sanitation, groundwater exploitation, etc. Accordingly, the following zones are categorised and presented in Table 6.59.

From the table (see Table 6.59), it is very clear that Bengaluru North, South and South-East fall under environmentally sensitive zones (ESZ). The causes which attribute to Bengaluru North as the most stressed or acute (as RED zone) which are as follows: the north zone accounts for (i) highest encroachment and (ii) overexploited zone of groundwater resources. In particular, for the south zone, the causes attributed are (i) highest groundwater extraction and the resultant depletion; (ii) existence of a large number of borewells; and (iii) south zone being the largest hazardous waste generator. Similarly, the south-east zone faces the problem of lowest supply of drinking water, and as a result, a majority of the residents are dependent on groundwater resources.

Table 6.59 Identification of environmentally sensitive zones in MBA

Zones	Overall scoring	Reasons for scoring
North	Red	Highest green belt encroachment Overexploited zone of groundwater resources
South	Red	Highest groundwater consumption/extraction and depletion More number of borewells Highest hazardous waste generator
Central	Orange	Green belt encroachment Air pollution
East	Orange	Severe shortage of drinking water Highest conversion of agricultural land to nonagricultural/industrial purpose
West	Orange	Lowest supply of drinking water Second highest groundwater consumption
South-east	Red	Lowest supply of drinking water

Bengaluru North is a combination of industrial areas like Peenya, academic institutes, large-scale industries and green belt areas. The zone includes Banaswadi, HBR Layout, Hebbal, Hennur, Jakkur, Jalahalli East and West, Peenya Industrial Area, Sanjeevani Nagar and Yeshwantpura. Meanwhile, areas surrounding Puttenahalli constitute a green belt area. The north zone has a vast area of eucalyptus and bamboo plantations and also horticultural and botanical gardens. This area is marked with various academic and research institutes and colleges. In addition, a few national and state-based industries are located here such as BHEL (Bharat Heavy Electronics Limited), BEL (Bharat Electronics Limited), Hindustan Liver and HMT (Hindustan Machine Tools) around Jalahalli area. The areas in the north zone has been encroached upon for various purposes by way of converting agriculture land to nonagricultural purposes. The study findings corroborate the fact that between 1984 and 2007, the green belt zone got drastically reduced to mere 262.5 sq.km (see Table 6.40), while the built-up area of the city has increased by nearly 62% (BMRDA 2009: 265). A study by Ramachandra and Mujumdar (2009) clearly reveals that the Revised CDP 2015 by BDA covers the entire 1279 km^2 of the BMA, comprising 565 km^2 of 'conurbation area' and 714 km^2 of green belt region. The CDP actually permits developmental activities in the zone, while extensive construction activities[91] have taken place in the green belt area.

Similarly, the south-east zone is a major destination for the IT/BT sector. The area under Mahadevapura zone now constitutes south-east zone consisting of Ramamurthy Nagar, Vijinanapura, KR Pura, Hudi, Devasandra, A Narayanapura, Vignana Nagar, Garudacharpalya, Kadugodi, Hagadur, Dodda Nekkundi,

[91] Conversion of Agricultural land to nonagricultural purposes amounts to the violation of the Karnataka Land Reforms Act and Karnataka Land Revenue Act (Ramachandra and Mujumdar undated).

Marathahalli, HAL Airport Ward and Varthuru. Well-established industries are spread across Mahadevapura and Sadaramangala (i.e south-east zone) including a eucalyptus plantation cover followed by the Whitefield green belt area. In addition, several lakes like Varthur Kere, Bellandur Kere, Yallamallappachetty Kere and KR Puram Kere drape this part of the city.

Cauvery water from BWSSB enters the city from the south zone, and as a result, its residents receive a good supply of drinking water. Yet, with the development of Kanakapura road for nonagricultural purposes, the groundwater extraction through borewells is reported very high in the south zone. Koramangala situated in the south zone has gradually developed into a commercial and cosmopolitan hub of Bengaluru City. It is one of the largest neighbourhoods besides being a highly sought after residential locality with wide tree-lined boulevards and a mix of luxury apartments, commercial structures and posh bungalows, hence one of the high-price-propertied areas in the city. Further, south zone is one of the highly commercialised areas with shopping centres, malls and restaurants and naturally with increasing traffic and population; the city faces a serious threat to its environment.

6.5 Environmental Conflicts in Bengaluru

Rapid urbanisation has exposed the city to a wide range of environmental challenges with respect to the management of natural resources and protection and conservation of ecosystems.

As a result, urban residents are often exposed to environmental disasters and conflicts – human and climate change induced. The impact of such conflicts and disasters is often disproportionately borne by the most vulnerable sections of the urban society like the urban poor, temporary settlers like migrants and low-income dwellers. Most often degraded or misused or overused natural resources and poorly managed ecosystems can lead to social conflicts over dwindling water, food or fuel resources, or encroachments of urban ecosystems can increase the exposure of urban residents to vulnerability in the form of urban floods or urban climate change (Table 6.60).

Table 6.60 Typology of environmental threats in urban areas

Urban issues	Typology of environmental threats/challenges
Encroachment of urban lakes	Urban floods
Poor management of waste	Mavallipura conflict Mandur
Increase in temperature	Urban heat islands

Table 6.61 Urban floods in Bengaluru Metropolitan Region (BMR): causal factors and key impact on environment and ecology

Causes	Effects	Environmental and ecological impact
Urbanisation and urban sprawl	Demographic and spatial explosion Increase in in-migration	Changes in land use pattern Encroachment of land
Unplanned urban growth	Haphazard development of the city Inadequate provision for main drains Development below the flood level Construction of new services (water, telecom, electricity) Reclamation of water bodies like lakes and tanks for various developmental activities	Increase in informal settlements or slums alters the topography prone to flash floods and increase the frequency of floods Encroachment of open running drains Causes gradual reduction in waterway capacity due to encroachments Obstructs the flow of drains/nalas thereby causing floods Loss of water network for storage leading to floods
Growth of industries	Increase in the generation of hazardous waste Reduction in storm water capacity Water bodies infested with weeds lead to unstable water ways Dumping of solid waste and other hazardous waste Litter, dumping waste from debris, sediments	Causes blockage and pollution in the drainage system significantly increasing surface floods Loss of vegetation eroded riparian zones Causes siltation of downstream water ways and blockage of drains causing floods

Source: Ramachandra and Mujumdar (2009: 66)

6.5.1 Urban Floods in Bengaluru

Urban floods are a consequence of susceptibility to climate change and related disasters in cities. Increased temperatures and severe precipitation rates are other factors that can cause greater incidents of urban floods (Satterthwaite et al. 2007: 15; UNHSP 2011: 169). A rapid urbanisation with an emphasis on economic growth has cumulative environmental consequences such as poor quality of air and water, high levels of traffic congestion, poor quality of built-up environment in informal settlements/slums, emergence of urban sprawls due to drastic changes in the land use pattern, greenhouse gas emissions and the generation of hazardous waste (Satterthwaite et al. 2007: 12; Gupta and Nair 2011: 1648) (Table 6.61).

There are about 36 flood-prone regions within Bengaluru Metropolitan Region (BMR) (Ramachandra and Mujumdar 2009: 43–44).

1. The two most significant factors responsible for an increase in urban floods are (i) land use changes like encroachments and conversion for residential, industrial and other commercial purposes and (ii) loss of water bodies. This reduces the absorbing capacity and interconnectivities in the undulating terrain of the city. An intense

urbanisation inevitably results in the transformation of natural topography and alterations in the hydrological flow and shifts in the chemical and biological set-up of the storm water drains (Ramanchandra and Mujumdar 2009: 55).

2. Similarly, the loss of water bodies like lakes and tanks in the name of developmental activities has put undue stress on the local ecology. In 1961, the lakes and tanks in the city numbered to 262. In a study, Ramachandra and Mujumdar (2009: 5–7), while carrying out a temporal analysis of water bodies in the city, observed a sharp decline of 58% in Greater Bengaluru (i.e. BMR) due to an increase in the built-up area (466%) over the period 1973–2007. There is also a considerable loss of open spaces (the built-up area which stood at 16% in 2000 is currently estimated to be around 23–24%). The study clearly shows a decline in the water bodies from 32.47% in 1973 to 1992 to 53.67% in 1973–2002 and to 60.83% in 1973–2007. This could be attributed to the use of lake catchment areas for residential, commercial, industrial and other developmental activities. In addition, they are used for dumping illegal waste by industries and households.
3. Increase in various types of pollution, particularly air pollution, due to an increase in the number of vehicles and consequent traffic congestion. This has resulted in an increase in the emission of greenhouse gases (CO2) creating urban 'heat islands' (a sudden rise in the local temperatures).
4. Due to an increase in in-migration to the city, the slum population has risen manyfolds. In 2006, the KSCB declared the presence of 542 slums. Now it is official that the slum population has been increasing in the city of Bengaluru unabatedly, from 473 in 2003 to 597 in 2013 (from 23% in 2001 to 30–40% by 2013). According to official records, 13.86 lakh out of 84.25 lakh people in Bengaluru (i.e. 16.45%) live in slums (CED 2011: 3; Kumar 2013). But unofficial records state that the city hosts more than 1500 slums, i.e. every third person lives in slums (Rao 2011). It is apparent that the concentration of such a huge migrant population in the city has put a considerable stress on the availability of land, and as a result, they are induced to occupy private or vacant lands on the edges of running drains or open drains/nalas, thereby exposing themselves to floods during rains.
5. The poor maintenance of infrastructure can be attributed to urban floods. Most parts of the sewer lanes have been or illegal constructions and buildings encroached by migrants. Such activities have led to siltation/corrosion of sewers and diversion of sewers either for agricultural purpose or industrial activities, leading to flooding of the entire surrounding and low-lying areas.
6. The depletion of water bodies and discharge of sewerage and waste water into storm water systems lead to floods during rainy season. Also, the increasing volume of run-offs enhances the risk of urban flooding.
7. Lack of drainage upgradation and increase in land use change in the catchment from open area to high-density urban development activities inevitably lead to floods.

6.5.2 Ecological and Environmental Impact of Urban Floods

1. Urban floods inevitably lead to the destruction of human settlements and property. The situation is more serious in informal settlements with slums getting washed away during floods. The problem is accentuated by water stagnation and poor sanitation practices in the surrounding areas of slums, leading to serious health hazards especially among women and children.
2. Street flooding especially causes a severe distress to the commuters and the general public, leading to traffic congestion, and disrupts daily mobility causing loss of working hours and business opportunities.
3. Precipitation, intensity and duration are the most significant factors associated with urban floods that cause considerable damages to properties in terms of cost.

6.5.3 Mavallipura Waste Dumping Site[92]

The city population has reached 80 lakhs (2011) and is likely to reach 1 crore in the immediate future. The city generates about 5000 tonnes of solid waste daily (Saldanha et al. 2012; BBMP 2009). Studies have clearly shown that though eight landfill sites have been identified by BBMP for dumping, segregation and management of waste, none of them is functional, and some of them are even under litigation. Non-committal and poor governance have led to weak management practices with respect to solid waste in the city. As a result, BBMP has been illegally dumping municipal waste in and around the city and the surrounding villages posing serious ecological and environmental threats, besides affecting the health of local communities.

One such case is Mavallipura waste dumping site located at Mavallipura. BBMP has allowed the use of 32 acres of forest land for dumping municipal waste and the private operator Ramky Infrastructure Pvt. Limited. Over the years, the two landfill sites have accumulated waste exceeding 1500,000 tonnes, threatening the livelihood and health of the local communities. The landfill sites have been gathering/accumulating municipal waste from 2007 onwards, affecting nearly 12 village communities north of Bengaluru City near Yelahanka. The private operator has completely failed to comply with legal norms and procedures of waste dumping and segregation. The company dumps massive piles of waste, mostly unprocessed, into massive pits (over 40 feet deep) affecting pastures, local open wells, streams, lakes and tanks surrounding the region.

[92] The case of Mavallipura landfill site involving the local community's struggle has been documented by the Environmental Support Group (ESG) which is also a partner in the struggle.

6.5.4 *Environmental Effects of Illegal Dumping of Waste in Mavallipura*

(i) Illegal dumping of waste at Mavallipura site has immensely affected the health of local communities, triggering high levels of morbidity.
(ii) A study by ESG (Environmental Support Group), a NGO, has reported three deaths over a 3-month period due to kidney failure, asthma and jaundice in view of living close to the Mavallipura landfill.
(iii) The health condition of women, children and elderly seems to be deteriorating constantly with a variety of chronic and infectious diseases.
(iv) Morbidity levels of animals like sheep and cattle are reported extremely high as they are prone to drinking contaminated water and grazing polluted plants and grass. Particularly, dogs are reportedly infected, attacking both humans and animals near the landfill.
(v) The landfill site is very close to river Arakavathy, 5 km away, a major source of drinking water for the population of Bengaluru City which will contribute to ground water contamination.

6.5.4.1 Efforts of Local Communities to Close down the Landfill Site

The local community efforts were organised under Dalit Sangharsh Samiti (Samyojaka) and Environmental Support Group (ESG). Following consistent efforts over the years to close down the landfill site, KSPCB (Karnataka State Pollution Control Board) issued a notice to Ramky Pvt. Ltd. (in gross violation of the Environment Protection Act and Municipal Solid Waste Management Rules, 2000) with 7 days' time to close down the landfill site by August 2011–2012. Later, a PIL (Public Interest Litigation) was filed with the High Court of Karnataka, and this was followed by landmark decision by Mr. A.S. Sadashivaiah, IFS (Retd), KSPCB, directing the commissioner of BBMP to immediately stop and close the private operations under Ramky and the landfill site.

The community struggle of Mavallipura has pledged to continuously monitor the implementation of the decision and not to rest until the entire area is decontaminated and the affected families appropriately compensated.

6.5.5 *Mestripalya Case*[93]

In the name of development and urban growth, several water bodies which include lakes and tanks have been either encroached upon or used as sites for dumping hazardous and solid waste. Till 1960, there were reportedly about 262 major water

[93] Details of which were collected through discussions with Koramangala Residential Welfare Association.

bodies of different sizes in Bengaluru with fresh water. These have since declined to about 81 of which only 34 are recognised as live lakes. These water bodies are not suitable for living beings. Each of these has been turned into a virtual cesspool of filth and polluted sewage. The Government and the designated development agencies, while professing the need for 'sustainable development' through series of master plans and developmental plans, have, in fact, prevented sustainable development. But communities with their determined efforts across the city of Bengaluru have quietly crusaded to retrieve some of the water bodies and open spaces in Bengaluru. The retrieval of Mestripalya Tank, in Koramangala, is a case in point.

Mestripalya Lake, located in an area occupying 17+ acres next to Mestripalya Village in Jakkasandra, has been part and parcel of more than 300 odd acres of land acquired by the Government for developing what is now called Koramangala Layout.

This measures a total of about 17+ acres and is divided into two parts: (a) 11.2 acres of tank bed and (b) 6.2 acres of adjoining area. Koramangala-Challaghatta Valley is a major system through which run-off water flows via a chain of water bodies. In the 1980s, a rapid growth of population resulted in an abnormal demand land which, in turn, unfortunately resulted in an unscrupulous conversion of land into built-up areas. Water bodies and valleys were the prime spots for conversion and illegal occupation. In the process, Challaghatta Lake, Chinnagara Lake, Koramangala Lake and vast tracts of low-lying areas were encroached upon with some even converted officially by the Government (parts of Koramangala Layout and ST Bed now have replaced Chinnagara Lake). Alarmed by a rapid and unscrupulous conversion of environmentally sensitive areas, a high-level committee chaired by N Lakshman Rau was constituted by the Government in 1988, which, in turn, made some very significant recommendations. However, while accepting the recommendations, and gazetted by way of a government notification (with a few exceptions), no effective mechanism was put in place to enforce the recommendations. Further, in the original layout plan of Koramangala Layout, on the basis of which land had been acquired, both these areas were marked as 'open spaces'. After BDA took over from CITB (and till May 2009), it showed the area as 'Open/CA'. In 1995 CDP of BDA, the entire area was classified as 'public/semipublic'. Consequently, in one corner of the area, a telephone exchange was set up. In 2004, the remaining area of the tank bed was brought under the supervision and care of the Forest Department (Lake Development Authority) to be developed as 'tree park' in accordance with Lakshman Rau Report of 1988. As a result, Koramangala continued to lose its vital lung space and environmentally sensitive areas. Alarmed by the rapidly deteriorating situation, environmentally and socially active residents of Koramangala decided to fight the forces that were unscrupulously damaging their neighbourhood and thus started the struggle for restoration of Mestripalya Tank. 'Save Koramangala' and 'The Koramangala Initiative' later joined by 'Citizens Action Group', with active and fulsome support of citizens living in the immediate proximity of Mestripalya Tank, spearheaded the crusade.

With a series of court cases ensued between the community of Koramangala and BDA (Bengaluru Development Authority) (Civil Appeal No. 971 of 2003), but following a writ petition, the High Court of Karnataka did not permit any construction activity in the existing open spaces besides directing BDA in 2001 not to allow any construction activities. Finally, an SLP in the Supreme Court in 2007 challenged the earlier court order on the ownership issue of the tank bed. The Supreme Court fixed November 30, 2009 as the date for final disposal, and the Government of Karnataka responded with its objections resulting in getting a favourable verdict from the Supreme Court. The residents quickly formed a group of technical professionals who, after a considerable research involving collection of an enormous amount of data, both ground level and historical, and interaction with several noted environmentalists and landscape architects, prepared a concept plan for the restoration of Mestripalya Tank.

6.5.5.1 Community Agitation Against Mandur Landfill

While the city of Bengaluru generates nearly 5000 tonnes of solid waste per day, out of which, around 1800 tonnes through 200 trucks is dumped into the landfill site at Mandur. The fact is, only 10–20% of the waste generated is treated, while the rest is sent to either Mandur or Mavallipura landfill site. Although the private companies are entrusted with the responsibility of scientifically disposing of the waste, the reality is they dump untreated waste that pollutes groundwater, in addition to affecting the health of villagers (Picture 6.24).

Photos: Chethan Shivakumar

Source: BBMP 2013

Picture 6.24 Mandur North: dumping site of unsegregated waste

Source: The Hindu (2014a, b)

Picture 6.25 Residents of Mandur staging a protest against the dumping of garbage by BBMP on June 2 2014

For more than 10 days, i.e. from June 2, 2014, communities residing near Mandur landfill erupted into an agitation against BBMP for dumping untreated garbage into the landfill. The communities formed vigilance squads for preventing the entry of garbage into Mandur. Initially, communities warned against the dumping of waste, giving 6 months to withdraw. But BBMP continued to pour piles of garbage into the Mandur landfill, seriously threatening the villagers in the vicinity (The Hindu 2014a, b). On June 3, 2014, all business establishments were closed and a rally organised against the dumping of waste into the landfill (Times of India 2014) (Pictures 6.25 and 6.26)

6.6 Green Initiatives/Activism in the City

Citizen groups can play a significant role in urban environmental stewardship, and network structure often influences function and performance. In Enqvist et al. (2014) paper on Bengaluru's citizen networks, he emphasise that citizen network functions as a platform that enables interaction between diverse interest groups and as a watchdog that monitors arks, lakes and trees to prevent further loss of fragmented urban ecosystems. He further in his study found that the network activities

Source: The Hindu (2014a,b)

Picture 6.26 Mandur dumping yard

are influenced by internal tensions between inclusiveness and efficiency and between internal and external legitimacies. Although core actors have central network positions, strong leadership or political alliances are not considered important; members instead prefer to emphasise transparency and democratic participation. This is important for monitoring Bengaluru's fragmented ecosystems and for raising public awareness and support. Findings indicate an urgent need to develop a comprehensive framework for urban environmental stewardship, to better describe potential roles of citizens in governance across diverse social, political and ecological conditions and during different periods of urban change.

6.6.1 Urban Agriculture: Organic Farming/Terrace Gardening/Vertical Farming

The Indian domestic market recorded Rs. 1000 crore (which is US $190 million) worth of organic produce in 2011–2012 (which is expected to grow further to US $1 billion by 2015). Most of the organic revolution has been taking place in cities in India. As per the estimates of a study conducted by Nina Osswald and Manoj Menon (2014), the total sale value of organic produce stands at Rs. 9.9 crore (US $2 million) in the case of Hyderabad City followed by Mumbai at Rs. 17.9 crore (US $3.4 million) and Bengaluru at Rs. 21.4 crore (US$ 4.1 million) (Osswald 2014). The city of Bengaluru is living up to its tag 'Garden City' by popularising urban agriculture. The efforts of residential communities, private individual organic farms and local NGOs are very prominent in promoting various urban agricultural initiatives. In fact, the city of Bengaluru can be referred as 'organic hotspot' for initiating organic movement in the country (Bhumik 2013). The city has more than 68 retail shops[94] (while Mumbai has 45 and New Delhi 36) selling organic produce[95] (Nandy 2011; Kumar 2012). A study conducted by Nina Osswald (2014) titled *Organic Food Marketing in Urban Centres of India* reports that Bengaluru has the highest number of organic food outlets in the country. The organic movement has been popularised since the 1990s by retired family members and NRIs who have returned from the USA and opted for producing their own agriculture produce, particularly vegetables on terrace or a piece of agricultural land. In fact, the garden city farmers have organised and initiated the concept of 'Oota From Your Thota' (OFYT) conducting various workshops and organic fairs at local levels on weekends. Such an organic movement has gained momentum, promoting general well-being and health of the communities, given the present chemical laden agricultural produce, particularly vegetables and milk. Many private individuals are engaged in popularising green visits' such as vocational or educational visits to their farms which are chargeable. Similarly, land is rented out for producing organic vegetables and fruits around the city. Various websites focused on organic produce are promoting retail shopping and awareness regarding organic produce.

6.6.2 Lake Adoption and Management by Communities

The city of Bengaluru was once rich in biodiversity consisting of water bodies such as lakes and tanks. But nearly 2/3 of it is lost to the urbanisation process such as developmental projects, land encroachment, land mafia and pollution. With the constitution of Lake Development Authority (LDA), there emerged the novel concept 'adopt-a-lake' scheme in 2004. Like-minded communities or private individuals can approach LDA for adopting lakes. This concept was mooted to protect and preserve wetlands and ecology of the city. The adopting agency or the individual has to adhere to the strict regulation and monitoring process by LDA. Besides, the adopting agency or communities have to bear the cost of restoration, rejuvenation and maintenance for a period of 5 years. Some of the core works like desilting, deweeding, fencing and diversion of sewerage can be carried out by the adopting agency. In addition, the agency can go ahead with its beautification besides and providing other amenities. Initially, the response to the LDA's adopt-a-lake scheme was lukewarm with only six lakes being adopted, namely, Sheelavanthakere, Challakere (in HRBR layout, adopted by the Royal Concord Education Trust),

[94] Most stores are concentrated in Jayanagar, J.P. Nagar and Bannerghatta Road in the south; in Rajajinagar, Malleshwaram and Dollars Colony in the north; and Whitefield, Indiranagar and Koramangala in the east (Bhumik 2013).

[95] The city each year conducts international organic trade fair earning business transaction worth Rs. 17.5 crore (for the year 2012) (Kumar 2012).

Kelaginkere (in Kaggadaspura, adopted by Bagmane Developers), Kundanahalli (adopted by Kalyani Developers), Mahadevapura (adopted by Abishek Developers) and Nayandahalli lakes (in Rajarajeshwari Nagar, adopted by the Rajarajeshwari Temple Trust). The adopting agency has to deposit 1% of the guidance value for the area in which the lake is located (a minimum of Rs. 10,000 and a maximum of Rs. 50,000 per acre) with the LDA (Yeshwanth 2007). In 2011, BBMP formally handed over the maintenance of Puttenahalli Lake to PNLIT[96] (Puttenahalli Neighbourhood Lake Improvement Trust) and Agara Lake Protection and Management Society (ALPMS), Kaikondrahalli Lake managed by MAPSAS Trust with the help of volunteers since August 2011 (2 years) for taking care of security, gardening and general maintenance (regular cleanups, de-weeding, etc.). The community trusts have been formed for the protection of lakes.

Despite plagued by many loopholes such as poor monitoring and implementation, the scheme has been successful in terms of protecting the rich heritage of lakes and tanks against encroachments, stopping sewerage flow and promoting awareness among local communities by involving them in the maintenance and preservation of lakes (ibid 2007). Meanwhile, local communities have initiated Save Bengaluru Lakes Trust (SBLT), sensitising people to the issues of preservation and protection of Bengaluru lakes (Yajaman 2013). Similarly, web pages[97] have been developed to save lakes in the city.

6.6.3 Role of NGOs and Community-Based Organisations (CBOs) in Green Initiatives

The city of Bengaluru is home for more than 20 NGOs working in various areas such as water supply, sanitation, urban poor, land and housing and green issues. Some of the NGOs specifically working on environmental and ecological issues in the city include Environment Support Group (ESG), Go Green, Keep it Clean, Rainwater Club, Saashas, SayTrees, Hasiru Usiru and Indian Youth for Climate Network. These NGOs are focused on issues such as water conservation techniques like rainwater harvesting, protection of biodiversity through lake conservation and protection, solid waste management, segregation of domestic waste, recycling domestic waste, etc. Similarly, many residential welfare associations are working on various green issues at community levels like waste segregation[98] and waste

---

[96] PNLIT has been actively working with Bruhat Bengaluru Mahanagara Palike (BBMP) towards rejuvenation and maintenance of Puttenahalli Lake, JP Nagar 7th Phase, Bengaluru. Visit the following link for more details on the Puttenahalli Lake: (http://www.puttenahallilake.in/).

[97] Visit www.agaralake.org.

[98] Access the article on Konanakunte Ward No 195 which is known for 100% segregation. (http://bangalore.citizenmatters.in/blogs/community-matters/blog_posts/volunteer-to-make-segregation-at-source-happen-at-konanakunte-ward).

recycling at household and apartment levels. A Bengaluru-based environment group called Eco Club[99] started 'I Own a Tree' which allows a person to sponsor and own a tree for 2 years. The concept of 'Waste to Wealth' has been initiated through which ragpickers and pourakarmikas are trained in waste collection and disposal methods. ITC Limited has come forward to adopt 12 wards in the city under the same concept (Rao 2013). The government of Karnataka has initiated a novel programme called 'Kasa Muktha', particularly in the south to make Bengaluru City a 'garbage-free city' in 6 months (Mahesh 2013). The largest-scale week-long expo called 'Wake Up Clean Up' was held in Bengaluru (Freedom park) in February to bring together various stakeholders from all platforms to initiate an effective dialogue for addressing the waste problem in the city (Vinitha 2013). Also Malleswaram Residential Association has initiated 'zero-waste' workshops and segregation and recycling of household waste.

6.6.4 Environmental and Ecological Awareness in Educational Institutions

Environmental Information System (ENVIS)[100] has initiated 'Know Your Ecosystem', an environmental education programme (in 1997) for conservation of ecosystems, environmental management and sustainable planning education for schools. Its holistic, participatory approach and a combination of learning and action make it an ideal way for schools to embark on a meaningful path for improving the environments of schools and their local communities and also for influencing the lives of young people, school staff, families, local authorities, NGOs and many more. 'Know Your Ecosystem'is a participatory programme that provides an excellent opportunity for students to taking decisions to improve both the school and home environments. It strives to increase environmental awareness of students, staff and communities for improving school environments. Actions taken during the programme have also helped schools save on electricity and water consumption costs. The schools are given the opportunity to create links with other schools (rural, urban, etc.), giving them a chance to share environmental education-related ideas and creating a means for cultural exchange and language improvement.

[99] See article on 'I own a tree' campaign to save Bengaluru's greenery', November 19 2009, The Hindu.(http://www.thehindu.com/news/national/karnataka/i-own-a-tree-campaign-to-save-bangalores-greenery/article50524.ece).

[100] Information on ENVIS is available on http://envis.frlht.org/.

6.6.5 *Solid Waste Management: Bengaluru Is the First Metropolis in India to Have Comprehensively Adopted Segregation of Solid Waste at Source*

The city of Bengaluru is the first metropolis in India to have comprehensively adopted and implemented the management of municipal solid waste on the principle of segregation at source and composting of wet waste locally. The rationale underlying the initiative is to comprehensively comply with the progressive standards laid down in the Municipal Solid Waste (Management and Handling) Rules, 2000.

In response to a Public Interest Litigation (PIL) filed by Environment Support Group (ESG) (WP No. 46523/2012) and other connected PILs (in particular, the ones filed by Kavit Shankar, WP No. 24739/2012 and G. R. Mohan, WP No. 30450/2012), a series of unprecedented directions were issued by the Principal Bench of High Court of Karnataka regarding the handling and management of municipal solid waste in Bengaluru on November 22, 2012. The court directed that all municipal waste generated in the city be segregated at source. Further, both BBMP and the government of Karnataka were directed that the 'segregation and wet waste processing stations be located and made operational in the 28 assembly constituencies within two months and also that, in each of the 198 wards of the city segregation and wet waste processing stations be constituted with each ward having at least three segregation and wet waste processing stations'. Additionally, BBMP was directed to seek the assistance of NGOs and other private agencies for segregation and disposal of dry waste with wet waste to be transported to composting sites identified by the BBMP (ESG 2012; Business Standard 2012 (Picture 6.27).

6.6.5.1 Poor Implementation

The city generates around 3600 tonnes of waste a day, and with the waste piling up across the city, segregation of waste has been made mandatory since October 1, 2013; yet it is not implemented strictly. In fact, BBMP has failed to evoke a positive response from the city residents in terms of complying with the rule (Ramani 2013). Further, a poor awareness level among residents as well as pourakarmikas could be the most significant reason for non-compliance even as BBMP continues to dump mixed waste in the four landfills – Mandur, S. Bingipura, Lakshmipura and Terra Firma.

Meanwhile, BBMP has introduced a mandatory waste segregation pilot project in 34 wards besides training pourakarmikas regarding disposal methods. ITC Limited has come forward to adopt 12 wards of the city under the project 'Waste to Wealth', with ragpickers and pourakarmikas being trained in the collection and disposal of segregated waste (Rao 2013) (Picture 6.28).

To promote awareness and to evoke positive compliance among the city's residents, Chief Minister Siddaramaiah launched a pilot programme called 'Kasa

Picture 6.27 BBMP Pourakarmikas holding a rally in Bengaluru in support of waste segregation

Picture 6.28 'Kasa Muktha', Bengaluru

Muktha' on July 24, 2013. The aim of the 'Kasa Muktha' was to achieve NGOG (No Garbage on Ground). Apart from the programme and door-to-door awareness campaigns for the public, training was given to 250 pourakarmikas of the 22 wards for three days. RWAs and volunteers were assigned the job of monitoring and keeping records of 22 wards. The residents were motivated to follow 'one mane, 2 dabba, Mix Maada Beda'. But unfortunately, even the 'Kasa Muktha' programme has not been successful due to the lack of institutional and organisational support from the government as well as from the BBMP.

6.6.5.2 'Neralu': Reminiscing 'Garden City'

'Neralu festival' is a voluntary initiative of the community conducing and organising 'citizen-led urban tree festival' to create appreciation and awareness regarding the protection and conservation of greenery in the city. Many volunteer-driven activities are organised such as workshops on tree journaling and identification, documentary and films exploring local and global narratives on trees, photograph exhibition and mega 'hug-tree' campaigns.

6.6.5.3 Institutions Working on Various Aspects of Sanitation

Indian Institute of Cartoonists and Awareness of Sanitation Issues: *Sustainable Sanitation Alliance and Indian Institute of Cartoonists on Awareness Initiative to Improve Sanitation*

The Sustainable Sanitation Alliance (SuSanA) is an open international alliance with members who share a common vision on sustainable sanitation and are committed to understanding sustainable sanitation solutions. It is related to ground experiences and engages with a group consisting of practitioners, policymakers, researchers and academicians at different levels with the aim of promoting innovation and best practices in policy, programming and implementation. The Sustainable Sanitation Alliance (SuSanA) supports open dialogue through open discussion forum where shared learning can take place with questions, answers, opinions and experiences being shared among the community members. In Bengaluru, SuSanA, in collaboration with the Indian Institute of Cartoonists, conducted a program 'Let's talk about toilets' and displayed 50 pictures at the Indian Cartoon Gallery highlighting various issues related to sanitation problems in Bengaluru (Plate 6.7).

Pee Project

'Pee Project' is a new public form of art that aims at regulating/restricting public peeing in the open places through inspiration and motivation. Ms. Riya Rajan, a local artist from Bengaluru with two partners and visiting German street artists, has taken up this initiative. They have identified a group of older craftsmen, experts in

Plate 6.7 Cartoon on sanitation awareness

Plate 6.8 The Pee project – pasting posters

making lithographic movie posters, a product of an intensive, dying technique unique to south India. These posters are mounted on walls on the locations used as peeing spots (Plate 6.8).

The Ugly Indians

The Ugly Indians, an anonymous group of citizens of Bengaluru, works on improving sanitation in Bengaluru. They identify ugly spots and fix them and call it 'spot fixing'. As a way of restoring public places used as urinals, removing garbage and paan (betel nut leaf) stains on walls. The group members ensure that the spot fixed remains the same with a 15-day test by keeping a close watch on the place. If it remains clean, they feel convinced. Interestingly, most of the spots fixed by the group have remained clean. The group has beautified a footpath that was earlier a garbage dump and local urinal, by clearing obstacles for pedestrians, and cleaned, painted walls, planted plants and beautified the locations with flowers with the rationale that people do not dirty the clean footpath that people actually use. The group prohibits its members from speaking to the media, and even if they do speak, they do not reveal their identity. In their website – www.theuglyindian.com – they go by the viewpoint that the names of individuals are not important, but the results are. Their motto is 'Kaam chalu, mooh bandh' (start working, stop talking). They believe that the problem lies with the people, not the system. They are also focused on bringing about a change in the attitude of people towards sanitation. People are welcome to join the group as well. Similar initiatives have begun in other cities, drawing lessons from the Ugly Indian Initiative, like Mumbai Rising, Gurgaon Rising, Kolhapur Rising and Meerut Rising.

Newer Initiatives by National Government for Awareness and Education

These initiatives have also had an implication to Bengaluru and its involvement in improving sanitation governance.

One of the latest initiatives of the government is to create awareness by involving celebrities for the promotion of a social cause under various programs. Some such initiatives are:

- Promoting Nirmal Bharat Abhiyan (NBA) aimed at eradicating the practice of open defecation by 2017 by involving Vidya Balan, India's Brand Ambassador for Nirmal Bharat Abhiyan. NBA is intended to facilitate the fight against open defecation, through sensitising people on hygiene and thereby to help improve their health, livelihoods and security.
- Creating awareness on importance of hygiene and sanitation under 'Dettol Banega Swachh India' campaign through involving the Bollywood mega star Amitabh Bachchan. The company Reckitt Benckiser would spend Rs. 100 crore in the coming 5 years to reach out to 2000 villages under this initiative *(source: http://swachhindia.ndtv.com/ (accessed on Jan 16 2015).*
- Promoting cleanliness across schools in the country under National School Sanitation Initiatives involving Bollywood actor Aamir Khan. It is a joint programme of the Ministry of Urban Development (MoUD), Ministry of Human Resource Development (MHRD), Central Board for Secondary Education (CBSE) and the GTZ (Germany). The main aim of the project is to generate awareness among school children, teachers, principals, administrative staff and parents regarding two key issues – (i) need for promoting hygiene and its impact on health and environment and (ii) importance of waste segregation.
- Improving children's lives under the Total Sanitation Campaign in India, Pakistan, Sri Lanka, Bangladesh and Nepal involving Indian cricket legend Sachin Tendulkar, UNICEF Brand Ambassador.
- Swachh Bharat Abhiyan – under this initiative the Prime Minister Narendra Modi invited several public figures (Mridula Sinha, Sachin Tendulkar, Baba Ramdev, Shashi Tharoor, Anil Ambani, Kamal Hassan, Salman Khan, Priyanka Chopra, Team Tarak Mehta, Suresh Prabhu, Saina Nehwal, Arvind Gupta, Anand Mahindra and Ronnie Screwvala) to promote clean India and sanitation for all.
- The Ministry of Urban Development has a separate link for Swachh Bharat Mission and its updates wherein celebrities, politicians and religious leaders have taken up the challenge to create Swachh Bharat at local levels. Several of the celebrities who have taken up the challenge update the various initiatives taken up so far and provide comments and suggestions. This has acted as an important platform to promote the programme. These campaigners of the

programme also update the initiatives taken through other social networking sites like Facebook, Twitter, etc. that are more accessible to the larger section of society.

- *Swachh Bharat Apps*. Several independent app developers showed interest to develop adds using mobile technologies. An article was published by the *Times of India* on how 'Desi' companies beat Facebook in 'Swachh' apps race.
- *Swachh Bharat Short Film. Phd* students from NIT Rourkela made a short film on Swachh Bharat with the message that Swachh Bharat should not be a 1-day event but a lifetime to achieve the goal.

6.7 Annexure

Table 6.62 Basic status of lake rejuvenation: water recharge before and after rejuvenation by 2014 (Part I)

Status of Lakes	Konasandra Lake	Sompura Lake	Kothanur Lake	Thalaghattapura Lake	Doddabidarakallu Lake	Jakkur-Sampigehalli Lake
Survey no	15 of Chudenapura 16 of Hemmigepura	11 and 12 of Sompura	54 of Kothanur	73 of Thalaghattapura	125 of Doddabidarakallu and 6 of Nagasandra	55 of Yelahanka, Amanikere, 15,16,17,23 of Sampigehalli, 13 of Agrahara
Catchment area	128.00 ha	93 ha	89.90 ha	217.45 ha	457.00 ha	806 ha
Storage spread area	12.09 ha	5.37 ha	6.03 ha	4.33 ha	11.97 ha	–
Water spread area	12.36 ha	5.52 ha	6.53 ha	4.66 ha	11.97 ha	56.60 ha
Area of lake	15.11 ha	7.48 ha	7.375 ha	7.85 ha	16.05 ha	65.05 ha
Shoreline length	1820.70 ha	1377.00 m	1514.00 m	1039.00 m	1680.00 m	5461.34 m
Maximum depth of the lake	1.4 m	3.20 m	4.00 m	5.00 m	3.80 m	2.70 m
Average depth of the lake	2.70 m	2.20 m	2.50 m	3.20 m	2.20 m	1.25 m
Number of inlets and outlets	2 inlets, 1 outlet	1 inlet 1 outlet	1 inlet 1 outlet	2 inlets 1 outlet	1 inlet 1 outlet	5 inlets 2 outlets
Storage capacity before and after rejuvenation	191.15 million ltrs (before) 280.00 million ltrs (after)	65.214 million ltrs (before) 140.00 million ltrs (after)	115.20 million trs (before) 240.00 million ltrs (after)	118.40 million ltrs (before) 170.40 million ltrs (after)	124.00 million ltrs (before) 320.00 million ltrs (after)	697.05 million litres
Length of bund	437.98 m	260.00 m	410.00 ,	279.00 m	422.00 m	818.15 m

Source: Dept of Town and Country Planning, BDA, Bangalore

Table 6.63 Basic status of lake rejuvenation: water recharge before and after rejuvenation by 2014 (Part II)

Status of lakes	Rachenahalli Lake	Venkateshpura Lake	Madhavaram Lake	Ullal Lake	Malathahalli Lake	Kommagatta Lake	Ramasandra Lake
Survey no	69 of Rachenahalli, 61 of Dasarahalli, 82 of Jakkur	12 of Venkateshpura, 37 of Sampigehalli	48 of Madhavaram, 32 of Thirumalapura, 98 of Doddabidarakallu, 21 of Chikkabidarakallu	93 of Ullal	101 and 26 of Malathahalli, 6 of Gidadakonenahalli	3 of Kommagatta and 46 of Ramasandra	159 of Ramasandra, 36 of Kenchanapura, 37 of Kanalli
Catchment area	850 ha	108 ha	1250 ha	342 ha	618.08 ha	553.00 ha	1629.00 ha
Storage spread area	–	–	–	–	–	–	–
Water spread area	44.35 ha	1.98 ha	23.45 ha	8.97 ha	23.15 ha	13.25 ha	48.62 ha
Area of lake	51.86 ha	4.40 ha	25.69 ha	9.83 ha	23.51 ha	13.82 ha	54.68 ha
Shoreline length	4908.87 m	723.00 m	2613.00 m	1634.00 m	2732.065 m	1830.80 m	4075 m
Maximum depth of the lake	3.30 m	4.90 m	3.63 m	3.15 m	3.0 m	4.2 m	6.61 m
Average depth of the lake	1.70 m	2.33 m	3.30 m	2.91 m	2.1281 m	2.51 m	3.20 m
Number of inlets and outlets	4 inlets 2 outlets	1 inlet 1 outlet	2 inlets 2 outlets	2 inlets 1 outlets	4 inlets 1 outlets	2 inlets 1 outlets	4 inlets 1 outlets
Storage capacity before and after rejuvenation	760.10 million ltrs	88.86 million ltrs	470.00 million ltrs	242.62 million ltrs	441.966 million ltrs	303.34 million ltrs	1539.00 million Ltrs
Length of bund	539.00 m	723.00 m	438.00 m	537.22 m	436.42 m	360 m	660.00 m

Source: Dept of Town and Country Planning, BDA, Bangalore

References

Abide. (2010). *Plan Bengarulu 2020: Bringing back to Bengaluru of Kempe Gowda's dreams*. Bangalore: Agenda for Bengaluru Infrastructure Development Task Force.

Achanta, P. (2013). *Time to "clear the air" in Bangalore, 20th July, Citizen Matters*. Access the link for more details (http://bangalore.citizenmatters.in/blogs/volunteer-diary/blog_posts/can-we-clear-the-air-in-bangalore)

Anand, B.K., Raju, K.V., Praveen, N., Deepa, N., Shashidharan, H.L., Latha, N., and Manasi S. (2006). Urban water supplies dependency on groundwater: initial findings from South Indian cities. In Raju, K.V. (eds) 'Elixir of life: the socio-ecological governance of drinking water', Bangalore: Book for Change.

Angad, A. (2012). City should focus on sewage treatment, 60% unused: CSE report, July 6th, Citizen Matters. http://bangalore.citizenmatters.in/articles/4321-cauvery-is-too-far

Anil, U. (2013). ITC to set up waste processing units in Bangalore, June 30th, Business Line. http://www.thehindubusinessline.com/companies/itc-to-set-up-waste-processing-units-inbangalore/ article4866268.ece

Ashwini, Y.S. (2012). Packaged water no safe bet: Health Department study, August 8. http://www.deccanherald.com/content/269993/packaged-water-no-safe-bet.html

Balakrishna, P. H. B. (2013). Decentralized composting of municipal solid waste in Bengaluru city – an overview. *International Journal of Research in Engineering and Technology, 3*(6), 260–263.

Balasubramanian, V. (2013). Bangalore water crisis, the big hoodwink, 5th March, DNA. http://www.dnaindia.com/analysis/report-bangalore-water-crisis-the-big-hoodwink-1807556

BBMP. (2006). City development plan for Bangalore: urban infrastructure and governance. Jawaharlal Nehru National Urban Renewal Mission (JNNRUM), Bangalore: Bruhat Bangalore Mahanagarapalike.

BBMP. (2009). Urban infrastructure and governance, revised city development plan: JNNRUM. Bangalore: Bhruhat Bangalore Mahanagarapalike.

BBMP. (2012). Integrated solid waste management, Department of Environment Cell. Bangalore: Bruhat Bangalore Mahanagarapalike.

BBMP. (2013). *Evaluation of technology for processing existing waste at seven landfill sites of BBMP, Bangalore technical committee recommendations on EOI application REPORT*. Bangalore: Bruhat Bangalore Mahanagarapalike. Access the link for the report.

BDA. (2007). *Revised master plan – 2015: Vision document* (Vol. 1). Bangalore: Bangalore Development Authority.

BDA. (2010). Revised Master Plan 2015: Revised Zoning of Land Use and Regulations, Volume III, Bangalore: Bangalore Development Authority.

Beary, H. (2005). BBC News, Bangalore, January 31.http://news.bbc.co.uk/2/hi/south_asia/4222521.stm

Bhattacharya, B. (2005). *Yet, life's in tatters*, Down to Earth, pp. 42.

Bhattacharya, P. (2013). Pollution weakening lung functions, 8th September, *The New Indian Express*. Access the link (http://www.newindianexpress.com/cities/bangalore/Pollution-weakening-lung-functions/2013/09/08/article1773857.ece#.UvC__7QuIgU).

Bhumik, K. (2013). Organic flavour of the year, 29th December, The Hindu. http://www.thehindu.com/features/metroplus/society/organic-flavour-of-the-year/article5512202.ece

BMRDA. (2009). *Bangalore metropolitan region revised structure plan 2031*. Bangalore: Bangalore Metropolitan Regional Development Authority.

BWSSB, KUIDFC, AusAID and World Bank. (2014). *Strategic action planning for revival of Bangalore Lakes*, First Mulit-Stakeholder Workshop, 4th April, Bangalore. Access the link for further details (http://www.kuidfc.com/website/webpage.nsf/6dfb1eea694920ff65256e2c00360da2/996f23d0f04c0f0465257a4f003baae2/$FILE/Multi-Stakeholder%20Workshop%20on%20Revival%20of%20Bangalore%20Lakes-%20KUIDFC1.0.pdf)

CED. (2011). Vulnerability assessment of urban marginalised communities: a pilot study in Bangalore slum areas, Bangalore: Centre For Education & Documentation (CED).

Centre for Sustainable Development. (2008). *Extension of water supply to BBMP area by BWSSB, Feasibility Report*. Bangalore: CDS.

Centre for Sustainable Development. (2009a). *Study on noise pollution in Bangalore*. Bangalore: Centre for Sustainable Development.

Centre for Sustainable Development. (2009b). *A lifestyle change approach: Assimilating best energy consumption practice, study report*. Bangalore: Centre for Sustainable Development.

Centre for Sustainable Development. (2010). *A lifestyle change approach: Assimilating best water consumption practices, study report*. Bangalore: CSD.

Centre for Sustainable Development. (2012). *Environment report card 2012: Bangalore*. Bangalore: Centre for Sustainable Development.

Centre for Sustainable Development. (2013). *Karnataka state action plan on climate change*. Bangalore: CSD.

CGWCB. (2008). *Ground water information booklet*. Bangalore Urban District, Karnataka, Bangalore: Central Ground Water Control Board.

Chanakya, H. N., Ramachandra T. V., and Shwetmala (2010). Small-scale decentralized and sustainable municipal solid waste management potential for Bangalore anchored around total recycle and biomethanation plants, Draft Paper, Centre for Ecological Science and Centre for Sustainable Technologies, Bangalore: Indian Institute of Science.

Chanakya, H.N., Ramachandra, T. V., and Shwetmala. (n.d.). Towards a sustainable waste management system for Bangalore, Centre for Sustainable Technologies, Bangalore: Indian Institute of Science, Bangalore, India.

Charan, C. S (2011). Alight from metro, hop on to bicycle, 24th June, *The Times of India*. Access the link (http://articles.timesofindia.indiatimes.com/2011-06-24/bangalore/29698917_1_public-bicycle-stations-space)

Chaturvedi, A. (2014). Cash-crunch hits BBMP's tree-planting drive, 28th May. *Bangalore Mirror*

Chhakchhuak, R. & Jain, N. (2014). Pulse of the Bangalore commuter, 7th February, *Deccan Herald*. Find the link (http://www.deccanherald.com/content/325818/pulse-bangalore-commuter.html)

Chandrashekar. (2013). TBI water heroes: presenting a strong case for open wells, June 10. http://www.thebetterindia.com/7297/tbi-water-heroes-chandra-shekar-presenting-a-strong-case-for-openwells/# sthash.aDX8Sg9V.dpuf

Citizen Matter. (2013). *Where does Bangalore get its water from?*, Citizen Matter Desk. Access the following link (http://bangalore.citizenmatters.in/articles/5306-understanding-bengalurus-water-sources-cauvery-groundwater-borewells-rainwater)

CSE. (2013). Pesticides in bottled water. New Delhi: Centre for Science and Environment. http://www.cseindia.org/pesticides-in-bottled-water-532

Dayalapwar, N. (n.d.). *Wastew.d.ater reuse in megacities-emerging trends in Bangalore city, a report*. Bangalore: Symbiosis Institute of International Business.

Deccan Herald. (2013). Bangalore third in E-waste generation in the country, November 13. http://www.deccanherald.com/content/368875/bangalore-third-e-waste-generation.html

Deepthi, M. R. (2012). Glass in high-rise buildings making Bangalore hotter, March 20th. www.dnaindia.com

DNA. (2011). *Over half of Bangalore's sewage flows into storm water drains, lakes: Report*, 15th March, DNA. Access (http://www.dnaindia.com/bangalore/report-over-half-of-bangalores-sewage-flows-into-storm-water-drains-lakes-report-1520311)

DNA. (2012). *Level of noise pollution in Bangalore is alarming*, 20th November, DNA. Visit the link (http://www.dnaindia.com/bangalore/report-level-of-noise-pollution-in-bangalore-is-alarming-1766987)

EMPRI and TERI. (2011). *Karnataka state action plan on climate change, draft report*. Environmental Management and Policy Research Institute and The Energy and Resources Institute (TERI): Bangalore.

Enqvist, J., Maria, T., & Örjan, B. (2014). Citizen networks in the Garden City: protecting urban ecosystems in rapid urbanization. *Landscape and Urban Planning, 130*, 24–35.

ESG. (2012). Environmental Support Group Press release. November 24th 2012. http://www.esgindia.org/education/teaching-resources/press/karnataka-high-court-directs-bangalore-b.html

GOK. (2003). State of Environment Report, Karnataka 2003-04. Bangalore: Department of Forest, Ecology and Environment, Government of Karnataka.

GoK. (2009). *State of environment report, Bangalore 2008*. Bangalore: Department of Forest, Ecology and Environment and Government of Karnataka.

GOK. (2011). State of Environment Report, Karnataka 2011-12. Bangalore: Department of Forest, Ecology and Environment, Government of Karnataka.

Gupta, A. K., & Nair, S. S. (2011). Urban floods in Bangalore and Chennai: risk management challenges and lessons for sustainable urban ecology. *Current Science, 100*(11), 1638–1645.

Hedge, G. V., & Chandra, S. (2012). Resource availability for water supply to Bangalore City, Karnataka. *Current Science, 102*(8), 1102–1104.

Herald Deccan. (2013). Bangalore third in E-waste generation in the country, November 13. http://www.deccanherald.com/content/368875/bangalore-third-e-waste-generation.html

Hindustan Times. (2013). How air and water pollution plagues Indian cities, 1st December, *Hindustan Times*. Follow the link (http://www.hindustantimes.com/india-news/how-air-and-water-pollution-plagues-indian-cities/article1-1157717.aspx).

Honaganahalli, P. (2014). Bangalore's ambient air quality is under threat, 4th February, *Deccan Herald*. Visit (http://www.deccanherald.com/content/56221/F)

IBNLIVE.com. (n.d.). Bangalore activist on the road to plastic success story. http://ibnlive.in.com/news/bangaloreactivist- on-the-road-to-plastic-success-story/46501-3.html

Jayashree, N. (2009). Bangalore has become heat island, January 9th. www.timesofindia.indiatimes.com

Jain, P. (2013). A mere 500 public toilets in Bangalore, 19th November, DNA. Visit the following link (http://www.dnaindia.com/bangalore/report-a-mere-500-public-toilets-in-bangalore-1921462)

Kaminska, A (2008): Review of 'Towards an Urban Aesthetics: The Aesthetics of Human Environments, Broadview Press'. https://public.journals.yorku.ca/index.php/public/index

Kamath, V. (2014). Garbage crisis triggers fear of dengue epidemic in city, 6th June, *Bangalore Mirror*.

Kehle, N., Mahadevia, D., & Parasher, A. (2008). Centralised and decentralised solid waste management in Bangalore. In Mahadevia (Ed.), *Solid waste management in Indian Cities: Status and Emerging Practices*. New Delhi: Concept Publications.

Kidiyoor, S. (2013). Forget tap water, even packaged water may not be safe in Bangalore, February 18. www.dnaindia.com

Kumar, S. B. R. (2012). Bangalore has the highest number of organic outlets, November 21, The Hindu. visit http://www.thehindu.com/todays-paper/tp-national/tp-karnataka/bangalore-has-the-highest-number-of-organicoutlets/ article4117845.ece

Kumar, S. K. (2013). The big fat truth about public toilets, 21st October, *The New Indian Express*. Visit the link (http://www.newindianexpress.com/cities/bangalore/The-big-fat-truth-about-public-toilets/2013/10/21/article1846059.ece)

Kushala, S. (2013). Arkavathy Layout yields bounty to the farmers who lost land, 2nd September, The Hindu. http://www.bangaloremirror.com/bangalore/others/Arkavathy-Layout-yields-bounty-to-the-farmers-who-lostland/ articleshow/22256917.cms

Lvovsky, K. (1998). Economic costs of air pollution with special reference to India. http://siteresources.worldbank.org/PAKISTANEXTN/Resources/UrbanAir/Economic+costs+of+air+pollution+KL.pdf

Mahesh, A. (2013). Three reasons why Kasa Mukta failed, 18th December, Citizen Matters. http://bangalore.citizenmatters.in/blogs/ideas-at-work/blog_posts/three-reasons-why-kasa-mukta-failed

Manasi, S. (2013). Emerging trends in E- waste management – status and issues: a case study of Bangalore city, Working Paper series 297, Bangalore: Institute for Social and Economic Change.

Manasi, S., & Umamani, K. S. (2013). Water conservation in urban areas- a case of rain water harvesting initiative in Bangalore city. In S. Nautiyal et al. (Eds.), *Knowledge systems of societies*

for adaptation and mitigation of impacts of climate change. Germany: Springer-Verlag Berlin Heidelberg.
Manasi, S., and Umamani, K.S. (2014). Biomedical waste management: issues and constraints: A Case Study of Bangalore City, Working Paper Series, Bangalore: Institute for Social and Economic Change.
Manasi. S., Latha N., & Hemalatha B. R. (2016). *Sanitation challenges of the poor in urban and rural settings: Case studies of Bengaluru city and rural north Karnataka*, UNU – MERIT Maastricht University Working Paper Series, 2017-026, ISSN 1871-9872.
Mathur, A., & Dilip Da, C. (2006). *Deccan traverses: The making of Bangalore's terrain*. New Delhi: Rupa & Co..
Mishra, S. (2006). Urban development and climate change: a case study of Bangalore. http://www.qub.ac.uk/research-centres/RespondingtoClimateChangeIndia- UKPerspectives/Filestore/Theme2/Filetoupload,239583,en.pdf
MOEF (2009). State of Environment Report, 2009 India, New Delhi: Ministry of Environment and Forestry, Government of India.
Mondal, S. (2013). Land acquisition policy leaves Karnataka's owners fuming, 26th April, The Hindu. http://www.thehindu.com/news/cities/bangalore/land-acquisition-policy-leaves-karnatakas-ownersfuming/ article4656614.ece
MoUD. (2008). *Study on traffic and transportation policies and strategies in urban areas in India, Final Report*. New Delhi: Ministry of Urban Development.
Mukherjee, S. (2014). Stop drilling, there's no groundwater left, 13th March, *Times of India*. Access the link (http://timesofindia.indiatimes.com/city/bangalore/Stop-drilling-theres-no-groundwater-left/articleshow/31914653.cms)
Nagendra, H., & Gopal, D. (2010). Tree diversity, distribution, history and change in urban parks: studies in Bangalore, India. *Urban Ecosyst*. https://doi.org/10.1007/s11252-010-0148-1.
Nandy, P. (2011). Stores that offer organic food around Bangalore, 1st September, DNA. http://www.dnaindia.com/lifestyle/report-stores-that-offer-organic-food-around-bangalore-1582323
Navya, P.K. (2011). Lying unwatched: Bengaluru's 1067 parks and playgrounds, Citizen Matters. http://bangalore.citizenmatters.in/articles/2734-parks-playgrounds-bbmp
Navya, P. K. (2012). Where does Bangalore's power come from?, 6th November, *Citizen Matters*. Find the link (http://bangalore.citizenmatters.in/articles/4648-bescoms-power-sources).
Osswald, N. (2014). India: organic taking root in the cities. http://www.organicmarket. info/web/Organic_Retail_Trade/General/India:_Organic_taking_root_in_the_cities/207/210/0/15262.html
Pavan, M.V. (2013) Plastic waste will soon pave Bangalore roads, November 5th, Times of India. http://articles.timesofindia.indiatimes.com/2013-11-05/bangalore/43693043_1_plastic-waste-bbmp-dry-waste
Prasad, K. (2008). Where rights are watered down. The Hindu, December 24. http://www.thehindu.com/todays-paper/tp-national/tp-karnataka/Where-rights-are-watereddown/ article15367005.ece
Prasad, A., and Khan, S. (2013). Bangalore IT investment region plan hit by rising land price, 20th December, The Economic Times. http://articles.economictimes.indiatimes.com/2013-12- 20/news/45417943_1_land-acquisition-acquisition-process-acquiring-land
Raju, K. V., Manasi, S., & Latha, N. (2008). *Emerging ground water crisis in urban areas: A case study of Ward no. 39, Bangalore City, working paper series 196*. Bangalore: Institute for Social and Economic Change.
Ramachandra, T.V., and Bachamanda, S. (n.d.). Environmental auditing: management of municipal solid waste. http://www.ces.iisc.ernet.in/biodiversity/sahyadri_enews/newsletter/issue24/article1.htm
Ramachandra, T. V., & Bachamanda, S. (n.d.). *Environmental audit of municipal solid waste management, Energy and Wetlands Research Group, Centre for Ecological Sciences*. Bangalore: Indian Institute of Science.
Ramachandra, T. V., & Mujumdar, P. P. (2009). Urban floods: Case study of Bangalore. *Disaster & Development, 1*(2), 1–97.

Ramachandra, T. V., & Kumar, U. (2010). Greater Bangalore: emerging heat island. *GIS for Development, 14*(1), 86–104.

Ramachandra, T. V., Aithala, B. H., & Sannab, D. D. (2012). Insights to urban dynamics through landscape spatial pattern analysis. *International Journal of Applied Earth Observation and Geoinformation, 18*, 329–343.

Ramani, C. V. (2013). Bangalore is obviously more 'power'ful, 9th April, *The Hindu*. Access the link for more details (http://www.thehindu.com/news/cities/bangalore/bangalore-is-obviously-more-powerful/article4595615.ece)

Rao, S. (2011). Every 3rd B'lorean lives in sub-human slum. March 17th, Times of India. https://timesofindia.indiatimes.com/city/bengaluru/Every-3rd-Blorean-lives-in-sub-humanslum/ articleshow/7724063.cms

Rao, S. (2013). Finally, Bangalore Development Authority breaks silence on denotified land, 19th June, The Hindu. http://articles.timesofindia.indiatimes.com/2013-06-19/bangalore/40068854_1_bda-arkavathy-layout-siteallottees- denotified-land

Ravi, P. (2005). Karnataka inches forward in water harvesting, December 17th. http://www.indiatogether.org/2005/apr/env-karwater.htm

Ravindra, A. (1996). *Urban land policy: study of metropolitan city*. New Delhi: Concept Publishers.

Ray, A. (2010). Lack of planners turns Bangalore into a heat island. Saturday, September 2010. www.dnaindia.com

Ray, A. (2012a): Off metro, take cycle, 2nd July, *The Times of India*. Access the following link (http://articles.timesofindia.indiatimes.com/2012-07-02/bangalore/32507637_1_cycles-urban-land-transport-mg-road)

Ray, A. (2012b). Many Bruhat Bangalore Mahanagara Palike areas not to get Cauvery water, 18th September, *The Indian Express*. Visit the link (http://articles.timesofindia.indiatimes.com/2012-09-18/bangalore/33924840_1_cauvery-water-bwssb-chairman-gaurav-gupta-greater-bangalore-water)

Ray, A. (2013). Old pipelines are bursting at the seams, 26th November, *Times of India*. Visit (http://articles.timesofindia.indiatimes.com/2013-11-26/bangalore/44485956_1_old-pipelines-sewage-water-drainage-network)

Rohith, B. R. (2012). Now, have your say in City's future. May 9th, Deccan Herald. http://www.deccanherald.com/content/248364/now-have-your-say-citys.html

Rohith, B. R. (2013). Treated wastewater goes down the drain, September 13, Times of India. http://articles.timesofindia.indiatimes.com/2013-09-13/bangalore/42039914_1_wastewater-treatment-stpstreated-water

Saldanha, L. F., Rodriques, W., Narayanan, S., Shekar, N., Srinivas, and Yadav, A. (2012). Guiding Principles to resolve the current Garbage Management Crisis of Bruhat Bengaluru Mahanagara Palike, Bangalore: Environment Support Group. http://www.esgindia.org/campaigns/press/mayor-bangalore-dumpsmavallipura-landfi.html

Satterthwaite, D., Huq, S., and Pelling, M. (2007). Adapting to climate change in urban areas: the possibilities and constraints in low- and middle-income nations, Human Settlement Discussion Paper Series: Theme: Climate Change and Cities-I. London: International Institute for Environment and Development (IIED).

Sawkar, R. H. (2012a). *Rainwater harvesting and ground water development in greater Bengaluru region*. Bengaluru: Geological Society of India.

Sawkar, R. H. (2012b). Evaluation of surface, ground and sewage water for sustainable supply of potable water to Bengaluru. *Journal of Geological Society of India, 80*, 877–880.

Schenk, H. (2001). Living in India's Slums: A Case Study of Bangalore. IDPAD, New Delhi: Manohar.

Sharma, K. (n.d.) *Why public toilets get clogged*. Access the following link (http://infochangeindia.org/urban-india/cityscapes/why-public-toilets-get-clogged.html)

Shekar, S., & Shah, M. (2006). *Bench marking Bangalore's public services: What the third citizen report card reveals*. Bangalore: Public Affairs Centre.

Sherwood institute. (n.d.). Better Lakes, Better Bangalore: recapture the health, heart and heritage of the 'Garden City'. www.sherwoodinstitute.org/bangalore
Siraj, M. A. (2014). Our cities are decaying, 31st may. *The Hindu.*
Smitha, K.C. (2006). Urban Governance and Bangalore Water Supply and Sewerage Board (BWSSB). In K.V. Raju (ed) 'Elixir of Life: The Socio-Ecological Governance of Drinking Water', Bangalore: Book for Change.
Suresh, T.S. (2001). An Urban Water Scenario: A Case Study of the Bangalore Metropolis, Karnataka, India, IAHS Publication No. 268, Netherlands: IAHS.
Sridevi, H., Shreejith, K., & Ramachandra, T. V. (2014). Comparative analysis of greenhouse gas emissions from major cities of India, international journal of renewable energy and. *Environmental Engineering, 2*(1), 38–43.
Srinivasan, S. (2013). Bangalore cycle day: One step closer to Cyclovia?, 23rd October, *The Alternative*. Access the following link (http://thealternative.in/environment/bangalore-cycle-day-one-step-closer-to-cyclovia/)
Srivatsa, S. (2008). Bangalore's hunger for land threatens forest patches too. 22nd August, *The Hindu*.Access (http://www.hindu.com/2008/08/22/stories/2008082254410700.htm).
Subramanian, N. (2014). Appreciating the City's Green Cover: 'Neralu' is aimed at reminiscing about the Garden City, *The Hindu*, dated 5th February 2014
Subramanyan, S. K. (2013). Massive project to fix water leaks underway. *The Citizen Matters*. Access the link (http://bangalore.citizenmatters.in/articles/4979-bwssb-project-to-check-water-theft-and-leakage)
The Hindu. (2011). Shrinking green cover, rising temperatures, 12th January, *The Hindu*. Aisit (http://articles.timesofindia.indiatimes.com/2011-01-12/bangalore/28358895_1_global-warming-forest-survey-forest-report)
The Hindu. (2012a). No road-cutting charge to get water connection in 54 wards, 8th August, *The Hindu*. Access the link (http://www.thehindu.com/todays-paper/tp-national/tp-karnataka/no-roadcutting-charge-to-get-water-connection-in-54-wards/article3740052.ece)
The Hindu. (2012b). Bangalore thirsts for better water supply, 16th March, *The Hindu*. Aisit the link for more details (http://www.thehindu.com/todays-paper/tp-national/tp-karnataka/bangalore-thirsts-for-better-water-supply/article3000806.ece)
The Hindu. (2012c). Bangalore has potential to save 10% of power generated, 4th May, *The Hindu*. Find the link for more details (http://www.thehindubusinessline.com/economy/bangalore-has-potential-to-save-10-of-power-generated/article3384631.ece)
The Hindu. (2013a). *Bangalore among 14 cities with high levels of pollution*, 23rd March. Access the link (http://www.thehindu.com/news/national/karnataka/bangalore-among-14-cities-with-high-levels-of-pollution/article4538834.ece)
The Hindu. (2013b). Public toilets in Bangalore to go PPP way, 30th October, *The Hindu*. Access the following link (http://www.thehindu.com/news/cities/bangalore/public-toilets-in-bangalore-to-go-ppp-way/article5286275.ece)
The Hindu. (2014a). *Mandur observes bandh, continues protest against garbage dumping*, 2nd June. Follow the link (http://www.thehindu.com/news/cities/bangalore/mandur-observes-bandh-continues-protest-against-garbage-dumping/article6075579.ece)
The Hindu. (2014b). *Another garbage crisis looms*, 2nd June, Access the link for more details (http://www.thehindu.com/news/cities/bangalore/another-garbage-crisis-looms/article6072586.ece).
The New Indian Express. (2013). *Bangaloreans lack knowledge on noise pollution, says HC*, 8th October.Access the link (http://www.newindianexpress.com/cities/bangalore/Bangaloreans-lack-knowledge-on-noise-pollution-says-HC/2013/10/08/article1824592.ece#.UvDCyLQuIgU)
The Times of India. (2014). *Mandur residents dig their heels in as garbage crisis brews*, 3rd June. The following link for more details (http://timesofindia.indiatimes.com/city/bangalore/Mandur-residents-dig-their-heels-in-as-garbage-crisis-brews/articleshow/35970805.cms)
Thipaiah, P. (2009). *Vanishing Lakes: A study of Bangalore City, Monograph series* (Vol. 17). Bangalore: Institute for Social and Economic Change.

Umamani, K.S., and Manasi, S. (2013). Water Conservation in Urban Areas- A Case of Rain Water Harvesting Initiative in Bangalore city, Working Paper Series, Bangalore: Institute for Social and Economic Change.

UN-HABITAT. (2010). State of Asian Cities. Japan: United Nations Human Settlement Programme (UN-HABITAT).

UNHSP. (2011). *Cities and Climate Change: Global Report on Human Settlements 2011*. London: Earthscan.

Vaidyanathan, V. & King, R. A. (2011). Institutional analysis of urban transportation in Bangalore, Conference paper presented at the 4th Annual International Conference on Next Generation Infrastructures, Virginia Beach, Virginia, November 16–18, 2011

Varshney, R. (2012). BBMP bylaws to dispose waste give hopes to many waste management players of Bangalore and Koramangala, October 12, *Economic Times*. Visit (http://articles.economictimes.indiatimes.com/2012-10-12/news/34413109_1_waste-management-wet-waste-household-hazardous-waste)

Venkataraman, R. (2000). Lakes in our City: How they are linked to the life of People. Bangalore: Department of Forest, Government of Karnataka.

Vinitha.(2013).Bangaloregetstogetherfor'clean-up'fest,3rdFebruary,CitizenMatters.http://bangalore.citizenmatters.in/articles/4879-bangalore-looks-at-optimal-waste-management-solutions

Vishwanath, S. (2007). Sustainable water for Bangalore, The Hindu, May 26. www.thehindu.com

Vishwanath, S. (2010). Every layout must have a sewage treatment plant, May 29th, The Hindu. http://www.hindu.com/pp/2010/05/29/stories/2010052950210300.htm

Vyas, H. (2014). Rs 50 crore loss to BWSSB every year, 21st January, *Bangalore Mirror*. Follow the link for more details (http://www.bangaloremirror.com/bangalore/cover-story/Rs-50-crore-loss-to-BWSSB-every-year/articleshow/29162048.cms)

Walkabilityasia (2013): Bangalore kicks off cycle day, 28th October, *Citizen Matters*. Follow the link for more details (http://walkabilityasia.org/2013/11/11/bangalore-kicks-off-cycle-day/)

World Bank. (2000). *Natural Resource Management Strategy*, Eastern Europe and Central Asia, Technical Paper No. 485, September.

World Bank. (2006). *World Development Report, 2006*. New York: Oxford University Press.

Yajaman, A. M. (2013). Save Bangalore Lakes Trust formed, first workshop scheduled, 18th October, Citizen Matters. http://bangalore.citizenmatters.in/blogs/like-a-lake/blog_posts/save-bangalore-lake-trust

Yasmeen, A. (2013). BWSSB to launch UFW project in city, 5th October, The Hindu. Access (http://www.thehindu.com/todays-paper/tp-national/tpkarnataka/bwssb-to-launch-ufw-project-in-city/article5202899.ece)

Yasmeen, A. (2005). Plastic waste to be used for asphalting roads, September 21st, The Hindu. http://www.hindu.com/2005/09/21/stories/2005092117280300.htm

Yousaf, S. (2012). Karnataka govt notifies land to set up IT park: Karnataka govt notifies land to set up IT park, 31st May. The Livermint: Wall Street Journal. http://www.livemint.com/Companies/t39AQppVx6YNRVKUVoDnjM/Karnataka-govt-notifies-land-to-set-up-ITpark.html

Chapter 7
Local Institutions and Governance Structures for Environmental Governance: Critical Issues and Emerging Options

Abstract This chapter examines how the dynamics of institutions at the local level triangulates and interfaces, affecting the management of environmental resources for an effective environmental governance. Locating institutional developments at the local level provides insights into the areas of convergence and divergence particularly when it is an acknowledged fact that the institutions at multilevels are often complex, evolving and incomplete. It is argued that institutions facilitate the resolving environmental conflicts. The case of Bengaluru Metropolitan Area (BMA) has been selected to demonstrate how diverse institutions and governance structures at the local level impact the regulating of environmental governance. The study has focused on how the environment and resource management issues at the local level contribute to environmental governance. The study has followed an interdisciplinary approach, deriving methods from both environmental economics and political science to explore and analyse the role of institutions and governance structures in environmental governance.

Keywords Environmental governance · BMA · Institutions · Urban planning · Urban services

7.1 Introduction

An environmental governance model cannot be conceived without a coherent institutional integration. There is no denying the fact that a two-way relationship between environment and institutions brings into sharp focus the inextricable link between environment and sustainable development. Environmental governance is largely concerned with the 'processes', i.e. a sum of organisations, policy instruments, rules, norms and procedures and actors which regulate the environmental management and protection (Najam, Papa and Taiyab 2006). The governance process provides an institutional framework within which the civic public realm is managed (Halfani 1997: 147) and creates favourable conditions for other actors to undertake economic and social activities. Similarly, Rhodes (1996) defined governance as

K. V. Raju et al., *Urban Environmental Governance in India*, The Urban Book Series, https://doi.org/10.1007/978-3-319-73468-2_7

'self-organizing networks characterized by interdependence between organizations, continuing interactions between network members and a significant degree of autonomy from the state'. Similarly, Plumptre and Graham (2000) explain governance as how governments and other organisations interact, how they interrelate to citizens and how decisions get taken in an increasingly complex world. According to UN-HABITAT, urban governance constitutes the process by which individuals, both public and private, plan and manage their common affairs, encompassing formal institutions, informal arrangements and the social capital of citizens. It includes three principal groups of actors (i.e. government, private sector and civil society) and the ongoing and complex process of harmonising their competing priorities. From an institutional perspective, 'environmental governance is best understood as the establishment, affirmation, or change of institutions to resolve environmental conflicts over environmental resources' (Paavola 2007). Further, it is argued that institutions facilitate the resolving of environmental conflicts through striking at a particular balance between conflicting interests by way of reaffirming, establishing or redefining entitlements (Paavola 2007: 95). The assessment, therefore, suggests that the 'choice of environmental governance institutions is a matter of social justice rather than economic efficiency, demanding a greater emphasis on public participation as the foundation of their political legitimacy' (Paavola, undated).

Governance structures or institutional arrangements are mechanisms for coordinating economic transactions for implementing informal constraints (customs, traditions, norms) and formal rules (constitutions, laws, property rights) defined by the institutional environment (Davis and North 1971; Young et al., undated). Institutions are simply defined as 'rules of game' (North 1990: 3), 'rules as the "humanly devised constraints that structure political, economic, and social interaction"' (North 1991: 97). But Paavola (2005) argues that 'they also create entitlements by defining the rights, liabilities and powers of involved agents'. Institutions assume many forms such as formal or informal, private or public and centralised or decentralised (North 1981 in King and Toffel 2007: 2). Similarly, Young et al. (1999) define that institutions give rise to social practices, assign roles to participants in these practices and govern the interactions among the occupants of various roles. Thus, the role of institutions is critical in solving or managing environmental problems.

The role of institutions in natural resource management is being increasingly recognised in the context of global environmental changes. The proliferation of urban institutional linkages has resulted in overlapping of institutions and sometimes conflicts, adversely impacting environmental governance (Najam, Papa and Taiyab 2006; Papa 2014: 143). In view of enormously daunting urban environmental challenges such as loss of biodiversity, degradation of green cover and overreaching issues of climate change, institutions at the local level are critical to addressing problems arising from human-environment interface. It is argued that 'environmental resource regimes are types of institutions' (Young et al. undated) emanating from a governance process essentially for understanding the way in which varied legislative enactments and policies affecting environmental governance are planned and implemented. Interactions between multilevels of governance regimes and institutions have been recently recognised (Young 2002 in Paavola 2005). Therefore, environmental governance is a decision-making 'process' in terms of norms, rules, regulatory regimes, standards, procedures and policies essentially embedded in political context, institutionalised at global, national and local levels. In addition, varied measures such as rule of law, accountability, transparency and public participation may equally impact the management in addressing a wide range of emerging environmental challenges and impacts.

A vast literature on 'New Institutional Economics' (NIE) emphasises the role of institutions in the management of common property resources (Ostrom 1990; Ostrom et al. 1994; Keohane and Ostrom 1995; Young n.d). But studies have cautioned that NIE has not per se focused on environmental governance, excepting Elinor Ostrom's prolific work on the governance of common property resources (Ostrom 1990) and Oran Young's work on International Environmental Governance (Young 2002 in Delmas and Young). A study by Paavola (2005) clearly reveals that NIE offers a new dimension for the study of environmental governance in two ways, i.e. (i) transaction cost exists and influences environmental outcomes signifying the implications of an institutional design for governance outcomes, and (ii) NIE analysis of environmental problems is based on the concept of interdependence (Ostrom 1990 and Keohane and Ostrom 1995 in Paavola 2005) rather than externalities, which are critical to promote environmental governance.

Therefore, varied approaches and innovative methods are followed to address the most pressing urban environmental challenges and crises. But a significant chunk of literature emphasises a particular type of agent for promoting effective environmental governance such as state, market (private sector) and most recently civil society, i.e. NGOs, CBOs (community-based organisations) or local organisations. Therefore, the term 'environmental governance' indicates broad areas of institutions such as rules, regulatory processes, norms and standards and governance mechanism, participatory forums, capacity building, policy research (which includes technical assistance, seminars, workshops, training, conferences, etc.), networking, advocacy, communication, global campaigns, monitoring, issues of coordination, knowledge production and building and technical and scientific support. These governance measures are devised to address the multiple types and layers of worsening urban environmental problems and conflicts. This clearly denotes that relationship has been dispersed at several institutional levels, not hierarchical, and is more complex and contextual.

The purpose of this chapter is to explore how the hybrid and dynamics of institutions at local level triangulate and interface affect the management of environmental resources for effective environmental governance. Today a broad array of governance strategies are being practised, i.e. state, market (private) or civil society at different spatial and institutional scales which impact environmental governance. Locating such institutional developments at the local level provides insights into the areas of convergence and divergence particularly when it is an acknowledged fact that the institutions at multilevel are often complex, evolving and incomplete.

7.2 Institutions and Environmental Governance in Urban Areas

A plethora of institutions characterise the urban setting in India and are affecting environmental policy and law. The interplay takes place through a broad spectrum of actors towards a multilevel, multi-actor governance system ranging from bottom-up to top-down in the hierarchy such as local, national, international and global setting the character by each actor or player with prevailing practices and set the goals to be achieved. Often the nature of interaction is conflict in the nature context of diverse political, legal and financial aspects. Such an interaction between polycentric institutions can produce both positive and negative implications for environment governance. The multilevel institutional framework often results in mixed policy instruments influenced by local level economic and voluntary measures.

The case of Bengaluru Metropolitan Area (BMA) has been selected to demonstrate how diverse institutional mechanisms and programmes affect each other at all levels of social organisation (global, national, state and local) and significantly impact and bring environmental governance in the process. The study, in particular, has focused on how the environmental and resource management issues contribute to environmental governance. The research has focused on various issues related to environment, local ecology and resources affected by both formal and informal institutions at the local level. The focused environmental issues at the local level range from deficient drinking water supply, overexploitation of groundwater, poor waste management, loss of biodiversity, land use changes, diminishing water bodies, climate change, to pollution (air/water), etc. The study has followed an interdisciplinary approach, deriving methods from both environmental economics and political science to explore and analyse the role of institutions and their impact on environmental governance.

Environmental governance is critical in the context of rapid urbanisation with a concomitant population explosion. Besides the resultant depletion of natural resources and environmental degradation, institutional fragmentation in the urban setting further aggravates environmental governance.

The framework (see Fig. 7.1) helps draw specific inferences on the role of each local body in the management of (natural) resources as well as explore the dynamic relationship between human beings, environment and natural resources on the one hand and institutions governing at the global, national, state and local levels on the other. In particular, the study seeks to address the following issues:

1. To explore the dynamics of institutions (both formal and informal) in terms of influencing the management of environment and resources at the local level
2. To examine the role of local institutions through community-based interventions or 'urban commons' for creating new institutions in the context of a new urban set-up of the twenty-first century
3. To suggest, if possible, a model for institutional integration at the local level for promoting environmental governance

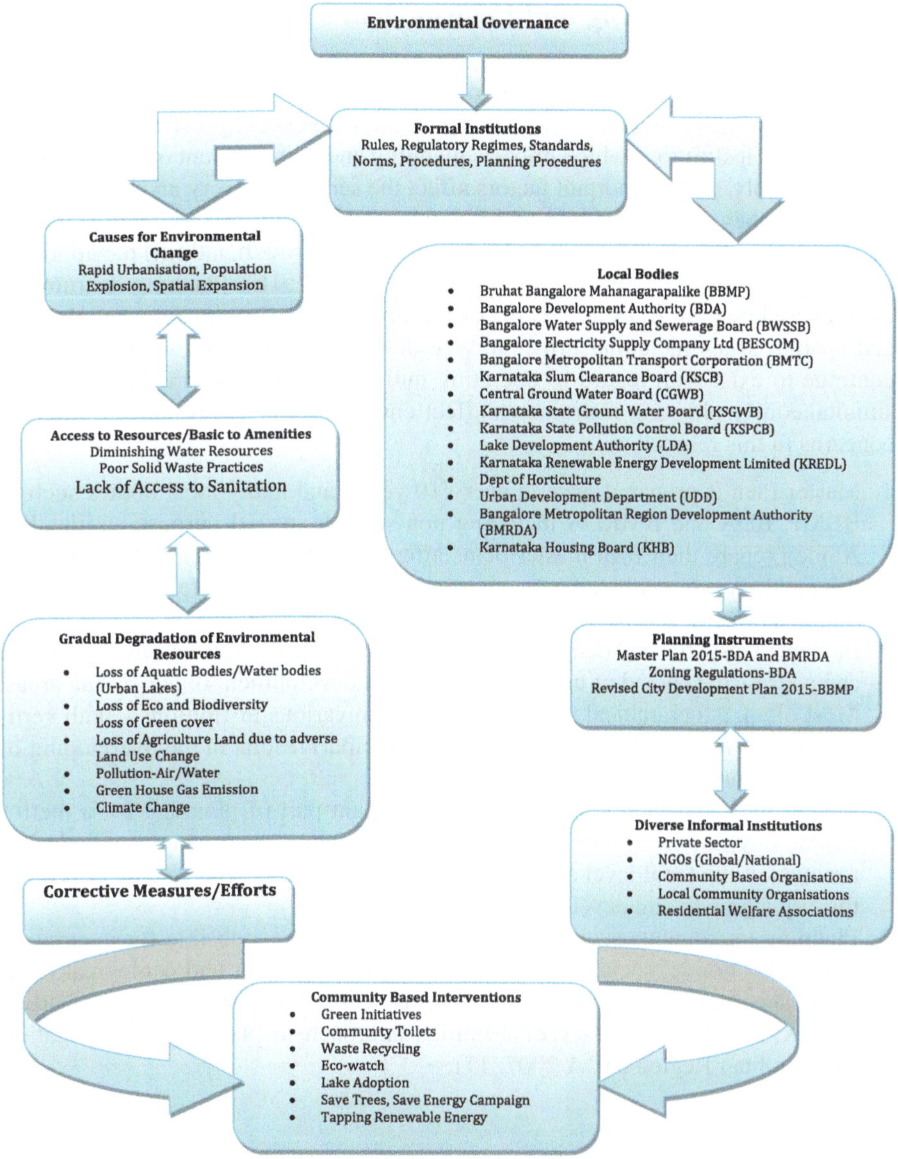

Fig. 7.1 Framework for analysing institutions at the local level for environmental governance

7.2.1 Urban Planning and Environment

Various institutions govern the city of Bengaluru's growth as well as service delivery. The figure (see Fig. 7.1) clearly demonstrates the presence of multiple agencies and institutions which impacts planning and management of service delivery. Essentially, three important factors affect the service delivery and environment in the city; they include (i) role and functions of institutions (national, state and local), (ii) the planning instruments for regulating the growth and (iii) the approach towards a long-term city development plan (BDA 2007: 16). Planning instruments are prepared by both BDA and BMRDA for a period of 10 years aimed at physical and spatial issues including urban service delivery. As a result, many loopholes continue to exist with regard to operating multiple agencies and institutions, and simultaneously planning instruments affect environmental governance. Some key concerns in this respect are:

1. Master Plan is prepared once in every 10 years, and many local bodies such as BBMP, BDA and BMRDA including non-state or special purpose entities like Abide prepare their own master plans affecting service delivery. Consequently, this can have a direct bearing on environment and quality of life in the city while implementing at administrative, technical and operational levels.
2. In view of a spatial and demographic expansion, the city lacks a holistic plan that can address the spread of urban sprawl and the development of peri-urban areas.
3. Most often, a hierarchical relationship between various institutions – both vertical and horizontal – (municipal and nonmunicipal) results in the overlapping of functional jurisdiction and conflicts.
4. Management of environment forms a significant part of planning for a metropolitan city like Bengaluru, yet provisions relating to the management of environment at the local level are not visible in the planning instruments governing the city's growth and development.
5. Planning instruments such as Master Plan 2015 are responsible for integrating spatial, economic, social, transportation and environment and ecological planning. This includes town planning, land use control and management of urban fringes. But there is a lack of a unified approach to planning for Bengaluru Metropolitan Region (BDA 2007: 17).

7.3 Urban Services and Institutions

The subsequent section presents key institutional challenges associated with the management of natural and environmental resources in Bengaluru Metropolitan Area (BMA). An analysis of institutional barriers to strengthening the environmental governance at the local level is presented with regard to certain selected sectors like water supply, land use management, power supply and urban lakes.

7.3.1 *Management of Water Resources at the Local Level*

For the city of Bengaluru, diverse institutions (both formal and informal), through an interface, provide drinking water[1] to citizens. Primarily, Bengaluru Water Supply and Sewerage Board (BWSSB)[2] is responsible for providing drinking water and sewerage facilities in the city. Prior to 2005, BWSSB was in charge of the erstwhile Bengaluru City consisting of 100 wards; post 2005, through a notification for the creation of Bruhat Bengaluru Mahanagara Palike (BBMP), the city was merged with 7 CMCs (City Municipal Councils) and 1 TMC (Town Municipal Council) and 110 villages. Although river Cauvery Phase III (130 MLD) is the main source for providing drinking water to the city, BWSSB is dependent on groundwater resources (100 to 200 mld) also for supplying water. Although groundwater resources are taken care of by the Central Ground Water Board (CGWB) and Karnataka Ground Water Board (KGWB), BBMP also drills borewells within the city.

With the merger of urban periphery and suburban areas, the urban sprawl has come to exert an enormous strain on BWSSB in terms of providing drinking water. To quench the thirst of a booming population of the city, BWSSB has now commissioned Cauvery Phase IV with the financial assistance from JBIC (Japanese Bank for International Cooperation). Yet, less than 10 per cent of the population of the newly merged areas is served with drinking water facility.

At present, according to the officials of the BWSSB,[3] an additional ten TMC of water is required to service the additional areas of the city. As a result, a large proportion of the population in the urban periphery is dependent on groundwater resources. Such high dependency on groundwater has inevitably led to an unabated tapping of groundwater resources nearly to a saturation point.[4] An illegal groundwater mining is rampant in the periphery region of the city. It has been estimated that 282 mld of groundwater is extracted from more than 7000 borewells. Further, an estimated 120 mld of water is drawn from nearly 80,000 private borewells in the city. Such an enormous extraction accounted for 28 per cent of the total drinking water supply for 2001–2002 (GoK, 2009: 85).

One of the major problems associated with water resources in the city of Bengaluru is the presence of multilevel institutions governing water resource (see Fig. 7.2). Some of the major institutional problems are:

1. A weak enforcement of regulatory norms particularly in the urban setting affects the governance of water resources. There is no compliance with respect to laws, regulations and norms governing water resources in the city. The city is witnessing an unauthorised/illegal use of groundwater. Therefore, there is a need for devising stringent regulations and laws to contain illegal groundwater usage. For example, the residents of Bengaluru periphery region indulge in a rampant

[1] Refer Chapter V for complete details on BWSSB and its organisational structure.

[2] BWSSB is a parastatal body constituted in 1964.

[3] Interview with officials of BWSSB on April 11, 2014.

[4] Interview with Regional Director, Central Ground Water Board (CGWB) on April 28, 2014.

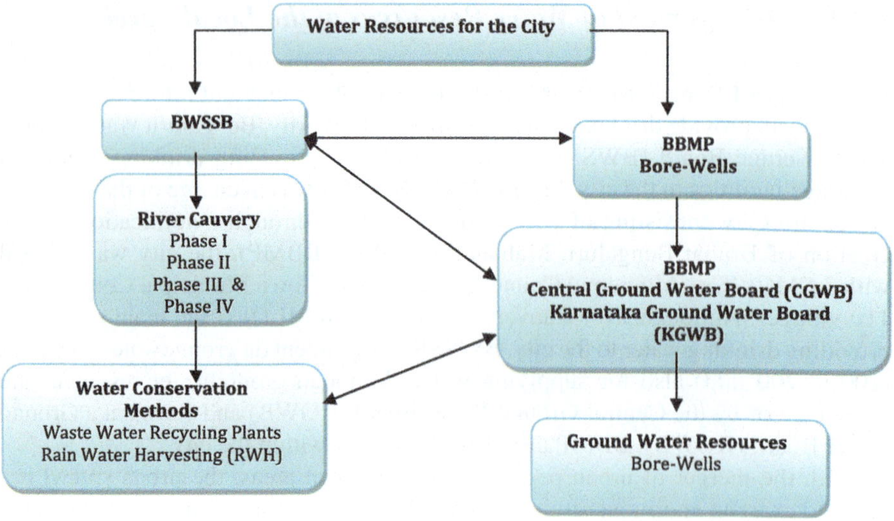

Fig. 7.2 Institutional dynamics of water supply delivery

exploitation of groundwater resources. Neither BWSSB nor Karnataka Ground Water Board nor Central Ground Water Board has maintained a data bank on the number of borewells or illegal water drawn from groundwater resources. Similarly, even though the city of Bengaluru has made installation of rainwater harvesting structures mandatory for new buildings, compliance levels are very low.

2. There exists a lot of political apathy in that local level political representatives like MLAs, MPs and ward level councillors interfere, blatantly violating regulations by way of promoting groundwater mafia, particularly in the peripheral region of Bengaluru.
3. Data regarding groundwater is scattered among multiple organisations. For instance, BBMP has a certain number of additional borewells, while private borewells are also very rampant in the periphery region. The absence of consolidated data sources on the number of borewells dug or operational or defunct borewells has further complicated the effective water management of water resources in the city.
4. A weak monitoring by multiple organisations has resulted in an ineffective resource management. The unauthorised tapping of groundwater is a very serious issue for the city.
5. Although BWSSB water recycling plants are one of the best models in Asia, treated water is not used for domestic or residential purpose; rather it is restricted only to nondomestic purpose. Both social and economic factors such as conservative attitude of households and income are attributed to the non-use of treated water for domestic or residential purpose.

6. At the field level, BWSSB lays pipe lines for new connections. This is done without a prior notice to BDA or seeking the concurrence of other departments like BDA or BESCOM. The already congested roads and lanes complicate the laying of pipe lines, and further digging of roads on their own affects the pedestrian mobility besides damaging the surroundings of the environment.

7.3.2 Integration of Institutions for Water Management

Institutional coordination is essential as far as sector-wise local bodies are concerned. The issues include the linkages of legal and regulatory issues, norms and practices. Integration of urban institutions is one of the biggest problems, affecting effective environmental governance. Integration should occur at both the administrative and technical levels involving all the stakeholders.

From the above figure (see Fig. 7.3), it is very clear that institutional synergy is possible only with active participation of all the local bodies concerned. For an effective management of water resources, there is a need for integrated efforts on the part of agencies like BBMP, BWSSB, CGWB and KGWA in the following ways:

1. At the administrative level, regulations, rules and laws must be effectively enforced and implemented. For instance, while implementing building by-laws, water conservation measures such as RWH and the use of treated water initiated by BWSSB must be strictly enforced, i.e. before BWSSB approves of new water

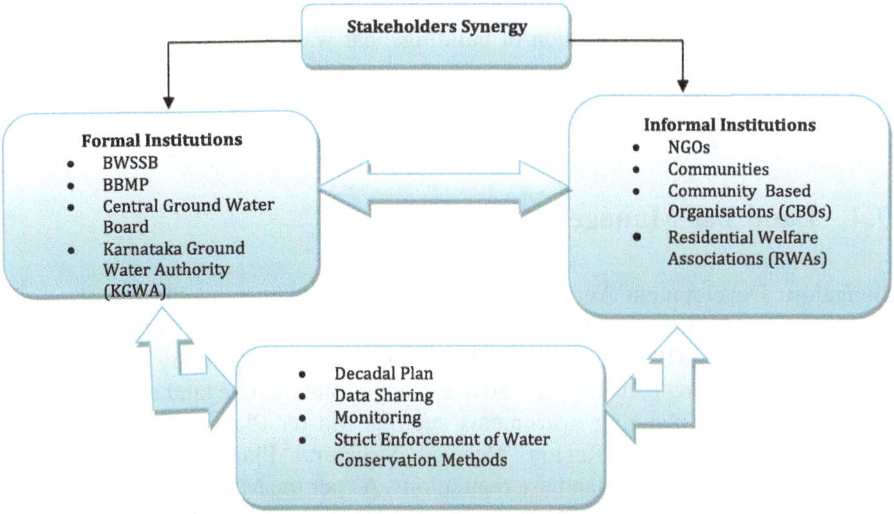

Fig. 7.3 Institutional integration for water resource management in the city

connections, builders must get the required certificate from BBMP to the effect that they have successfully installed RWH structure.

2. A decadal plan involving all the stakeholders with clearly defined rules, regulations and laws can help effectively manage the resources.
3. Data regarding water resources (i.e. the number of borewells dug, levels of water extraction, number of defunct borewells) must be consolidated at one place to facilitate the dissemination and sharing of information.
4. Networking between formal and informal institutions like community participation through community-based organisations (CBOs) and Residential Welfare Associations (RWAs) should be effectively encouraged for the monitoring and management of water resources such as waste water recycling and installing of rainwater harvesting structures (RWHs). These arrangements at the community level can assist water conservation such as groundwater recharge, reduction in the dependency on freshwater sources, etc. For instance, A.R. Shivakumar, 52, a senior scientist at the Karnataka State Council of Science and Technology (KSCST), a Bengaluru-based government body, is a vociferous proponent of rainwater harvesting in the city. Due to his sincere efforts, the Karnataka State Government effected an amendment to the Bengaluru Water and Sewerage Act, making rainwater harvesting compulsory for residences and offices with an area of more than 2400 sq. ft. in the heart of the city of Bengaluru. So far, only 45,000 cases have been recorded with respect to implementing RWH in the city.
5. There is a need for a strict implementation of regulations for promoting waste water usage, particularly for bulk supply and high-rise buildings. Treated water can be used for secondary purposes.
6. There is also a need for synergy of policy regarding both groundwater usage and waste water treatment.
7. Decentralised models of STP have been proposed for the city to be implemented in 100 wards. The usage of waste water must be made compulsory for various activities like the construction of buildings, car washing, etc. The use of STPs should be made mandatory, particularly for new constructions in the newly added areas and periphery region.

7.4 Land Use Management

Bengaluru Development Authority (BDA) is primarily responsible for planning, regulating, controlling, monitoring and facilitating urban development in Bengaluru Metropolitan Area (BMA), as per the Karnataka Town and Country Planning Act, 1961. BDA is also responsible for land use changes that include land use regulations and controls. But planning instruments such as Master Plan 2015 by BDA and Bengaluru Metropolitan Region, Revised Structural Plan 2031 prepared by BMRDA, are governed by land use regulations. As per the Master Plan 2015, based on Zonal Regulations, Local Planning Areas (LPAs) are created for strictly enforcing land use regulations besides partial powers being delegated to BBMP which is responsible for approving of building plans (as per the Karnataka Town and Country Planning Act 1961, Section 13D). As per zoning regulations of the Master Plan, environmental rules and laws are implemented and enforced.

From the figure above (see Fig. 7.4), it is very clear that multiple organisations interact in respect of land use management in the city of Bengaluru. As a result of an institutional fragmentation, some of the constraints being faced are:

1. Land use management has become one of the biggest problems as far as the city of Bengaluru is concerned. Especially, with the merger of 7 CMCs, 1 TMC and 110 villages, the conversion of agriculture land to nonagricultural purposes has become a major constraint in the urban periphery region. Besides, blatant violations of land use regulations and the conversion of residential layouts into commercial purposes are becoming increasingly rampant in the core areas of the city.
2. As a result, green belt area has shrunk drastically over a period of time, i.e. from 800 sq. kms in 1972–400 sq. kms. As per the zoning regulations, green belt region does not appear in the Master Plan.
3. The enforcement of land use regulations occurs at two levels: (i) BDA and (ii) BBMP. While BDA is the sole authority for approving of layout plans, BBMP is responsible for approval of building plans.
4. Planning instruments such as Bengaluru Master Plan 2015 and Bengaluru Metropolitan Region, Revised Structure Plan, 2031, have further complicated the enforcement of land use regulations. Each of these Master Plans has its own administrative jurisdiction for implementing the set objectives, complicating further the delivery of services. Planning instruments such as Master Plans governing the city's development complicate the existing land management scenario. Many such Master Plans are prepared by each local body such as BDA, BMRDA, BESCOM, BWSSB, etc., defining their own regulations and norms for the city. The result is, besides wasting enormous funds in the preparation of individual sets of Master Plans, none of the stakeholders consult each other. This results in the duplication of work and overlapping, leading to conflicts.

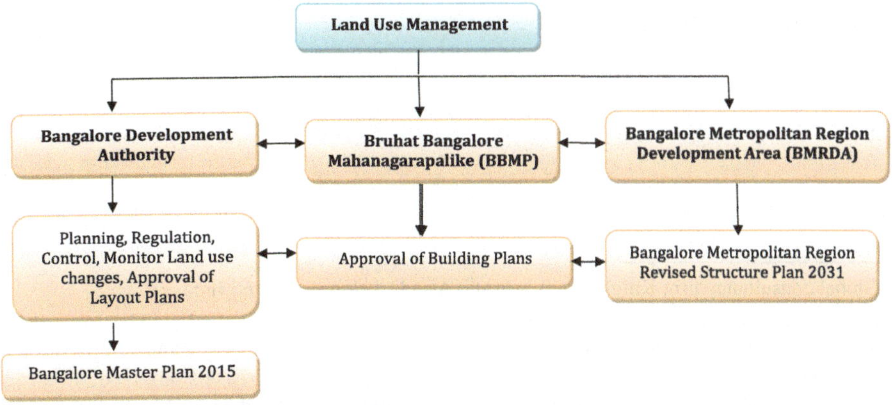

Fig. 7.4 Institutions governing land use changes

5. BDA has completely failed to regulate the land use pattern in the city. As a result, land prices have skyrocketed with private builders manipulating the land use pattern. For example, a study conducted by a private consultancy[5] firm reveals that Prime Asia Development Land Index for Bengaluru and the National Capital Region (NCR) shows a growth of 26.1 per cent and 24.9 per cent, respectively (Times of India, 31st May, 2014).[6]
6. Meanwhile, Karnataka Industrial Area Development Board (KIADB), an independent statutory authority, is in charge of developing industrial areas and regions. Unfortunately, land pockets, particularly, surrounding the Bengaluru rural and periphery region, are being indiscriminately converted from agricultural to nonagricultural purposes without seeking the required approval from BDA.[7] As a result, land pockets are notified by KIADB without complying with the BDA regulations and norms.
7. Similarly, land acquisition is a major problem affecting agricultural lands in Bengaluru rural areas, thereby affecting the local ecology and environment. Most often, de-notification is done without a prior approval from BDA. In the name of de-notification, much of the agricultural land is used for nonagricultural purposes.
8. The question is what needs to done to regulate the booming new development activities leading to congestion, traffic jams and pollution. Although both BBMP and BDA are responsible for the overall development of the city as also for executing the existing rules, violations are rampant. A weak regulation and implementation of the Karnataka Town and Country Planning Act, 1961, has further complicated the land management in the city.

7.4.1 Integration of Land Use Management

At present, a proposal has been mooted by BDA for integrating nearly 20 departments at the local level. A common data model[8] is suggested across the departments and boards at the city level.

[5] Global consultancy firm Knight Frank's Prime Asia Development Land Index.

[6] Refer Times of India (2014): Bengaluru is second in land price appreciation, 31st May, 2014. Access the (link http://timesofindia.indiatimes.com/business/india-business/Bangalore-is-second-in-land-price-appreciation/articleshow/35811716.cms.)

[7] Interview with EOI, Dept. of Town and Country Planning, BDA, April 26, 2014.

[8] Interview with EOI, Dept. of Town and Country Planning, BDA, April 26, 2014.

7.5 Management of Power Supply

The state of Karnataka embarked upon major power sector reforms from 1999 onwards. As a result, Karnataka Electricity Board (KEB) was dissolved and Karnataka Power Transmission Corporation Limited (KPTCL) constituted. This was followed by the constitution of Karnataka Electricity Regulatory Commission (KERC) in November 1999. Meanwhile, one of the four distribution companies formed to distribute power in Karnataka happens to be Bengaluru Electricity Supply Company Limited (BESCOM). BESCOM, responsible for power distribution in eight districts of Karnataka, covers an area of 41,092 sq. kms. With a population of

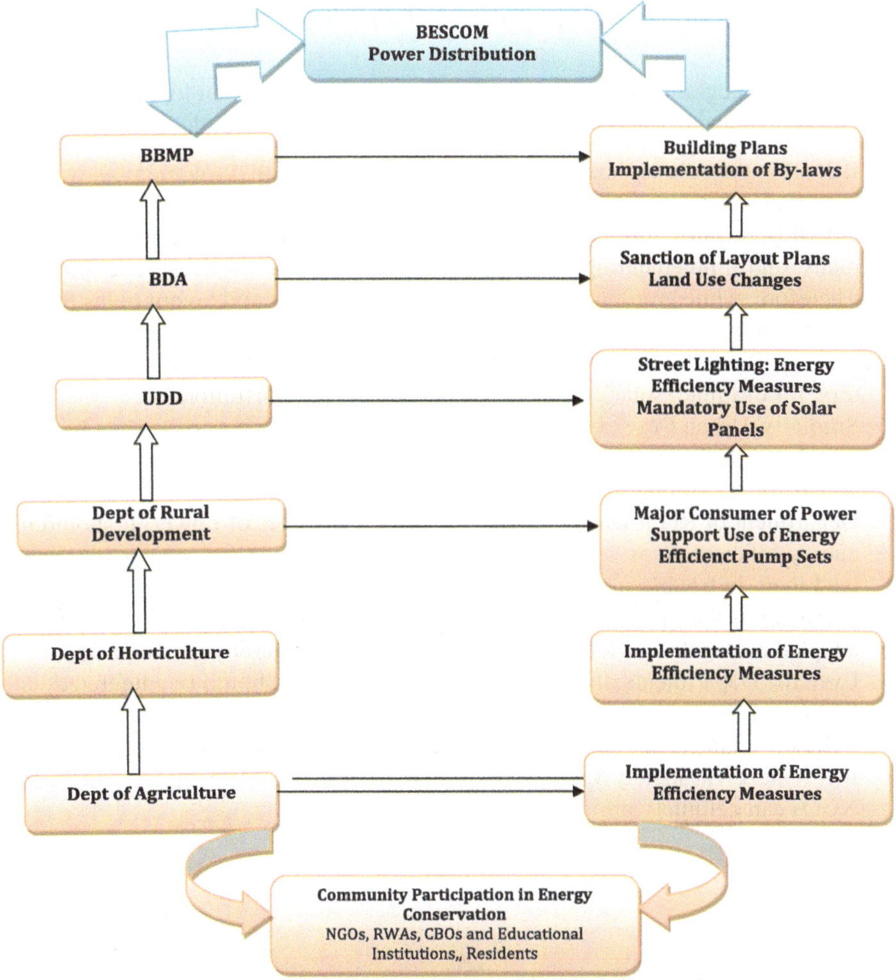

Fig. 7.5 Institutional arrangement for power supply

over 207 lacs. The company has three operating zones – Bengaluru Metropolitan Area Zone, Bengaluru Rural Area Zone and Chitradurga Zone.

BESCOM being the sole distributor of power to entire Bengaluru Metropolitan Area (BMA) has to interact with many urban local bodies such as BBMP, BDA, UDD, etc. There is an urgent need for the integration of BESCOM with various urban local bodies. Particularly, there is an ample scope for the integration of BESCOM with BBMP and UDD. The emphasis here is to reduce the peak hour load and to promote an efficient lighting system in the city and if possible, to provide 24 h power supply to the entire city. At present, the utilisation capacity stands at 2000 megawatts, but it should be substantially reduced to 500 megawatts (Fig. 7.5).[9]

7.5.1 *Issues in the Integration of Institutions with BESCOM*

Institutional integration cannot happen without integrating urban local bodies at the organisational level. The issues that need to be considered include the following:

1. BBMP is the sole authority for granting permission for implementation of by-laws and sanctioning building plans for new constructions in the metropolitan city of Bengaluru. Similarly, BDA is in charge of granting permission for layout plans and implementing zoning regulations related to land use changes. Therefore, while granting permission for building plans and layout plans for new constructions, these local bodies should strictly enforce energy efficiency measures; particularly they should make the use of energy-efficient power mandatory, for example: solar energy for residential power consumption.
2. Similarly, Urban Development Department (UDD) is in charge of street lighting. The department must strictly enforce the use of energy-efficient bulbs and solar panels for generating power.
3. Department of Rural Development is a major consumer of power. It should use energy-efficient pump sets. Farmers must understand that free power supply can land them in trouble in the long run. In view of free power supply, farmers are overusing power, partly out of negligence. For instance, flood irrigation model adopted at Bagalakote reduces wastage of water and thereby saves energy.
4. Even the Departments of Horticulture and Agriculture should be integrated with BESCOM to promote awareness regarding energy efficiency usage.
5. Policy-level decisions can help make the use of solar energy mandatory.
6. Involving civil society organisations like RAWs and local organisations like NGOs can definitely promote energy conservation.

[9] Interview with Chairman, BESCOM, Corporate Office, Bengaluru, April 24, 2014.

7.6 Management of Urban Lakes

A rapid urbanisation process is transforming the city of Bengaluru from metro to 'Mega city'[10] (as observed by Justice N.K. Patil). The worst hit of all is the disappearance of Bengaluru lakes. This could be attributed to an unplanned urban development and has led to encroachments and pollution (caused by dumping garbage and diverting untreated sewerage in to pristine lakes).

Urban lakes constitute an ecological heritage of the city's landscape. Preservation and management of urban lakes is the responsibility of various urban local bodies, as presented below (Fig. 7.6).

In the ongoing urbanisation process, the management and protection of aquatic bodies, particularly urban lakes, have become a daunting task. Some of the institutional constraints involved include the following:

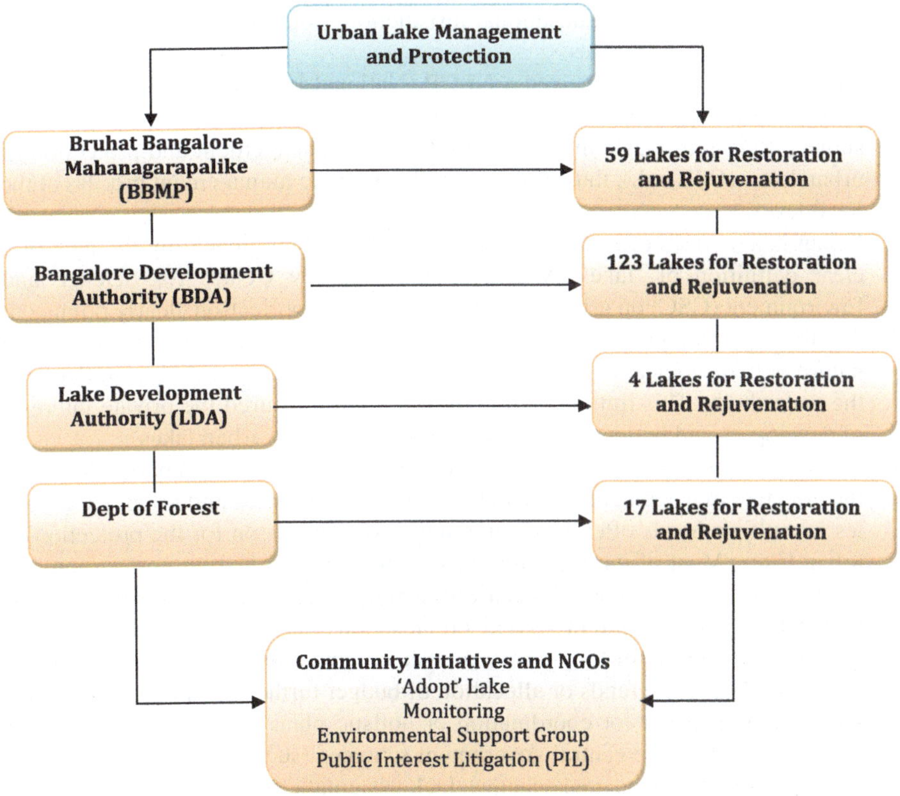

Fig. 7.6 Institutional arrangement for management of urban lakes

[10] An observation by Justice N.K. Patil on the deteriorating status of Bengaluru Lakes.

1. Particularly in a million (it is already 9.5 million) plus city like Bengaluru, the task of managing urban lakes is a complex process, given the fragmented institutions and multi-tasking by multiple agencies which include central and state level policies and programmes through Ministry of Forest, Ministry of Environment and Forest (MoEF), urban local bodies, development authorities and boards besides special purpose agencies like Lake Development Authority (LDA), etc.
2. Although special purpose agencies have created for protection, restoration and rejuvenation of urban lakes, for instance, Lake Development Authority (LDA), Bengaluru City continues to suffer from inadequate staff, funds and organisational strength.
3. The most significant of them is the lack of separate acts or laws for the protection and restoration of lakes in the city. As a result, numerous local bodies such as BBMP, BWSSB, BDA and LDA have their own set-ups for the protection of lakes under their respective jurisdiction. However, the problem is often overlapping institutional and organisational jurisdictions complicate the protection and conservation of urban lakes.
4. Although MoEF has conceived a NLCP (National Lake Conservation Plan) to promote a holistic approach towards lake conservation, as water being a state subject, the centre has a limited authority in the management and protection of urban lakes (CSE n.d.), thus further complicating the management process at the local level.
5. Urban lake management and protection is further complicated by the lack of a clear definition of 'lake'. A study carried out by Centre for Science and Environment (CSC) on urban lake management and protection clearly points out that the definition given by NCLP does not take into consideration the changing nature of hydrological criteria in that it limits itself to only the size and depth of the water body. This limited definition has in fact favoured the private interests of developers and builders in terms of encroachment of urban lakes.
6. The lack of institutional and organisational synergy and coordination hampers the effective management of urban lakes. Various agencies and urban local bodies at multiple levels often lack communication and vision for the protection of urban lakes. Most of the plans and programmes for protection, restoration and rejuvenation of lakes at individual levels are mostly short term in nature. Besides, both NGOs and community-based organisations such as RWAs lack a clear vision and direction for long-term management and monitoring of urban lakes.
7. The lack of separate funds or allocation of budget further complicates the long-term planning either for coordination or holistic plans for the management of urban lakes. Both the central government schemes (such as JNNRUM) and the state government-sponsored schemes lack systematic budgetary allocations for managing urban lakes.

7.6.1 *Seeking Autonomy for Lake Development Authority*[11]:

For the effective protection and management of lakes, according to the officials[12] of LDA, (i) LDA should be made the sole authority; (ii) both BBMP and BDA must relinquish their management rights on city lakes and transfer the same to LDA. Particularly, political representatives BBMP tend to influence the restoration efforts, and more importantly they happen to be the main encroachers of lakes surrounding the city. To achieve this, recently a proposal has been submitted to the government of Karnataka to pass the Karnataka Lake Development Authority Bill, 2014. If the bill is passed, LDA will become the sole authority with all the required funds, staff and infrastructure to manage and protect the city's lakes.

In its efforts to retain autonomy over lakes, LDA has drafted Lake Development Authority Bill, 2014, defining powers and functions so that it can prosecute those found polluting or encroaching upon lakes in the city. The bill is now with the Department of Law, Government of Karnataka (GoK) for approval; further it must be approved by state legislative assembly as well. The bill is intended to sufficiently empower the LDA to (i) seize vehicles and machinery and (ii) confiscate property and remove encroachments. As per the draft bill, any offender can be arrested by the designated officers of LDA without warranty. Besides, the violators can be fined (Rs.10,000 to Rs.20,000) or imprisoned (3–5 years) (Gandhi 2014).

The bill effectively prohibits (i) the use of lakes for any purpose other than storaging or impounding of water; (ii) construction/occupation of any structure on lake land, blocking inflow or outflow; (iii) construction of any commercial, recreational or industrial complexes or houses; (iv) dumping of debris or municipal solid waste; (v) discharging of untreated waste; (vi) construction of roads, bridges or any other structures; (vii) breaching of bunds, waste; and (viii) indulging in any other act which is detrimental, directly or indirectly, to lakes.

Further, a model DPR (Draft Project Report) has been commissioned under LDA involving national level tenders. This DPR will be a model for the management and protection of the city's lakes under LDA. The DPR will consist of all the lake management issues such as (i) recreational activities, (ii) community participation, (iii) maintenance, (iv) revenue generation, (v) lake advisory committee, etc.

7.7 Management of Urban Sanitation (Public Toilets)

According to the 74th Constitutional Amendment Act 1992, water supply and sanitation services for domestic, industrial and commercial purposes are one of the 18 functional responsibilities of ULBs (Urban Local Bodies) (Smitha 2011). Matters related

[11] Information collected from Lake Development Authority (LDA), Office, Bengaluru, on April 24, 2014.

[12] Interview with Chief Executive Officer, LDA, Bengaluru, April 24, 2014.

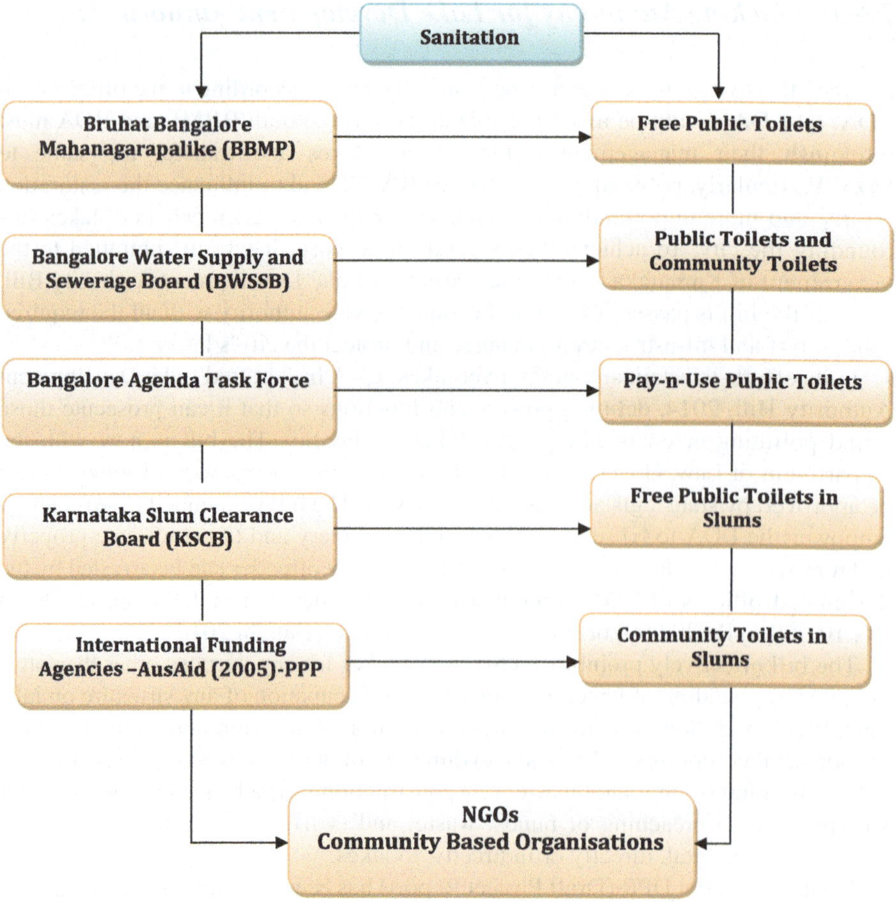

Fig. 7.7 Institutional arrangement for urban sanitation

to sanitation are subject to Article 56 of the Indian Constitution, resting with the state. The state is expected to fulfil water supply and sanitation services through (i) their own departments, (ii) state level boards and corporations, (iii) statutory and nonstatutory bodies at the city level and (iv) urban local bodies (Ruet et al. 2002 in Smitha 2011).

Urban sanitation constituting public toilets and storm water drains is exclusively managed by Bruhat Bengaluru Mahanagara Palike (BBMP) in Bengaluru City. However, household sewerage connections are managed by Bengaluru Water Supply and Sewerage Board (BWSSB). As part of corporate efforts, Bengaluru Agenda Task Force (BATF), 2005, also funds the construction of pay-n-use public toilets under 'Nirmal' programme. Similarly, Sulabh Souchalaya, an NGO, is also involved in the construction and maintenance of pay-n-use public toilets. Besides, a public-private partnership (PPP) scheme/project has been forged with the support of *BWSSBs AusAid* (2005) *funding* for the construction and maintenance of community toilets in three pilot study areas in Bengaluru City (Fig. 7.7).

7.7.1 *Institutional Constraints Involved in the Management of Urban Sanitation*

At the institution level, though norms and regulations do not conflict over urban sanitation, yet many agencies and organisations involved are complicating matters at the operational and functional levels. Some of the constraints involved include the following:

1. Public toilets managed and provided by City Corporation are highly inadequate and are often poorly maintained in the city.
2. Most often, due to the lack of adequate resources from City Corporation, many agencies such as BWSSB and Corporate Sector like BATF have funded the construction and maintenance of community and public toilets, particularly in the slums of the city. Therefore, a poor institutional financing has affected local bodies in terms of undertaking an effective implementation of schemes related to urban sanitation (Smitha 2011).
3. Many international donor agencies such as JIBC and AusAid have stepped up their funding for PPP model of construction of community toilets in the slums of Bengaluru. But the efforts are very limited and selective. For example, in 2006, only three slums were selected on a pilot basis, namely, Cement Huts, Sudamanagar and Chandralayout for implementation of AusAid-supported project (under PPP model) for providing water connection and UGD and constructing community toilets in the city of Bengaluru. But BWSSB has completely failed to replicate and upscale the projects in other slums of the city (Smitha 2011). Such an involvement of multiple agencies undertaking similar and often overlapping functions has led to conflicting agendas during operation and maintenance. Most often the efforts are not holistic and cannot be replicated in terms of promoting a complete access to public toilets by slum-dwellers.
4. The lack of coordination between City Corporation and parastatal bodies like BWSSB affects the implementation of meaningful, long-term sustainable plans for urban sanitation.

Therefore, a greater challenge is the lack of a strong institutional convergence at the local level for mediating the jurisdictional conflicts so as to promote sustainable development in the area of urban sanitation.

7.8 Conclusion

Institutions and governance structures at the local level are often dynamic, multilayered structures, with a range of stakeholders (across sectors and agents), and their interrelationships strongly influence environmental actions and their outcomes. Many bottlenecks and impediments exist in the formulation and implementation of governance measures such as policies and programmes, affecting the environment of the city. The context-specific issues and challenges vary across different sectors such as resources, monitoring, finance, stakeholder integration, transparency and accountability. Therefore, improving the governance process by identifying the context-specific challenges helps promote better environmental outcomes. Effective environmental governance is, therefore, largely dependent on an integrated institutional, regulatory and legal framework and the capacity of the governing authorities in providing an 'enabling environment'.

Government ought to play a more analytical, political and managerial role than the technical role they have played in the past. Amos (1989) states that certain forms of decentralisation have been characterised by the central state devolving the responsibility of delivering services to local government without ensuring that this tier of government has the capacity to execute this task and, hence, proposes new role demands skills require a rethink with appropriate training programmes for employees. Delivery of urban services is a vehicle for determining how the states, national and local level, distribute services more equitably. Decentralisation of government functions and evolution of strong local government is possible with allocation of necessary resources required to deliver urban basic services. The relationship between national and urban governance stems from the structures that underpin power and authority.

At the local (urban) level, decentralisation brings good governance through increasing popular decision-making. Popular participation is desirable and essential in improving the operation of urban services, making them relevant to local needs, increasing local commitment and involvement in provision of services. Governance must grapple with the number of problems in association with this rationale it adopts. It is essential to establish what sort of participation is intended in a particular situation which individuals and groups are to participate and degree of participation. There is little literature on urban governance with case studies to illustrate issues raised by Boeninger (1991) with respect to dimensions of governance indicating technical, political, institutional and cultural dimensions.

New urban governance is an effective system of 'multilevel governance' with well-defined spheres of government (national, regional and local) based on appropriate decentralisation policies that aims to construct a balanced and collaborative system of well-managed cities and improved urban-rural linkages so that no city or territory is left behind (OECD 2013). Given the various challenges and contexts, there is uncertainty and change in global public health problems, impacts of climate change, inequalities in cities, youth booms, vulnerable ageing population, unprecedented migration, crime and insecurity and conflicts over resources.

Given the vulnerability of cities, new governance should be informed, flexible, innovative, forward-looking and open to continuous learning for intelligent governance. Increasing complexity of urban governance multilevel systems and multilevel stakeholder challenges given the diversity in local conditions, new urban forms, coexistence of intermediary cities, small towns, rapidly growing cities and shrinking cities should be addressed holistically. Urban policy is required to meet specific characteristics in the context of absent or inadequate decentralisation besides budget constraints given the newer challenges of limited revenue-raising capacities and ineffective decentralisation, poor planning and economic inefficiencies. Ineffective legal and institutional frameworks add on by creating disconnect between legal, administrative and fiscal frames. There prevails lack of clarity in distribution of responsibilities across levels of government, contradictory regulations, rigid frameworks to address changing situations and urban dynamics like the title deeds for the urban poor (Hakku patra in Karnataka slums), industrial policies and apartment policies. Metropolitan challenges wherein more than 500 cities worldwide exceeded threshold of 1 million inhabitants and having physically grown beyond administrative boundary economies have become more globalised, attracting flow of funds, capital and migrants. Metropolitan cites have centralised administration or several municipalities with significant disparities and spatial segregation across neighbourhood governance reforms compounded with intense political debates/controversies. Inequity and exclusion among women, youth, elderly, minorities and urban poor seem prominent leading to social imbalances, friction, violence and political instability.

Insufficient monitoring and evaluation of urban policies complexed with poor data affects informed decisions. National governments do not promote involvement of local governments/stakeholders in definition, implementation and monitoring of urban and regional policies/plans' weak frameworks for service delivery partnerships. Legal frameworks like the tendering, contracts, weak and unimplemented projects discourage foreign business investment. PPP is difficult to undertake. Rapid technological change and new technological developments offer unforeseen possibilities in business, citizens and public actors. Public actors find it difficult to understand and regulate with little understanding on what needs to be regulated and what should be left to the market and how best to protect common good questions concerning surveillance, personal data protection, privacy, etc. and continue challenge for urban governance. Enabling institutions through local governments and integrated approach and participation by evolving key processes in planning, implementation and monitoring is important. Putting mechanisms in place is important to facilitate participation and national governments need to promote effective decentralisation and participation. Besides, defining roles and responsibilities more clearly to avoid tensions and establishing legal and institutional frameworks, establishments of National urban forums, judiciary and legislative branches strengthens the processes. It is important to accommodate at all levels by setting rules, regulations for new urban agenda, scope of interpreting and defining principles, values and rights, supranational governments and intergovernmental cooperation, endorsing global guidelines for good urban governance, encouraging cross border intermunicipal cooperation and sharing good practices. Civil society functioning relationship between government institutions and civil society requires mediators and facilitators. Educational and knowledge institutions about urban issues and closing the gap between academia and local govt realities by building strong links between private sector CSR, PPP, Corporate philanthropic organizations through innovation and facilitation initiates a holistic approach. Financial institutions and international development agencies like banks, insurance companies, etc. also play an important role in improving governance. Private sector incentives with public goals and incentivisation to adopt sustainable practices are to be promoted. International devt agencies through funding for sustainable infrastructure in sanitation, water, transport, etc. Policy design, implementation for improved governance and strong multilevel governance frameworks strengthen decentralisation processes, promote integrated national urban and territorial policies, reinforce metropolitan governance and promote participation and equity online networks. It is also required to have active RWAs; NGOs strengthen capacity building for urban governance through in-house training and outsourcing. Enabling digital era governance is relevant to bring about a significant improvement in governance. Both nationally and state-wise e-governance department is playing a significant role in improving governance.

References

Amos, A. F. (1989). The occurrence of hawks bills along the Texas coast In: Proceedings of the North Annual Workshop on Sea Turtle Conservation and Biology Feb 7-11, 1989, Jekyll Island, GA, NOAA-JM-NMFS-SEFC- 232, Miami FL.

BDA. (2007). *Revised master plan –2015: Vision document, Bangalore* (Vol. 1). Bangalore: Bangalore Development Authority.

Boeninger, E. (1991). Governance and Development in Proceedings of the World Bank Annual Conference on Development Economics, World Bank, Washington, DC.

CSE. (n.d.). *Protection and Management of Urban Lakes in India*. New Delhi: Centre for Science and Environment.

Davis, L., & North, D. C. (1971). *Institutional change and American economic growth*. New York: Cambridge University Press.

Gandhi, Divya (2014). Draft Bill gives 'toothless' LDA powers to prosecute polluters, dated 5th June, The Hindu.

GOK. (2009). State of Environment Report, Bangalore 2008, Bangalore: Department of Forest, Ecology and Environment and Government of Karnataka.

Halfani. (1997). Marginality and Dynamism: Prospects for the Sub-Saharan African City. In M. A. Cohen, B. A. Rable, J. S. Tulchin, & A. M. Garland (Eds.), *Preparing for the urban future, global resources and local forces*. Washington: Woodrow Wilson Centre Press.

King, Andrew and Michael W. Toffel (2007). Self-regulatory Institutions for Solving Environmental Problems: Perspectives and Contributions from the Management Literature, Working Paper 07–089. Access the following link for full paper: (http://www.hbs.edu/faculty/Publication%20Files/07-089.pdf).

Keohane, R., & Ostrom, E. (1995). *Local commons and global interdependence:Heterogeneity and cooperation in two domains*. London: Sage.

Najam, A., Papa, M., and Taiyab, N. (2006). Global environmental governance: a reform agenda, International Institute for Sustainable Development.

North, D. C. (1981). *Structure and Change in Economic History*. New York: London.

North, D. C. (1990). *Institutions, institutional change, and economic performance*. Cambridge: Cambridge University Press.

North, D. C. (1991). Institutions. *Journal of Economic Perspectives, 5*(1), 97–112.

OECD. (2013). Environment as a Glance, OECD indicators.

Ostrom, E. (1990). *Governing the commons: the evolution of institutions for collective action*. New York: Cambridge University Press.

Ostrom, E. R., & Gardener, J. W. (1994). *Rules, games and common pool resources*. Ann Arbor: University of Michigan Press.

Paavola, J. (2005). *Environmental conflicts and institutions as conceptual cornerstones of environmental governance research, CSERGE working paper EDM 05–01, Centre for Social and Economic Research on the global environment*. UK: University of East Anglia.

Paavola, J. (2007). Institutions and environmental governance: A reconceptualization. *Ecological Economics, 63*, 93–103.

Papa, M. (2014). Environmental foreign policy: Crossovers among levels of governance. In P. G. Harris (Ed.), *Handbook of global environmental politics* (pp. 138–149). New York: Routledge.

Plumptre, T., and Graham, J. (2000). Governance in the new millennium: challenges for Canada, Ottawa, ON: Institute on Governance.

Rhodes, R. A. W. (1996). The new governance: Governing without government. *Political Studies, 44*, 652–667.

Ruet, J., Saravanan, V. S., & Zevah, M. H. (2002). The water and sanitation scenario in Indian metropolitan cities: resources and management in Delhi, Calcutta, Chennai, Mumbai, Occasional Paper No. 6 Centre de Science Humanities, New Delhi.

Smitha, K. C. (2011). *Urban governance and service delivery in India: A study of water supply and sanitation in Bangalore, unpublished thesis*. Bangalore: Institute for Social and Economic Change.

Young, O. R. (2002). *The institutional dimensions of environmental change: Fit, interplay and change*. Cambridge, MA: The MIT Press.

Young, Oran R., Leslie A. King, and Heike Schroeder (1999): Institutions and Environmental Change: Principal Findings, Applications, and Research Frontiers, Access the link (http://www.eci.ox.ac.uk/research/climate/cop07/downloads/schroeder-summaryfor-policymakers.pdf).

Chapter 8
Environmental Governance in Bengaluru

Abstract Rapidly urbanising cities in Asia have two defining features, i.e. demographic expansion and urban concentrated economic growth and, secondly, scattered forms of urbanisation creating islands of 'agglomeration, suburbanisation and urban sprawl' threatening environmental management (Dahiya, B. (2012). 21st century Asian cities: Unique transformation, unprecedented challenges. *Global Asia, 7*(1), 98–104.). Environmental governance in India is challenged by an unprecedented urbanisation, demographic explosion, environmental degradation coupled with multi-level institutional jurisdictions and actors. More importantly, urban setting in India is often characterised by an overwhelming rural-urban migration, population growth, economic growth-related industrialisation, urbanisation-induced growth of slums that severely impinge on the natural resources, thereby adversely affecting the environment.

8.1 Introduction

Rapidly urbanising cities in Asia have two defining features, i.e. demographic expansion and urban concentrated economic growth and, secondly, scattered forms of urbanisation creating islands of 'agglomeration, suburbanisation and urban sprawl' threatening environmental management (Dahiya 2012). Environmental governance in India[1] is challenged by an unprecedented urbanisation,[2] demographic explosion, environmental degradation coupled with multi-level institutional jurisdictions and actors. More importantly, urban setting in India is often characterised by an overwhelming rural-urban migration, population growth, economic growth-related industrialisation, urbanisation-induced growth of slums that severely impinge on the natural resources, thereby adversely affecting the environment.

We explored the status of 'environmental governance' in Bengaluru Metropolitan Area (BMA). In the context of a rapid urbanisation process, the city has exploded both in terms of population and spatial transformation. Preliminary observations with respect to environmental governance in the city of Bengaluru clearly point to compounding environmental challenges and a poor urban planning. The burgeoning population of the city has led to varied problems such as poor access to basic ameni-

[1] 'Metropolitan cities host nearly 29 per cent of Asia's urban population' (Dahiya, 2012: 99).

[2] India's urban population is the second largest in the world after China (Rahman et al., 2009: 182).

K. V. Raju et al., *Urban Environmental Governance in India*, The Urban Book Series, https://doi.org/10.1007/978-3-319-73468-2_8

ties (water supply, sanitation, waste management, urban mobility, power supply, etc.), loss of biodiversity, high levels of pollution, depletion and contamination of groundwater resources and degradation of ecosystems. In fact, the city has reached the threshold levels of environmental exploitation and degeneration, clearly away from sustainable path. The situation is further accentuated by a weak institutional dynamics (poor regulations, flouting of norms and standards) at the local level.

The study, therefore, presents an overview of the state of environment prevailing in Bengaluru Metropolitan Area (BMA). The study has relied on secondary data collected from local bodies, boards, corporation and NGOs. Further, personal interviews and focus group discussions were conducted with key administrators, bureaucrats and officials of the local bodies for eliciting their opinions and perceptions on key areas influencing environmental governance. The report has been broadly divided into eight chapters based on issues and areas affecting environmental governance. Further, the study presents situational analysis of what has been done towards promoting environmental governance (reactively), how it is done, why it is done, what are the intricacies and constraints in promoting environment governance, so far how the city of Bengaluru has achieved its success, whether there exists any further scope for future and how to manage environmental governance proactively. In this context, the underlying systems, in particular, the role of institutions and governance structures are effectively discussed.

The study reviewed the literature on environmental governance from global and national perspectives. The purpose is to examine how the recent urban transformation has impacted the urban environmental governance and also to explore how different approaches and methods are applied in the context of global experience. The global urban experience clearly suggests that over half of the world's population[3] lives in Asian cities (Dahiya 2012: 98), posing a serious threat to the management of ecology and environment. Two most significant attributes of Asian urbanisation are (i) demographic expansion and an undue emphasis on economic growth and (ii) ineffective urban planning. However, environmental resources and their management have never kept pace with a rapidly evolving urbanisation and economic transformation of Asian cities. Many cities in Asia have emerged as 'knowledge economies', 'command-and-control' points such as Beijing, Shanghai, Bengaluru, Mumbai, Hyderabad, Hongkong, Seoul, Singapore, Tokyo, etc. Particularly, Bengaluru and Mumbai are aspiring to become international financial centres emulating the success of Tokyo and Singapore models. As a result, Asian cities are facing a unique set of challenges related to environment and climate change (Dahiya 2012: 101). In the quest for economic growth, cities in Asia are on the verge of environmental degeneration in terms of environmental pollution, unregulated natural resources, such as water, loss of biodiversity and ecosystems and unabated urban pollution (Hust 2005). The situation is further accentuated by urban poverty, for instance, the number of urban poor increased from 107 to 125 million between 1993 and 2002. Similarly, the Asian cities alone have hosted 505 million

[3] Between 1990 and 2010, Asian urban population increased by over 754 million people, equal to the combined population of the United States and the European Union (Dahiya 2012: 98).

slum-dwellers, more than half of the world's slum population by 2010 (Dahiya 2012: 100). As per 2011 census, every sixth urban resident lives in slums.[4] In other words, nearly 17 per cent of the urban population lives in slums. Further, the census declares that all-India slum population could be 27.5 per cent by 20,011 (Shrinivasan 2013). One of the chief characteristics of Asian urbanisation process is the emergence of 'informality' (UN-HABITAT 2010: 87). Urban slum-dwellers are significantly drawn to unregulated jobs with meagre incomes which further aggravate inequality and deprivation at the city level.

8.2 The City of Bengaluru: Geography and Environment

The city of Bengaluru is one of the fastest-growing cities of the twenty-first century. The city grew spatially from 69 kms in 1949 to 741 sq.kms by 2007 (Sudhira et al. 2007: 380). The city is growing in all directions unabated, exerting a severe stress on access to local resources, particularly water supply, sanitation, urban land, transportation, health and education. Due to the absence of a defined natural boundary, such a continuous expansion of the city well into the peripheral areas has led to the near destruction of agricultural lands and natural ecosystems (BDA 2007: 11; Ravindra 2012:78). Besides, an increase in urban poverty has further complicated access to resources. At present, one of the greatest challenges for the local governments is the promotion of a sustainable environmental governance.

8.2.1 Key Environmental Challenges

Some of the key environmental challenges faced by the city of Bengaluru include (i) the pace of urbanisation and economic growth has not translated into the actual capacity of local governments in terms of catering to the increasing demands for basic amenities. As a result, the local governments are unable to deal with the emerging environmental and ecological challenges; (ii) an increasing rural-urban migration has led to an increase in urban poverty which in turn has exerted an enormous pressure on the environment; (iii) a rapid expansion of the city is tremendously affecting the 'ecological footprint', a phenomenon that impacts the environment; and (iv) the lack of an effective urban planning has made the city more vulnerable to climate change and its impacts.

[4] Census (2011) defines a slum as 'residential areas where dwellings are unfit for human habitation' because they are dilapidated, cramped, poorly ventilated, unclean or any combination of these factors which are detrimental to the safety and health.

8.2.1.1 Institutions and Governance Structures

The study clearly illustrates how the institutions and governance structures directly confront man-made management systems which are designed to impact the environment. The metropolitan cities in India are hubs of varied institutional jurisdictions cutting across actors and sectors. In addition, environmental laws and legal pronouncements are interpreted as the desired levels of governance, and they represent the legal thresholds for environmental governance. Institutional complexity and the entwining of legislations, mandates, norms and regulations at both the spatial and sectoral levels (Lele et al. 2010: 15) further complicate the process of planning and management of environmental governance from national to local levels. As a result, there is the lack of a long-term strategic environmental plan and management that can further weaken the governance.

8.2.1.2 Local Institutions and Governance Structures in Bengaluru

The modus operandi of urban governance in the city intersects at two levels:

(i) administrative structures and 'processes' such as rules, norms, regulations and implementation of policies and programmes are often supported by grants from the central and state governments as also financial aid from international agencies and donors; and (ii) the city planning is supported by Master plans, clearly mandating zoning codes, building by-laws and planning schemes governing urban development. A preliminary investigation of the situation in Bengaluru metropolitan clearly demonstrates that the city is governed by multiple institutions cutting across sectoral departments (both central and state), regional authorities, special-purpose agencies and local bodies. Therefore, a broad array of institutional and governance strategies being practiced clearly create problems in discharging of their obligations at both the operational and management levels. There is an apparent lack of institutional coordination and synergy for promoting an effective environmental governance. The lack of funds and finance also presents serious impediments in terms of causing weak institutional capacity. In the view of complexity and multi-scalar character of institutions, local level bodies fall short of addressing most of the pressing environmental challenges. The study clearly highlights the emerging hybrid modes of governance structures across the centre vs state and market vs community partnerships. Therefore, the study has tried to explore the significance of institutions at different spatial scales with respect to environmental governance, mainly focusing on the emerging hybrid forms of governance structures.

8.2.1.3 Emerging Environmental Governance: Models and Approaches

A rapid urbanisation is the singular factor affecting the world. The contours of urbanisation are often found to be in conflict with natural resources and the built-in environment. A highly haphazard urbanisation process across most of the Asian metropolitan cities has led to consequences that can no longer be ignored. Some of the pressing environmental challenges are global climate change, ecosystem degradation, greenhouse gas emissions, urban poverty, impoverished living standards, increase in built-up area, loss of biodiversity, etc. The study reviews on how different approaches and models are being followed for attaining 'sustainability' so as to offer solutions in terms of effectively addressing multiple environmental issues and conflicts. These sustainable models and approaches are often driven by particular agents such as the centre, state and civil society-based actors such as NGOs and local communities across various sectors. However one of the striking factors is that none of the metropolitan cities in India qualifies as a 'sustainable' city model. An assessment of these 'sustainable models' across the world reveals that they constitute the emerging trends influencing and shaping the environmental governance across the cities.

8.2.1.4 Environmental Management in Bengaluru: Trends and Status of Environmental Resource Use in Bengaluru Metropolitan Area (BMA)

An assessment of environmental management in Bengaluru clearly points to the reality that spatially the city is growing by leaps and bounds. The city of Bengaluru happens to be one of the top 10 mega cities of India,[5] accounting for 15 per cent of GDP, 8 per cent of population and just 0.1 per cent of land area (IIHS 2011: 9). Further population growth and migration from rural to urban peripheral areas have led to growth of suburban and urban sprawls. Also, the industrial growth of the city has a significant impact on the nature and pattern of urbanisation in the city. The current urban growth and development plans are conditioned by the emergence of urban corridors such as Bengaluru-Mysore Infrastructure Corridor (South-West Bengaluru), Bengaluru International Airport (North Bengaluru), Information Technology (IT) Corridor (South-East Bengaluru), Bengaluru Metro Rail and location of IT/BT industries (East and South Bengaluru). Such an exponential growth of the city, coupled with the inability of the local bodies to provide necessary infrastructure and basic amenities, has inevitably led to an environmental chaos. As any other metropolitan city in India, Bengaluru also faces the same problem of an unbridled urban expansion, leading to depleting water sources, inadequate housing, poor transportation, lack of urban mobility, deficient sanitation, erratic power supply, etc. Thus, it is evident that the city lacks a long-term vision for planned urban development and infrastructure provision which, in turn, has inevitably led to plethora of environmental problems at varying scales and degrees. This raises the serious question of sustainability of the city and its impact on ecology and environment. Moreover, an increase in urban poverty due to the mushrooming of slums in the city

[5]According to a study carried out by Indian Institute for Human Settlements (2001), top ten mega cities include Greater Mumbai, New Delhi, Calcutta, Chennai, Bengaluru, Hyderabad, Ahmedabad, Pune, Surat and Jaipur.

is literally threatening the environmental fabric of the city. Most often slum-dwellers bear the brunt of environmental degradation in terms of lack of access to potable water, sanitation-toilets and sewerage facilities, land, health care and education.

A qualitative assessment of the state of infrastructure in the city of Bengaluru is succinctly profiled in the following sectors:

- Water Supply: The supply of water to the city is 840 MLD, while the demand being 1050 MLD at the supply rate of 140 lpcd. Correspondingly, the demand and supply gap[6] amounts to 210 MLD and 560 MLD (BDA 2007: 18; CSD 2010). The over all leakage and wastage of drinking water varies from 30 to 40 per cent (Smitha 2006: 400). At present, water distribution systems are available for only 226 sq.kms of BMP area, while the newly merged areas of 7 CMCs and 1 TMCs have less than 10 per cent of individual connections from BWSSB (BDA 2007: 47). While the Revised Master Plan 2015 prepared by BDA claims that 30 per cent of groundwater is tapped to fulfil the city's water requirements through about 5850 borewells and 15,180 public taps (BDA 2007: 47).
- Groundwater Management: A heavy reliance on groundwater resources as an alternative source of supply of potable water to the city residents constitutes one of the defining features of environmental deterioration. The total supply from private borewells for 2008 amounts to 170 MLD, while 200 MLD for 2012 (CSD 2012: 16). There is phenomenal growth of borewells which has increased from 5000 to around 4.08 lakh in the last three decades drawing 750 MLD (Sawkar 2012: 15) which is 3.7 times more than the recharge from the city's annual rainfall of 900 mm (Balasubramanian 2013).
- Sanitation: Access to toilets – As per the data available with BBMP,[7] there are merely 587 public toilets for an estimated population of 96.2 lakh out of which only 200 public toilets are functional, which means that for every 19,000 population, only one toilet is available. Studies have shown that 17 per cent of the slum-dwellers (i.e. every third in recognised slums and 700,000 in unrecognised slums) in the city do not have access to in-house toilets and that they usually rely on public or shared toilets or open spaces (GoK 2009: 94).
- Waste Water Management: At present, the core city area of 226 sq.kms (i.e. erstwhile BMP) is completely connected to sewerage network, while the newly merged areas of greater Bengaluru are yet to be adequately covered by sewerage network.
- Storm Water Drainage: Storm water drains under BMP cover 240.40 kms, while under greater Bengaluru, the area covered has increased to 869.83 kms covering all the 8 zones (CSD 2012: 18). Drainage is one of the major areas of concern (BDA 2007: 18).

[6] Access http://www.rainwaterharvesting.org/Crisis/Urbanwater-scenario.htm for details on demand and supply gap presented in Figure 2.

[7] BBMP is obliged to ensure access to public toilets as per the Karnataka Municipal Corporation Act, 1976 (The Hindu 2013).

- Solid Waste Management: The solid waste generation, treatment and disposal has been a great concern, seriously impacting the ecology and environment. The per capita generation of municipal solid waste has gone up from 2500 tonnes to 5000 tonnes per day over a span of 10 years between 2000 and 2010 (BDA 2007: 18; Varshney 2012; BBMP 2012). Although the city of Bengaluru has made segregation of domestic waste at source mandatory since 2011, it is yet to be practised due to lack of awareness, aggravating the environmental concerns. In addition, municipal waste mixed with hazardous, plastic and toxic waste seriously threatens the biodiversity of the city comprising water bodies such as lakes, tanks and acquifers.
- Urban Land Utilisation: The city of Bengaluru has grown from 226 sq.kms to 800 sq. kms incorporating 7 CMCs, 1 TMC and 111 villages (Greater Bengaluru region/Bruhat Bengaluru). The state environment report, Bengaluru (2009), clearly presents that urban land use has drastically changed due to rapid urbanisation and growth of urban sprawl. Urban land has been put to intense economic activities, from 284 km^2 in 1990 to 740 km^2 by 2007. Consequently, the built-up area has increased substantially while reducing wetlands and water bodies. It is estimated that the city green cover has been declining at an average rate of 30 sq.kms per annum. The built-up area has increased from 180.42 sq.kms to 301.27 sq.kms (61.6 per cent). The city has lost 23.7 per cent of water bodies at an annual rate of 1.7 sq.kms; similarly, it has lost 20.30 per cent of vegetation (BMRDA 2009: 265).
- Urban Transport: The city has grown from a population of 26 lakhs in 1971 spread over 250 sq.kms, at present city has blown up more than 9 million population spread over 1000 sq.kms (Vaidyanathan and King 2011). However, demographic and spatial expansion is not commensurate with the delivery of amiable transport services. The city is gradually experiencing a vehicular explosion from 400,000 in 1987 to 2.3 million by 2005 (BDA 2007: 36). Concomitantly, the vehicular GHG emissions have increased at alarming levels in the city.
- Energy/Power Supply: With a vertical growth of the city, the demand for power has increased enormously. At present, the city consumes nearly 30 per cent of the total power generated in the state with negligible savings. Over the years, the power consumption has grown by 14 per annually (ibid), resulting in a constant demand-supply gap.

Table 8.1 Action plan for urban environment governance

Issues	Constraints	Action required	Who should do it	How they should do it	Indicators for monitoring	How to measure monitoring
Institutions	Weak enforcement of rules, regulations, norms and standards Lack of effective compliance and deliberate violations of environmental rules, regulations, norms and values Lack of institutional interface at both administrative and technical levels among the stakeholders Lack of coordination between central and state level policies and programmes which intersect with local level by-laws, zoning regulations defined by master plans	To improve the coordination and networking for promoting effective of environmental institutions Effective enforcement of rules, regulations, norms and standards Promoting clear and simplified procedures Integration at both administrative and technical levels involving all the stakeholders	Centre and state and local bodies Among local bodies	Providing with an effective and strong coordinating role within the local system on environmental matters	Transparency and planning Strict enforcement of regulations, norms and standards	Level of planning and reporting sector-wise and at regional levels Extent of regular, widely accessible and comprehensive publication on environment Extent of promoting incentives and disincentives for adhering to environmental norms regulations and standards Extent of linkages established across sectors (stakeholders) for implementing regulations, norms and standards both at administrative and technical levels

Governance structure	Multiple actors/players/agencies at different scales and degrees complicate the operationalisation and management of environmental objectives Lack of awareness among political representatives regarding environmental issues Municipalities lack separate structures, mechanisms or instruments to promote environmental governance Lack of integrating demand management principles including legal, economic and social into efficient service delivery, adversely impacts the environment	'Tripartite' partnership between formal/government agencies, NGOs and community will certainly promote environmental governance Discharging environmental functions and obligations within the framework of intergovernmental synergy/interface Equipping political representatives particularly local officials with environmental protection and awareness about conservation Clearly identifying issues and priorities and detailed strategies for the implementation of environmental management and protection	Formal/government agencies (centre, state and local bodies)	Institutionalising partnership and engagement between government, civil society (NGOs) and community Motivating local officials through increased funding, staff and political support Municipalities should create separate structures, mechanisms or instruments into their processes for effective promotion of environmental governance Establish mechanisms and procedures within each sector for drawing expertise and views for design, implementation and evaluation across the sectors (functional departments) Effectively integrating demand management principles into the service delivery	Agenda setting Effective assessment and planning across sectors Data sharing among the stakeholders.	Extent of implementing both short-term and long-term plans/agenda across sectors for assessing performance on environmental management Extent of standards adhered to across thematic areas or sector-wise standards Extent of systematic data collection, data gap analysis, data generation and data synthesis

(continued)

Table 8.1 (continued)

Issues	Constraints	Action required	Who should do it	How they should do it	Indicators for monitoring	How to measure monitoring
	Lack of centralised database/sources for a realistic assessment of environmental resources (sectoral studies)	Creation of a centralised database/ source on environmental resources (sectoral studies)				Extent of regular maintenance and updation of a centralised database/source (centralised updation and regional data source on sectoral studies) on environmental resources
Infrastructure management	Lack of mainstreaming the management of infrastructure impacting environmental decisions Conflict between delivery of services and the protection of environment Lack of an effective strategy to respond to the environmental challenges and problems Lack of data on performance to effectively support decision-making	Mainstreaming information on infrastructure management to effectively incorporate environmental concerns into decision-making Strategising the departments to effectively respond to the environmental challenges and problems	Both state level and local bodies	Building resilient infrastructure management across sectors at the local level Effective strategising across departments to delineate clearly environmental responsibilities and discharging obligations	Mainstreaming planning processes across sectors. Periodic assessment of environmental resources (across sectors). Creation of a database/ databank on performance to support policy interventions.	Extent of periodic submission of state of environment reports across sectors Extent of clearly delineating environmental rules and regulations and discharging obligations Extent of regular maintenance of a data bank on performance to support policy initiatives

Operation	Lack of coordination within institutional framework leads to dispersed environmental activities at the operational level Lack of physical maintenance of natural and environmental resources affects operations	To strengthen coordination within the framework of environmental governance by integrating environmental activities at the operational levels	Centre, state and local governments		Conceiving specific tools, techniques and instruments for physical maintenance of environmental resources Integrating monitoring at all levels among stakeholders	Extent of specific tools techniques, and instruments implemented for physical maintenance (particularly quality forecasting) of environmental resources Extent of reports generated and implemented across sectors (on both positive and negative aspects environmental impacts) Extent and level of integrating monitoring among the stakeholders

(continued)

Table 8.1 (continued)

Issues	Constraints	Action required	Who should do it	How they should do it	Indicators for monitoring	How to measure monitoring
Accountability	Lack of accountability among governance structures to effectively implement their environmental obligations and to respond to environmental challenges Lack of provisions or mechanisms through which consumers of services can directly hold the service-providing agency accountable to the cost or quality of services provided Lack of a periodic impact assessment across sectors on quality, quantity and adequacy	Promoting accountability across sectors within their jurisdiction to effectively adhere to environmental obligations Constituting accountability mechanisms through which consumers can directly hold the service provider accountable to the cost or quality or quantity of services rendered	Local bodies across sectors	strengthening monitoring mechanisms for enhancing their accountability and promoting environmental governance Creating mechanisms for advocacy, promoting of environmental protection and conservation	Strict compliance mechanisms across sectors on quality, quantity and adequacy of services -Constituting regional monitoring systems Creation of comprehensive database on the state of environment across sectors Submission of regular impact assessment reports across sectors (regarding reaching targets)	Extent of penalties imposed on violation of environmental norms and standards Extent of submission of periodic assessment reports on benchmark indicators to identify gaps, initiate action to improve and create incentives for further improvements Extent of a comprehensive database created for an analysis of environmental compliance and state of environment across sectors

						Level of specific needs and use clearly defined across sectors Extent of regular impact assessment reports (both onsite and project related) generated across sectors to effectively meet targets Extent of record keeping at the sectoral and regional levels
Technology	Lack of an effective use of technology for sustainable management of infrastructure	Effective use of technological solutions such as GIS for the sustainable management of infrastructure	Local bodies across the sectors	Mandatory use of technology – (GIS) for the sustainable management of infrastructure	Periodic assessment of data on integrating the use of technology (GIS) for management of services across sectors	Extent of periodic collection, data analysis and dissemination of information (for tracking operational indicators, identifying performance gaps and incentivising the improvements) across sectors using GIS.

(continued)

Table 8.1 (continued)

Issues	Constraints	Action required	Who should do it	How they should do it	Indicators for monitoring	How to measure monitoring
Education and capacity building	Lack of awareness and education for community participation in the environmental management of local resources (especially on environmental rules, regulations and norms) Lack of capacity and confidence-building measures among the staff complicates the management of environment Lack of synergising environmental considerations into innovation and research programmes across sector.	Undertaking effective education and public awareness campaigns and outreach activities on environmental rules, regulations, standards and norms for promoting community participation as 'urban commons' or community stewardship of local ecology and environment	Local bodies vs community	Incentives for promotion of sustainable communities/ partnerships such as neighbourhood/RWAs/NGOs Promoting training programmes and workshops for building capacity and confidence in terms of knowledge and skills among the staff for promoting an effective delivery of services Mainstreaming the dissemination of information on environmental protection and management	Promoting capacity building programmes for the staff across sectors for encouraging (to elicit the compliance among the stakeholders)	Levels of incentives like regular appreciation and support for stakeholders participation Extent of outreach activities (education and awareness) created for participation of the communities in the management of environmental resources Extent of training and capacity and confidence-building programmes on environmental management Extent of integrating and implementating of environmental considerations into research and innovations

Urban planning	Haphazard urban planning Unabated urban sprawl Urban poverty Demand and supply gap of basic amenities	Development, implementation and enforcement of environmental by-laws need to be improved Incorporating strict environmental concerns/sustainable building programmes into the implementation of local by-laws, zoning regulations Municipalities should develop sustainable environmental vision for the city Creating a separate vision plan for effectively curbing urban poverty and environmental degradation	Local bodies-city corporation	Conceiving and implementingcomprehensive urban sustainable plans for each Metropolitan City in India Integrating national environmental priorities into local sustainable development projects Conceiving and implementing decadal plans across sectors	Efficient delivery of services across sectors	Extent of timely and adequate delivery of services across sectors like water, sanitation, etc. Extent of access to services by all the sections of urbansociety Extent of strict implementation of by-laws, incorporating environmental conservation methods and measures Extent of integration of national environmental priorities into the local sustainable development projects

(continued)

Table 8.1 (continued)

Issues	Constraints	Action required	Who should do it	How they should do it	Indicators for monitoring	How to measure monitoring
Finance	Gap between actual allocation and implementation of the budget Lack of a separate sustainable budget across sectors Lack of separate budget for education and capacity building programmes for the staff Lack of updation on consumer database, asset register and record on quantity of resources consumed or revenue accounts (base line information)	Promoting separate sustainable budgets for environmental protection and conservation functions	Centre, state and local bodies (across sectors)	Allocating a separate budget for ensuring environmental compliance in prospect functional areas for example, provision of services Allocating separate budget for promoting outreach, education and awareness programmes for the staff and communities Allocating finance for training programmes and capacity building measures for strengthening the staff in the management of environment	Finance spent on environmental management Finance spent on outreach and education promotion programmes Finance spent on training and capacity building programmes	Extent of the actual budget spent on environment management Extent of the actual budget spent on outreach, education and awareness programmes Extent of the actual budget spent on training and capacity building programmes on environmental management Extent of base line information available on assets maintained, consumer, quantity of resources used and available and revenue accounts

References

Balasubramanian, V. (2013). Bangalore water crisis, the big hoodwink, 5th March, DNA. Access the following link (http://www.dnaindia.com/analysis/report-bangalore-water-crisis-the-big-hoodwink-1807556) for more details.

BBMP. (2012). Integrated solid waste management, Department of Environment Cell, Bangalore: Bruhat Bangalore Mahanagara Palike.

BDA. (2007). *Revised master plan –2015: Vision document*. Bangalore: Bangalore Development Authority.

BMRDA. (2009). *Bangalore metropolitan region revised structure plan 2031*. Bangalore: Bangalore Metropolitan Regional Development Authority.

Centre for Sustainable Development. (2010). *A lifestyle change approach: Assimilating best water consumption practices, Study Report*. Bangalore: CSD.

Centre for Sustainable Development. (2012). *Environment report card 2012: Bangalore*. Bangalore: Centre for Sustainable Development.

Dahiya, B. (2012). 21st century Asian cities: Unique transformation, unprecedented challenges. *Global Asia, 7*(1), 98–104.

GoK. (2009). *State of environment report, Bangalore 2008*. Bangalore: Department of Forest, Ecology and Environment and Government of Karnataka.

Hust, E. (2005). Problems of Urbanaization and Urban Govlernance in India. In Hust E and M Mann (eds.,), *Urbanization and governance in India: South Asia Studies*, New Delhi: Manohar Publications.

IIHS. (2011). *Urban India 2011: Evidence*. Bangalore: Indian Institute for Human Settlements.

Lele, S., Dubash, N. K., & Dixit, S. (2010). A structure for environmental governance: A perspective. *Economic and Political Weekly, XLV*(6), 13–16.

Rahman, A., Netzband, M., Singh, A., & Mallick, J. (2009). An assessment of urban environmental issues using remote sensing and GIS technique: An integrated approach. A case study: Delhi, India. In de Sherbiniin, A., A. Rahman, A. Barbieri, J.C. Fotso, and Y. Zhu (eds.). 'Urban population-environment dynamics in the developing world: Case studies and lessons learned'. Paris: Committee for International Cooperation in National Research in Demography (CICRED) (316 pages). Also available at (http://www.populationenvironmentresearch.org/workshops.jsp#W2007.

Shrinivasan, R. (2013). 17% of urban India lives in slums: Census, 22nd March, Times of India. Access the link for more details. (http://timesofindia.indiatimes.com/india/17-of-urban-India-lives-in-slums-Census/articleshow/ 19118219.cms).

Smitha, K. C. (2006). Urban governance and Bangalore water supply and sewerage board (BWSSB). In K. V. Raju (Ed.), *Elixir of life: The socio-ecological governance of drinking water*. Book for Change: Bangalore.

Sudhira, H. S., Ramachandra, T. V., & Balasubramanya, M. H. (2007). City Profile: Bangalore. *Cities, 24*(5), 379–390.

Sawkar, R. H. (2012). Evaluation of surface, ground and sewage water for sustainable supply of potable water to Bengaluru. *Journal of Geological Society of India, 80*, 877–880.

Vaidyanathan, V., & King, R. A. (2011). Institutional analysis of urban transportation in Bangalore, conference paper presented at the 4th annual international conference on next generation infrastructures, Virginia Beach, Virginia, November 16-18, 2011.

Varshney, R. (2012). BBMP bylaws to dispose waste give hopes to many waste management players of Bangalore and Koramangala, October 12, Economic Times. Visit (http://articles.economictimes.indiatimes.com/2012-10-12/news/34413109_1_waste-management-wet-waste-household-hazardous-waste).

UN-HABITAT. (2010). *State of Asian Cities*. Japan: United Nations Human Settlement Programme (UN-HABITAT).

Index

K. V. Raju et al., *Urban Environmental Governance in India*, The Urban Book Series, https://doi.org/10.1007/978-3-319-73468-2

H

I

The manufacturer's authorised representative in the EU is Springer Nature Customer Service Centre GmbH, Europaplatz 3, 69115 Heidelberg, Germany. If you have any concerns regarding our products, please contact ProductSafety@springernature.com

Printed and bound by CPI Group (UK) Ltd, Croydon, CR0 4YY
30/09/2026
02225662-0006